# Moving Shape Analysis and Control

## Applications to Fluid Structure Interactions

T0173867

# PURE AND APPLIED MATHEMATICS

## A Program of Monographs, Textbooks, and Lecture Notes

# MONOGRAPHS AND TEXTBOOKS IN
# PURE AND APPLIED MATHEMATICS

## Recent Titles

*I. Graham and G. Kohr*, Geometric Function Theory in One and Higher Dimensions (2003)

*G. V. Demidenko and S. V. Uspenskii*, Partial Differential Equations and Systems Not Solvable with Respect to the Highest-Order Derivative (2003)

*A. Kelarev*, Graph Algebras and Automata (2003)

*A. H. Siddiqi*, Applied Functional Analysis: Numerical Methods, Wavelet Methods, and Image Processing (2004)

*F. W. Steutel and K. van Harn*, Infinite Divisibility of Probability Distributions on the Real Line (2004)

*G. S. Ladde and M. Sambandham*, Stochastic versus Deterministic Systems of Differential Equations (2004)

*B. J. Gardner and R. Wiegandt*, Radical Theory of Rings (2004)

*J. Haluska*, The Mathematical Theory of Tone Systems (2004)

*C. Menini and F. Van Oystaeyen*, Abstract Algebra: A Comprehensive Treatment (2004)

*E. Hansen and G. W. Walster*, Global Optimization Using Interval Analysis, Second Edition, Revised and Expanded (2004)

*M. M. Rao*, Measure Theory and Integration, Second Edition, Revised and Expanded (2004)

*W. J. Wickless*, A First Graduate Course in Abstract Algebra (2004)

*R. P. Agarwal, M. Bohner, and W-T Li,* Nonoscillation and Oscillation Theory for Functional Differential Equations (2004)

*J. Galambos and I. Simonelli*, Products of Random Variables: Applications to Problems of Physics and to Arithmetical Functions (2004)

*Walter Ferrer and Alvaro Rittatore*, Actions and Invariants of Algebraic Groups (2005)

*Christof Eck, Jiri Jarusek, and Miroslav Krbec*, Unilateral Contact Problems: Variational Methods and Existence Theorems (2005)

*M. M. Rao*, Conditional Measures and Applications, Second Edition (2005)

*A. B. Kharazishvili*, Strange Functions in Real Analysis, Second Edition (2006)

*Vincenzo Ancona and Bernard Gaveau*, Differential Forms on Singular Varieties: De Rham and Hodge Theory Simplified (2005)

*Santiago Alves Tavares*, Generation of Multivariate Hermite Interpolating Polynomials (2005)

*Sergio Macías*, Topics on Continua (2005)

*Mircea Sofonea, Weimin Han, and Meir Shillor*, Analysis and Approximation of Contact Problems with Adhesion or Damage (2006)

*Marwan Moubachir and Jean-Paul Zolésio*, Moving Shape Analysis and Control: Applications to Fluid Structure Interactions (2006)

# MONOGRAPHS AND TEXTBOOKS IN PURE AND APPLIED MATHEMATICS

## Recent Titles

*I. Graham and G. Kohr*, Geometric Function Theory in One and Higher Dimensions (2003)

*G. V. Demidenko and S. V. Uspenskii*, Partial Differential Equations and Systems Not Solvable with Respect to the Highest-Order Derivative (2003)

*A. Kelarev*, Graph Algebras and Automata (2003)

*A. H. Siddiqi*, Applied Functional Analysis: Numerical Methods, Wavelet Methods, and Image Processing (2004)

*F. W. Steutel and K. van Harn*, Infinite Divisibility of Probability Distributions on the Real Line (2004)

*G. S. Ladde and M. Sambandham*, Stochastic versus Deterministic Systems of Differential Equations (2004)

*B. J. Gardner and R. Wiegandt*, Radical Theory of Rings (2004)

*J. Haluska*, The Mathematical Theory of Tone Systems (2004)

*C. Menini and F. Van Oystaeyen*, Abstract Algebra: A Comprehensive Treatment (2004)

*E. Hansen and G. W. Walster*, Global Optimization Using Interval Analysis, Second Edition, Revised and Expanded (2004)

*M. M. Rao*, Measure Theory and Integration, Second Edition, Revised and Expanded (2004)

*W. J. Wickless*, A First Graduate Course in Abstract Algebra (2004)

*R. P. Agarwal, M. Bohner, and W-T Li,* Nonoscillation and Oscillation Theory for Functional Differential Equations (2004)

*J. Galambos and I. Simonelli*, Products of Random Variables: Applications to Problems of Physics and to Arithmetical Functions (2004)

*Walter Ferrer and Alvaro Rittatore*, Actions and Invariants of Algebraic Groups (2005)

*Christof Eck, Jiri Jarusek, and Miroslav Krbec*, Unilateral Contact Problems: Variational Methods and Existence Theorems (2005)

*M. M. Rao*, Conditional Measures and Applications, Second Edition (2005)

*A. B. Kharazishvili*, Strange Functions in Real Analysis, Second Edition (2006)

*Vincenzo Ancona and Bernard Gaveau*, Differential Forms on Singular Varieties: De Rham and Hodge Theory Simplified (2005)

*Santiago Alves Tavares*, Generation of Multivariate Hermite Interpolating Polynomials (2005)

*Sergio Macías*, Topics on Continua (2005)

*Mircea Sofonea, Weimin Han, and Meir Shillor*, Analysis and Approximation of Contact Problems with Adhesion or Damage (2006)

*Marwan Moubachir and Jean-Paul Zolésio*, Moving Shape Analysis and Control: Applications to Fluid Structure Interactions (2006)

# Moving Shape Analysis and Control

## Applications to Fluid Structure Interactions

**Marwan Moubachir**

INRIA, France

**Jean-Paul Zolésio**

CNRS and INRIA, France

CRC Press
Taylor & Francis Group
Boca Raton  London  New York

CRC Press is an imprint of the
Taylor & Francis Group, an **informa** business

A CHAPMAN & HALL BOOK

CRC Press
Taylor & Francis Group
6000 Broken Sound Parkway NW, Suite 300
Boca Raton, FL 33487-2742

First issued in paperback 2019

ISBN-13: 978-1-58488-611-2 (hbk)
ISBN-13: 978-0-367-39128-7 (pbk)
Library of Congress Card Number 2005053196

---

### Library of Congress Cataloging-in-Publication Data

---

Mouchabir, Marwan.
    Moving shape analysis and control : applications to fluid structure interactions / by Marwan Mouchabir and Jean Paul Zolesio.
        p. cm. -- (Monographs and textbooks in pure and applied mathematics ; 278)
    Includes bibliographical references and index.
    ISBN 1-58488-611-0 (acid-free paper)
    1. Shape theory (Topology) 2. Fluid-structure interaction--Mathematics. I. Zolésio, J.P. II. Title. III. Series.

QA612.7.M68 2005
514'.24--dc22                                                                    2005053196

---

**Visit the Taylor & Francis Web site at**
**http://www.taylorandfrancis.com**

**and the CRC Press Web site at**
**http://www.crcpress.com**

*A Valérie, Olivia et Timothé.*
*A Monique.*

# Contents

# List of Figures

# *Preface*

## Objectives and Scope of the Book

This volume intends to provide a mathematical analysis of problems related to the evolution of two or three dimensional domains. This topic includes a number of engineering applications such as free surface flows, phase changes, fracture and contact problems, fluid-structure interaction problems for civil transport vehicles, civil engineering constructions such as stayed-cable bridges or tall towers, biomechanical systems, boundary tracking problems, computational vision . . . .

The main objective is to furnish various tools to handle the motion of a moving domain on the level of its intrinsic definition, computation, optimization and control. The efficiency of these tools will be illustrated within this volume on different examples in connection with the analysis of non-cylindrical partial differential equations, e.g., the Navier-Stokes equations for incompressible fluids in moving domains.

We shall concentrate our analysis on the Eulerian approach. Hence the evolution of a 3D domain is chosen to be described by the flow $\mathbf{T}_t$ of a non-autonomous Eulerian velocity field $\mathbf{V}(.,t)$ for $t \geq 0$. This flow can be defined in a strong or a weak manner depending on the space and time regularity of the moving domain. The first two chapters give all the details of the existence and uniqueness of the flow for both cases. These results are fundamental in order to deal with the arbitrary Eulerian-Lagrangian formulation or the level-set approach. The sensitivity of the flow mapping with respect to its associated velocity field $\mathbf{V}$ in the direction $\mathbf{W}$ is completly characterized by the transverse field $\mathbf{Z}$. The evolution of this transverse field is governed by a dynamical transport equation involving the Lie bracket $[\mathbf{V}, \mathbf{W}]$. Among other applications, these differentiation results form the basis in order to analyse the first and second-order trajectory variations of any general dynamical system. We furnish different tools that can prove appropriate in order to deal with general variational formulations involving integral functionals using either the tensorial or the joint time-space metric over non-cylindrical time-space domains also called tubes. As an example, we can cite the two ways of handling the tube perimeter constraint that yield to different expressions for the surface tension.

We will also revisit the notion of shape differential equations driving the evolution of a domain subjected to certain geometrical constraints. In the context of optimal control and inverse problems, this equation corresponds to the first-order optimality conditions associated to the minimization with respect to the velocity field **V** of a given Eulerian cost function. In the context of the level-set method, this shape differential equation is related to the Hamilton-Jacobi equation driving the optimal evolution of the level-curves.

The next chapters aim at providing illustration of these Eulerian evolution and derivation tools for the control of systems involving fluids and solids. Especially, we shall consider in chapters 5 and 6 the control of a fluid described by the Navier-Stokes equations inside a 3D bounded domain through the dynamical evolution of its boundary. The Eulerian approach is based on the introduction of the transverse field characterizing the sensitivity of the flow with respect to the velocity field perturbations. We state the structure of the optimality conditions using either the fluid state differentiability with respect to **V** or the min-max derivation principle providing instantly the adjoint state problem. Similar results are obtained using a Lagrangian description of the boundary motion. In this case, the sensitivity analysis is performed using non-cylindrical identity perturbations. An interesting application is the rigorous mathematical justification of transpiration boundary conditions on fixed boundary usually used by engineers as a first-order approximation of no-slip boundary conditions on a moving solid surface.

The end of the volume deals with boundary control of fluid-structure interaction systems. The idea is to minimize a given tracking functional related to the solid evolution, with respect to a fluid inflow boundary condition. The case of a rigid solid coupled with an incompressible fluid flow is based on the Eulerian approach for characterizing the partial sensitivity of the fluid with respect to the solid state variables. We obtain a new fluid-structure adjoint problem that allows the computation of a given cost function gradient. In case the rigid solid is replaced by an elastic solid with an arbitrary constitutive law, we use the Lagrangian framework and the non-cylindrical identity perturbations. Again we obtain a new fluid-structure adjoint system.

---

## Intended Audience

This book will be useful for researchers or graduate students with a background in applied mathematics who are interested in the control of complex systems involving a moving boundary. This include topics such as free surface flows, phase changes, fracture and contact problems, fluid-structure interaction problems for civil transport vehicles, civil engineering constructions such as stayed-cable bridges or tall towers, biomechanical systems, boundary tracking problems, computational vision ....

More generally, this book will be of great use for people who are interested in the optimization of complex partial differential systems since it includes various tools that can be used in different contexts. Some parts of the book can require some knowledge of advanced mathematics that are recalled in the appendix at the end of the volume.

xx

# Chapter 1

## Introduction

Shape Optimization was introduced around 1970 by Jean Céa [31], who understood, after several engineering studies [127, 12, 35, 110, 102, 83, 84, 7], the future issues in the context of optimization problems. At that time, he proposed a list of open problems at the French National Colloquium in Numerical Analysis. These new problems were formulated in terms of minimization of functionals (referred as open loop control or passive control) governed by partial differential boundary value problems where the control variable was the geometry of a given boundary part [103, 76]. From the beginning, the terminology *shape optimization* was not connected to the structural mechanical sciences in which elasticity and optimization of the compliance played a central role. Furthermore, these research studies were mainly addressed in the context of the numerical analysis of the finite element methods.

At the same time, there was some independent close results concerning fluid mechanics by young researchers such as O. Pironneau [123, 124, 78], Ph. Morice [107] and also several approaches related to perturbation theory by P.R. Garabedian [74, 75] and D.D Joseph [91, 92].

Very soon, it appeared that the shape controlof Boundary Value Problems (BVP) was at the crossroads of several disciplines such as PDE analysis, non-autonomous semi-group theory, numerical approximation (including finite element methods), control and optimization theory, geometry and even physics. Indeed several classical modeling in both structural and fluid mechanics (among other fields) needed to be extended. An illustrative example concerns a very *popular* problem in the 80's concerning the thickness optimization of a plate modeled by the classical Kirchoff biharmonic equation. This kind of solid model is based on the assumption that the thickness undergoes only small variations. Therefore, many pioneering works were violating the validity of this assumption, leading to strange results, e.g., the work presented in the Iowa NATO Study [85] stating the existence of optimal beams having *zero cross section* values.

In the *branch* which followed the passive control approach, we shall mention the work of G. Chavent [32, 34] based on the theory of distributed system control introduced by J-L. Lions [98]. Those results did not address optimization problems related to the domain but instead related to the coefficients inside the PDE. At that time, it was hoped that the solution of elliptic problems would be continuous with respect to the weak convergence of the coefficients.

It appeared that this property was not achieved by this class of problem[1]. At that point a main *bifurcation* arose with the homogenization approach [10] which up to some point was considered as a part of the *Optimal Design* theory.

The mathematical analysis of shape optimization problems began with the correct definition of derivatives of functionals and functions with respect to the domain, together with the choice of tangential space to the familly of shapes. Following the very powerful theory developed by J. Nečas [117], the role of bilipschitzian mapping was emphasized for Sobolev spaces defined in moving domainsbased on the Identity perturbation method [115, 106, 134]. Concerning the large domain deformationviewpoint the previous approach led to the incremental domain evolution methods [143].

After 1975, the second author introduced [145] an asymptotic analysis for domain evolution using classical geometrical flows which are intrinsic tools for manifolds evolutions and gave existence results for the so-called *shape differential equation* (see also [79]). At that period, applications focused more on sensitivity analysis problems than on asymptotic analysis of domains evolution. In 1972, A.M. Micheletti introduced in parallel [105, 104] a metric based on the Identity perturbation method thanks to the use of differentiable mappings, in order to study eigenvalues perturbation problems. The associated topology was extended by M. Delfour *et al.* [52] and turns out to be the same as the one induced by the continuity along flow field deformations [147].

The systematic use of flow mapping and intrinsic geometrythrough the fundamental role of the oriented distance function [47, 50] led to the revised analysis of the elastic shell theory [48, 49, 25, 26, 27, 28], of the boundary layer theory [3] or of the manifold derivative tools [53].

The use of both Bounded Variation (BV) analysis and the notion of Cacciapoli sets led to the first compactness method for domain sequences and several extensions to more regular boundaries were done through the use of different concepts such as *fractal boundaries, density perimeter* [23, 20, 21, 19] or *Sobolev domains* [50].

At that point, an other important *bifurcation point* in that theory occurred with the relaxation theory and the Special Bounded Variation (SBV) analysis which was particularly well adapted for image segmentation problems [6]. At the opposite, the capacity constraint for Dirichlet boundary conditions led to a fine analysis initiated in [18] and is still going on for cracks analysis.

The method of large evolution based on the flow mapping (known from 1980 as the *speed method* [150]) turns to be the natural setting for weak evolution of

---

[1]Indeed, in his thesis [33], G. Chavent referred to such a result to appear in a work by F. Murat [113]. That paper [111] appeared but as a counterexample to the expected continuity property. He showed on a one dimensionnal simple example that with weak *oscillating* convergence of the coefficients, the associated solution was converging to another problem in which the new coefficients were related to the limit of the *inverse* coefficients associated to the original problem [112, 114].

geometry allowing topological changes through the convection of either characteristic functions or oriented distance functions.

After this non-exhaustive review of the context in which the shape optimization analysis emerged, we shall concentrate on the particular framework of the present book.

## 1.1 Classical and moving shape analysis

The object of this book is the mathematical analysis of systems involving the evolution of the geometry. This is motivated by many important applications in physics, engineering or image processing. The classical shape analysis investigates the effects of perturbations of the geometry in terms of continuity, differentiability and optimization of quantities related to the state of a system defined in that geometry. In this case, the geometry is usually perturbed thanks to a map involving a scalar parameter usually referred to as a fictitious time. On the contrary, the moving shape analysis deals with systems that are intrinsically defined on a moving geometry. Hence, we shall deal with sensitivity analysis with respect to a continuous family of shapes over a given time period. In this context, if we consider the geometry in a space-time configuration, the moving shape analysis may also be referred to as a non-cylindrical shape analysis[2].

A first issue in this analysis is to model the evolution of the geometry. This is a common topic with the classical shape analysis. There exist many ways to build families of geometries. For instance, a domain can be made variable by considering its image by a family of diffeomorphisms parametrized by the time parameter as it happens frequently in mechanics for the evolution of continuous media. This way of defining the motion of domains avoids a priori the modification of the underlying topology. This change of topology can be allowed by using the characteristic function of families of sets or the level set of a space-time scalar function. We refer the reader to [51] for a complete review on this topic. In chapter 2, we shall deal with the particular problem of defining in a weak manner the convection of a characteristic function in the context of the *speed method* developed in [147].

In numbers of applications, we shall consider a state variable associated to a system which is a solution of a partial differential equation defined inside the moving domain over a given time period. Hence, we need to analyse the solvability of this non-cylindrical PDE system before going further. Here, again this topic has been already studied since it enters the classical shape analysis

---

[2]The notion of tube (non-cylindrical evolution domains) was also independently introduced by J.P. Aubin via the concept of *abstract mutations* [8].

problem while introducing a perturbed state defined in the moving domain parametrized by the fictitious time parameter. Furthermore, this solvability analysis has been performed in numbers of mathematical problems involving moving domains. Here, we refer the reader to [135, 51] for some particular results in the context of the classical shape analysis. We also refer to the extensive litterature concerning the analysis of PDE systems defined in moving domains, e.g., [96, 126, 132, 62, 130, 88, 128, 70, 100, 11].

Contrary to the last topic, very few references exist for the sensitivity analysis with respect to perturbation of the evolution of the moving geometry. Early studies have been conducted in [90, 151, 158, 141, 120, 43, 142, 2] for specific hyperbolic and parabolic linear problems. An important step was performed in [154, 155] where the second author of the present book established the derivative of integrals over a moving domain with respect to its associated Eulerian velocity. These results were applied in order to study variational principles for an elastic solid under large displacements and for the incompressible Euler equation. This work was generalized in [58, 59].

## 1.2    Fluid-Structure interaction problems

As already mentioned above, we shall apply the moving shape analysis on systems involving the coupling between fluids and solids.

A general fluid-solid model consists of an elastic solid either surrounded by a fluid (aircrafts, automobiles, bridge decks ...) or surrounding a fluid flow (pipelines, arteries, reservoir tanks ...). Here the motion of the interface between the fluid and the solid is part of the unknown of the coupled system. It is a free boundary problem that can be solved by imposing continuity properties through the moving interface (e.g., the kinematic continuity of the velocities and the kinetic continuity of the normal stresses). This model has been intensively studied in the last two decades on the level of its mathematical solvability [87, 54, 82, 39, 80, 131, 15, 9, 138, 41], its numerical approximation [89, 55, 119, 118, 122, 95, 67], its stability [73, 38, 64, 65] and more recently on its controllability [66, 109]. In this lecture note, we will restrict ourself to viscous Newtonian incompressible fluid flows described by the Navier-Stokes equations in space dimension two or three. The case of a compressible Newtonian fluid can be incorporated in the present framework with the price of a heavier mathematical analysis (solvability, non-differentiability around shocks ...).

Our goal is to solve inverse or control problems based on the previous general fluid-solid model. As an example, we think to decrease the drag of a car inside the atmospheric air flow by producing specific vibrations on its body using smart materials such as piezoelectrical layers. In this example, the control variable can be chosen as the electrical energy input evolution inside the piezoelectrical device and the objective is to decrease the drag which is a function of the coupled fluid-structure state ( the air and the body of the car) and this state depends on the control variable. In order to build a control law for the electrical input, we need to characterize the relationship between the drag function and the control variable on the level of its computation and its variations.

As an other example, we can think of the problem of aeroelastic stability of structures. Both authors have been dealing with such a problem in the context of the stability analysis against wind loads of bridge decks. In [108], it has been suggested that such a problem can be set as the inverse problem consisting in recovering the smallest upstream wind speed that leads to the worst bridge deck vibrations. In this example, the decision variable can be chosen as the upstream wind speed and the objective is to increase a functional based on the vibration amplitude history of the bridge deck during a given characteristic time period which is a function of the coupled fluid-structure state ( the wind flow and the bridge deck) which is also a function of the decision variable. Again, in order to recover the wind speed history, we need to characterize the relationship between the objective functional and the decision variable on the level of its computation and its variations.

In order to characterize the sensitivity of the objective functional with respect to the control variable, it is obvious that we need to characterize the sensitivity of the coupled fluid-structure state with respect to the control variable. Here we recall that the coupled fluid-structure state is the solution of a system of partial differential equations that are coupled through continuity relations defined on the moving interface (the fluid-structure interface). The key point towards this sensitivity analysis is to investigate the sensitivity of the fluid state, which is an Eulerian quantity, with respect to the motion of the solid, which is a Lagrangian quantity. This task falls inside the moving shape analysis framework described earlier. Indeed the fluid state is the solution of a system of non-linear partial differential equations defined in a moving domain. The boundary of this moving domain is the solid wall. Then using the tools developed in [59], it has been possible to perform in [58] the moving shape sensitivity analysis in the case of a Newtonian incompressible fluid inside a moving domain driven by the non-cylindrical Navier-Stokes equations.

All the previous results use a parametrization of the moving domain based on the Lagrangian flow of a given velocity field. Hence, the design variable is the Eulerian velocity of the moving domain, allowing topology changes while using the associated level set formulation. In [13, 14], the author used a non-cylindrical identity perturbation technique. It consists in perturbating the space-time identity operator by a family of diffeomorphisms. Then, this family is chosen as the design parameter. It is a Lagrangian description of the moving geometry, which a priori does not allow topology changes but which leads to simpler sensitivity analysis results which are still comparable with the one obtained by the non-cylindrical *speed method*. In [57], the authors came back to the dynamical shape control of the Navier-Stokes and recovered the results obtained in [58] using the Min-Max principle allowing to avoid the state differentiation step with respect to the velocity of the domain.

Now, we come back to the original problem consisting in the sensitivity analysis of the coupled fluid-structure state with respect to the control variable. Using the chain rule, the derivative of the coupled state with respect to the control variable involves the partial derivative of the fluid state with respect to the motion of the fluid-structure interface already characterized in [58, 57]. Hence, again using a Lagrangian penalization technique, already used and justified in [45, 46], it has been possible to perform in [109] the sensitivity analysis of a simple fluid-structure interaction problem involving a rigid solid within an incompressible flow of a Newtonian fluid with respect to the upstream velocity field. As already mentioned, this simple model is particularly suited for bridge deck aeroelastic stability analysis [121].

---

## 1.3   Plan of the book

The book is divided in eight chapters. Chapter 2 furnishes a simple illustration to some of the moving shape analysis results reported in the core of the lecture note. We deal with a simple inverse problem arising in phase change problems consisting in recovering the moving interface at the isothermal interface between a solid and liquid phase from measurements of the temperature on a insulated fixed part of the solid boundary. We use a least-square approach and we show how to compute the gradient of the least-square functional with respect to the velocity of the moving interface. It involves an adjoint state problem together with an adjoint transverse state, which is the novelty of the moving shape analysis compared to the classical one.

In chapter 3, we consider the weak Eulerian evolution of domains through the convection, generated by a non-smooth vector field $\mathbf{V}$, of measurable sets. The introduction of transverse variations enables the derivation of functionals associated to evolution tubes. We also introduce Eulerian variational formulations for the minimal curve problem. These formulations involve a geometrical adjoint state $\lambda$ which is backward in time and is obtained thanks to the use of the so-called transverse field $\mathbf{Z}$.

In chapter 4, we recall the concept of shape differential equation developed in [145],[147]. Here, we present a simplified version and some applications in dimension 2 which enable us to reach the time asymptotic result. Furthermore, we introduce the associated level set formulation whose speed vector version was already contained in [149].

In chapter 5, we deal with a challenging problem in fluid mechanics which consists in the control of a Newtonian fluid flow thanks to the velocity evolution law of a moving wall. Here, the optimal control problem has to be understood as the open loop version, i.e., it consists in minimizing a given objective functional with respect to the velocity of the moving wall. This study is performed within the non-cylindrical Eulerian moving shape analysis describe in chapters 2 and 3. We focus on the use of a Lagrangian penalization formulation in order to avoid the fluid state differentiation step.

In chapter 6, we introduce the Lagrangian moving shape analysis framework. It differs from the Eulerian one from the fact that the design variable is the diffeomorphism that parametrizes the moving geometry. The sensitivity analysis is simpler since it does not involve the transverse velocity field. We apply these tools in order to deal with the control of a Newtonian fluid flow thanks to the displacement evolution law of a moving wall.

Chapter 7 moves to inverse problems related to fluid-structure interaction systems. Here, we consider a 2D elastic solid with rigid displacements inside the incompressible flow of a viscous Newtonian fluid. We try to recover informations about the inflow velocity field from the partial measurements of the coupled fluid-structure state. We use a least-square approach together with a Lagrangian penalization technique. We derive the structure of the gradient with respect to the inflow velocity field of a given cost function. Using the Min-Max principle, the cost function gradient reduces to the derivative of the Lagrangian with respect to the inflow velocity at the saddle point. This saddle point is solution of the first order optimality conditions. We use non-cylindrical Eulerian derivatives to compute the partial derivative of the Lagrangian functional with respect to the solid state variables, involved in the optimality system.

Finally, in chapter 8 we extend the results of chapter 7, to the case of an elastic solid under large displacements inside an incompressible fluid flow. The main difference with the previous case is the use of a non-cylindrical Lagrangian shape analysis for establishing the KKT system. It forms the adjoint counterpart of the sensitivity analysis conducted in [66].

---

## 1.4 Detailed overview of the book

In order for the reader to have a simple overview of the book, we shall describe the different steps encountered while designing a complex fluid-structure interaction system. Indeed, let us consider a mechanical system that consists of a solid and a fluid interacting with each other. We would like to increase the performances of this system. These performances have to be quantitatively translated inside a cost function that we have to optimize with respect to some parameters that we will call the control variables. In the sequel, we will describe different control situations:

1. *Control of a fluid flow around a fixed body:* it consists in trying to modify the fluid flow pattern around a fixed body using a boundary control which can act for example by blowing or suctioning the fluid at some part of the solid boundary. The control law will be designed in order to match some efficiency goals using the minimisation of a cost functional.

2. *Shape design of a fixed solid inside a fluid flow:* in this case, the control is the shape of the body. We would like to find the best shape satisfying some geometrical constraints that will optimize some cost functionals. This problem is somewhat classical in the aeronautical field, but it requires some subtle mathematical tools that we will quickly recall.

3. *Dynamical shape design of a solid inside a fluid flow:* the novelty compared to the last item is that the shape is moving and we are looking for the best evolution of this shape that both satisfies some geometrical constraints and optimizes some cost functionals. This is a rather natural technique in order to control a fluid flow pattern, but still its design requires some new mathematical tools that will be sketched in this introduction and more detailed in the core of this lecture note.

4. *Control of an elastic solid inside a fluid flow:* this is the most complex and most realistic situation where both the fluid and the solid have their own dynamics which are coupled through the fluid-solid interface. Then, we would like to control or optimize the behaviour of this coupled system thanks to boundary conditions. The mathematical analysis of this situation uses the whole framework introduced previously. This is a challenging problem, both on the mathematical point of view and on the technological side. The goal of this book is to partially answer to some issues related to this problem.

## 1.4.1 Control of a fluid flow around a fixed body

Let us consider a fixed solid obstacle $\Omega \subset \mathbb{R}^3$ with boundary $\Gamma$ surrounded by a viscous Newtonian fluid. We shall restrict the analysis of this system to a control volume $D \in \mathbb{R}^3$ containing the obstacle $\Omega$ (see figure 1.1). The fluid domain is denoted by $\Omega^f \overset{\text{def}}{=} D \setminus \Omega$. The fluid is described by its velocity $\mathbf{u} : \mathbb{R}^+ \times \Omega^f \to \mathbb{R}^3$ and its pressure $p : \mathbb{R}^+ \times \Omega^f \to \mathbb{R}$. The flow is supposed to be incompressible and the state variables satisfy the classical Navier-Stokes equations,

$$\begin{cases} \partial_t \mathbf{u} + \mathrm{D}\,\mathbf{u} \cdot \mathbf{u} - \nu\,\Delta\mathbf{u} + \nabla p = \mathbf{f}, & (0, \tau) \times \Omega^f \\ \mathrm{div}\,\mathbf{u} = 0, & (0, \tau) \times \Omega^f \end{cases} \tag{1.1}$$

We must endow this system with boundary conditions such as far-field Dirichlet boundary conditions on $\partial D$ and non-slip boundary conditions on the solid boundary $\Gamma$

$$\begin{cases} \mathbf{u} = \mathbf{u}_\infty, \ \partial D \\ \mathbf{u} = 0, \quad \Gamma \end{cases} \tag{1.2}$$

We also add initial conditions,

$$\mathbf{u}(t = 0) = \mathbf{u}_0, \ \Omega^f \tag{1.3}$$

A classical issue consists in reducing the drag exerted by the fluid on the

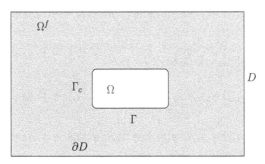

**FIGURE 1.1:** Fluid flow control

solid (e.g., car, aircraft, ship). The first step in this direction is to choose the type of control we would like to apply on this system to reach our objective. The boundary control consists in adding inside the solid a control device that modifies the boundary conditions on a part $\Gamma^c \subset \Gamma$ of its boundary,

$$\begin{cases} \mathbf{u} = \mathbf{g}, \ \Gamma^c \\ \mathbf{u} = 0, \ \Gamma \setminus \Gamma^c \end{cases} \tag{1.4}$$

Here the boundary control $\mathbf{g}$ can be either

- time-independent and in this case it can be viewed as a passive control strategy consisting in modifying, e.g., the friction of the surface.

- or time-dependent and in this case it can be viewed as an active control strategy consisting of a blowing/suction system that can be designed off-line (that means using an open-loop control) or on-line (that means using a feedback control law).

Let us describe quickly both situations since it will be useful for the rest of the book.

**The objective functional**

A common topic in the optimization and control field of PDE systems is the choice of appropriate cost functionals, i.e., meeting both our objectives and the mathematical requirements that guarantee the converge to at least one optimum parameter. This functional can depend both on the state variables $(\mathbf{u}, p)$ and on the control parameter $\mathbf{g}$. Let us choose this function as the work needed to overcome the drag exerted by the fluid on the solid boundary,

$$J_{drag}(\mathbf{u}, p) = \int_{(0,\tau) \times \Gamma} (\mathbf{u} - \mathbf{u}_\infty) \cdot \sigma(\mathbf{u}, p) \cdot \mathbf{n} \, d\Gamma \, dt \qquad (1.5)$$

where $\sigma(\mathbf{u}, p) \overset{\text{def}}{=} -p\,\mathbf{I} + \nu\, \varepsilon(\mathbf{u})$ stands for the fluid stress tensor, $\varepsilon(\mathbf{u}) \overset{\text{def}}{=} \frac{1}{2}(\mathbf{D}\,\mathbf{u} + {}^*\mathbf{D}\,\mathbf{u})$ stands for the fluid deformation tensor and $\mathbf{n}$ is unit normal field.

Another cost function is associated to the objective of steering the fluid along a given pattern $\mathbf{u}_d$ during a prescribed time,

$$J_{track}(\mathbf{u}, p) = \frac{1}{2} \int_{(0,\tau) \times \Omega^f} (\mathbf{u} - \mathbf{u}_d)^2 \, dx \, dt \qquad (1.6)$$

More generally, we can consider any cost functionals that are twice-differentiable with respect to their arguments.

### The control problem

Here we choose to work with a control variable defined as a velocity field on the solid boundary and our objective is to minimize a given cost functional,

$$\min_{\mathbf{g} \in \mathcal{U}} j(\mathbf{g}) \qquad (1.7)$$

where $j(\mathbf{g}) = J[(\mathbf{u}, p)(\mathbf{g})]$, $\mathcal{U}$ stands for an appropriate Hilbert space of the Sobolev type, typically $H^1((0,\tau) \times \Gamma^c)$ and the couple $(\mathbf{u}, p)(\mathbf{g})$ stands for a weak solution of the Navier-Stokes equations associated to the control $\mathbf{g}$,

$$\begin{cases} \partial_t \mathbf{u} + \mathbf{D}\,\mathbf{u} \cdot \mathbf{u} - \nu\,\Delta \mathbf{u} + \nabla p = \mathbf{f}, & (0,\tau) \times \Omega^f \\ \operatorname{div} \mathbf{u} = 0, & (0,\tau) \times \Omega^f \\ \mathbf{u} = \mathbf{u}_\infty, & (0,\tau) \times \partial D \\ \mathbf{u} = \mathbf{g}, & (0,\tau) \times \Gamma^c \\ \mathbf{u} = 0, & (0,\tau) \times \Gamma \setminus \Gamma^c \\ \mathbf{u}(t=0) = \mathbf{u}_0, & \Omega^f \end{cases} \qquad (1.8)$$

Our goal is now to furnish the first-order optimality conditions associated to the optimization problem 1.7. These conditions are very useful since they are the basis in order to build both a rigourous mathematical analysis and gradient-based optimization algorithms.

There exists two main methods in order to derive these conditions: the first one is based on the differentiability of the state variables with respect to the control parameter and the second one relies on the existence of Lagrangian multipliers.

## Sensitivity

Let us consider a control point $\mathbf{g} \in \mathcal{U}$, then the cost functional $j(\mathbf{g})$ is Fréchet differentiable with respect to $\mathbf{g}$ [1, 71] and its directional derivative is given by

$$\langle j'(\mathbf{g}), \mathbf{h} \rangle = \langle \partial_{(\mathbf{u},p)} J[(\mathbf{u}, p)(\mathbf{g})], (\mathbf{u}', p')(\mathbf{g}; \mathbf{h}) \rangle \qquad (1.9)$$

where $(\mathbf{u}', p')(\mathbf{g}; \mathbf{h}) \stackrel{\text{def}}{=} \frac{d}{d\mathbf{g}}(\mathbf{u}, p)(\mathbf{g}) \cdot \mathbf{h}$ stands for the directional derivative of $(\mathbf{u}, p)(\mathbf{g})$ with respect to $\mathbf{g}$.

In the case of the functional (1.6), we have

$$\langle j'(\mathbf{g}), \mathbf{h} \rangle = \int_{(0,\tau) \times \Omega^f} (\mathbf{u}(\mathbf{g}) - \mathbf{u}_d) \cdot \mathbf{u}'(\mathbf{g}; \mathbf{h}) \, dx \, dt$$

It can be proven by evaluating the differential quotient that $(\mathbf{u}', p')(\mathbf{g}; \mathbf{h})$ exists in an appropriate space and that it is the solution of the linearized Navier-Stokes system,

$$
\begin{cases}
\partial_t \mathbf{u}' + D\mathbf{u}' \cdot \mathbf{u} + D\mathbf{u} \cdot \mathbf{u}' - \nu \Delta \mathbf{u}' + \nabla p' = 0, & (0, \tau) \times \Omega^f \\
\text{div } \mathbf{u}' = 0, & (0, \tau) \times \Omega^f \\
\mathbf{u}' = 0, & (0, \tau) \times \partial D \\
\mathbf{u}' = \mathbf{h}, & (0, \tau) \times \Gamma^c \\
\mathbf{u}' = 0, & (0, \tau) \times \Gamma \setminus \Gamma^c \\
\mathbf{u}'(t = 0) = 0, & \Omega^f
\end{cases}
\qquad (1.10)
$$

Then the first-order optimality condition writes

$$\langle j'(\mathbf{g}), \mathbf{h} \rangle = 0, \quad \forall \mathbf{h} \in \mathcal{U} \qquad (1.11)$$

That means that the set of optimal controls is contained in the set of critical points for the cost function $j(\mathbf{g})$.

However, we would like to obtain an expression of this condition avoiding the direction $\mathbf{h} \in \mathcal{U}$. To this end, we introduce the adjoint variable $(\mathbf{v}, \pi)$ solution of the adjoint linearized Navier-Stokes system,

$$
\begin{cases}
-\partial_t \mathbf{v} - D\mathbf{v} \cdot \mathbf{u} + {}^*D\mathbf{u} \cdot \mathbf{v} - \nu \Delta \mathbf{v} + \nabla \pi = -(\mathbf{u} - \mathbf{u}_d), & (0, \tau) \times \Omega^f \\
\text{div } \mathbf{v} = 0, & (0, \tau) \times \Omega^f \\
\mathbf{v} = 0, & (0, \tau) \times \partial D \\
\mathbf{v} = 0, & (0, \tau) \times \Gamma \\
\mathbf{v}(t = \tau) = 0, & \Omega^f
\end{cases}
\qquad (1.12)
$$

Consequently, we are able to identify the gradient of the cost function as the trace on $\Gamma^c$ of the adjoint normal stress tensor, i.e.,

$$\nabla j(\mathbf{g}) = {}^*\gamma_{(0,\tau) \times \Gamma^c} \left[ \sigma(\mathbf{v}, \pi) \cdot \mathbf{n} \right] \qquad (1.13)$$

This formal proof provides the basic steps needed in order to build a gradient-based optimization method associated to the control problem (1.7).

An alternative approach consists in avoiding the derivation of the fluid state $(\mathbf{u}, p)$ with respect to the control $\mathbf{g}$ thanks to the introduction of a Lagrangian functional that includes not only the cost functional but also the state equation,

$$\mathcal{L}(\psi, r, \phi, q; g) = J(\psi, r) + \langle e(\psi, r; g), (\phi, q) \rangle \tag{1.14}$$

where $\langle e(\mathbf{u}, p; g), (\phi, q) \rangle$ stands for the weak form of the state equations (1.8), e.g.,

$$
\langle e(\psi, r; g), (\phi, q) \rangle = \int_{(0,\tau) \times \Omega^f} [-\psi \cdot \partial_t \phi + (\mathrm{D}\,\psi \cdot \psi) \cdot \phi - \nu\,\psi \cdot \Delta\phi
$$
$$
+ \psi \cdot \nabla q - r \operatorname{div} \phi] + \int_{(0,\tau) \times \Gamma^c} \mathbf{g} \cdot (\sigma(\phi, q) \cdot \mathbf{n})\, d\Gamma\, dt
$$
$$
+ \int_{(0,\tau) \times \partial D} \mathbf{u}_\infty \cdot (\sigma(\phi, q) \cdot \mathbf{n}) + \int_{\Omega^f} \psi(\tau) \cdot \phi(\tau) - \int_{\Omega^f} \mathbf{u}_0 \cdot \phi(0)
$$

Hence the control problem (1.7) is equivalent to the min-max problem,

$$\min_{g \in \mathcal{U}} \min_{(\psi, r)} \max_{(\phi, q)} \mathcal{L}(\psi, r, \phi, q; g) \tag{1.15}$$

For every control $\mathbf{g} \in \mathcal{U}$, it can be proven that the min-max problem

$$\min_{(\psi, r)} \max_{(\phi, q)} \mathcal{L}(\psi, r, \phi, q; g)$$

admits a unique saddle-point $(\mathbf{u}, p; \mathbf{v}, \pi)$ which are solutions of the systems (1.8)-(1.12). Finally the first-order optimality for the problem (1.15) writes

$$\partial_g \mathcal{L}(\mathbf{u}, p, \mathbf{v}, \pi; \mathbf{g}) = 0 \tag{1.16}$$

which turns out to be equivalent to (1.13).

Then, we can think to solve the optimality condition (1.11), using a continuous iterative method. Indeed let us introduce a scalar parameter $s \geq 0$, and a control variable $\mathbf{g}(s)$ that is differentiable with respect to $s$. Hence using the differentiability of $J(\mathbf{g})$, we get

$$J(\mathbf{g}(r)) - J(\mathbf{g}(0)) = \int_0^r \langle \nabla J(\mathbf{g}(s)), \mathbf{g}'(s) \rangle_{\mathcal{U}^*, \mathcal{U}}\, ds$$

Let us choose the control such that

$$\mathbf{g}'(s) + \mathbb{A}^{-1}(s) \nabla J(\mathbf{g}(s)) = 0,\ s \in (0, r) \tag{1.17}$$

where $\mathbb{A}$ stands for an appropriate duality operator, then the functional writes

$$J(\mathbf{g}(r)) - J(\mathbf{g}(0)) = -\int_0^r \langle |\nabla J(\mathbf{g}(s)|^2\, ds$$

That means that the control law (1.17) leads to a functional's decrease and is referred to as a continuous gradient based optimization method. Using a discretization of the parameter $s$ leads to a standard gradient-based method such as the conjugate-gradient or the quasi-Newton method depending on the choice of $\mathbb{A}(s)$.

## 1.4.2 Shape design of a fixed solid inside a fluid flow

We again consider the situation where a fixed solid is surrounded by a fluid flow. The shape control consists in finding the optimal shape of the solid that reduces some objective functional (e.g., the drag) under some perimeter, volume or curvature constraints. This optimization is an open-loop control since the shape of the obstacle is time-independent.

**The** *speed method*

In the previous section, we have been dealing with an optimization problem where the control belongs to a linear space. Here the space of shapes is no more a linear space and the associated differential calculus becomes more tricky.

Our goal is to build gradient-based methods in order to find the optimal shape, i.e., we would like to solve the following problem,

$$\min_{\Omega \in \mathcal{A}} J(\Omega) \tag{1.18}$$

In order to carry out the sensitivity analysis of functionals depending on the shape of the solid $\Omega$ , we introduce a family of pertubated domains $\Omega_s \subset D$ parametrized by a scalar parameter $0 \le s \le \varepsilon$. These domains are the images of the original domain $\Omega$ through a given family of smooth maps[3] $\mathbf{T}_s : \bar{D} \to \bar{D}$, i.e.,

$$\Omega_s = \mathbf{T}_s(\Omega), \Gamma_s = \mathbf{T}_s(\Gamma)$$

Two major classes of such mappings are given by :

- the identity perturbation method ([116, 125]),

$$\mathbf{T}_s = \mathbf{I} + s\,\boldsymbol{\theta}$$

  where $\boldsymbol{\theta} : \bar{D} \to \bar{D}$.

- the *speed method* [145, 147], where the transformation is the flow associated to a given velocity field $\mathbf{V}(s, \mathbf{x})$,

$$\begin{cases} \partial_s \mathbf{T}_s(\mathbf{x}) = \mathbf{V}(s, \mathbf{T}_s(\mathbf{x})), & (s, \mathbf{x}) \in (0, \varepsilon) \times D \\ \mathbf{T}_{s=0}(\mathbf{x}) = \mathbf{x}, & \mathbf{x} \in D \end{cases}$$

---

[3]Typically we have the following Lipschitz regularity assumptions :

$$\mathbf{T}(., \mathbf{x}) \in \mathcal{C}^1([0, \varepsilon]; \mathbb{R}^3), \forall \mathbf{x} \in D$$

$$\|\mathbf{T}(., \mathbf{x}) - \mathbf{T}(., \mathbf{y})\|_{\mathcal{C}^0([0,\varepsilon];\mathbb{R}^3)} \le C\|\mathbf{x} - \mathbf{y}\|_{\mathbb{R}^3}$$

$$\mathbf{T}^{-1}(., \mathbf{x}) \in \mathcal{C}^0([0, \varepsilon]; \mathbb{R}^3), \forall \mathbf{x} \in D$$

$$\|\mathbf{T}^{-1}(., \mathbf{x}) - \mathbf{T}^{-1}(., \mathbf{y})\|_{\mathcal{C}^0([0,\varepsilon];\mathbb{R}^3)} \le C\|\mathbf{x} - \mathbf{y}\|_{\mathbb{R}^3}$$

where $C > 0$.

In order for $\bar{D}$ to be globally invariant under $\mathbf{T}_s(\mathbf{V})$ we need to impose the following viability conditions,

$$\mathbf{V}(s, \mathbf{x}) \cdot \mathbf{n}(\mathbf{x}) = 0, \ \mathbf{x} \in \partial D$$

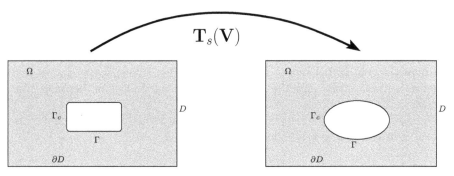

**FIGURE 1.2:** Shape deformation

Let us consider the family of functionals $J(\Omega_s)$ that depends on the shapes $\Omega_s$, e.g., the work to overcome the drag exerted by the fluid on the solid boundary,

$$J_{drag}(\Omega) = \int_{(0,\tau) \times \Gamma} (\mathbf{u} - \mathbf{u}_\infty) \cdot \sigma(\mathbf{u}, p) \cdot \mathbf{n} d\Gamma \ dt \tag{1.19}$$

This functional depends on $\Omega$ not only because it is an integral over the boundary $\Gamma$, but also because it involves the solution $(\mathbf{u}, p)$ of the Navier-Stokes system,

$$\begin{cases} \partial_t \mathbf{u} + \mathbf{D}\, \mathbf{u} \cdot \mathbf{u} - \nu\, \Delta \mathbf{u} + \nabla p = \mathbf{f}, & (0, \tau) \times \Omega \\ \operatorname{div} \mathbf{u} = 0, & (0, \tau) \times \Omega \\ \mathbf{u} = \mathbf{u}_\infty, & (0, \tau) \times \partial D \\ \mathbf{u} = 0, & (0, \tau) \times \Gamma \\ \mathbf{u}(t = 0) = \mathbf{u}_0, & \Omega \end{cases} \tag{1.20}$$

that depends on $\Omega$.

To perform our sensitivity analysis, we choose to work in the framework of the *speed method*[4]. We define the Eulerian derivative of the shape functional $J(\Omega)$ at point $\Omega$ in the direction of the vector field $\mathbf{V} \in \mathcal{V}$ as the limit,

$$dJ(\Omega; \mathbf{V}) = \lim_{s \searrow 0} \frac{J(\Omega_s(\mathbf{V})) - J(\Omega)}{s}$$

---

[4] which leads, at least for the first order terms, to the same results as the identity perturbation framework [51].

where $\mathcal{V}$ is a linear space[5]. If this limit exists and is finite $\forall\, \mathbf{V} \in \mathcal{V}$ and the mapping

$$\mathcal{V} \to \mathbb{R}$$
$$\mathbf{V} \mapsto dJ(\Omega; \mathbf{V})$$

is linear and continuous, then the functional $J(\Omega)$ is said to be shape differentiable.

Actually if $J(\Omega)$ is shape differentiable, then its Eulerian derivative only depends on $\mathbf{V}(0)$ and there exists a distribution $\mathbf{G}(\Omega) \in \mathcal{D}(D; \mathbb{R}^3)'$ that we call the shape gradient such that

$$dJ(\Omega; \mathbf{V}) = \langle \mathbf{G}(\Omega), \mathbf{V}(0) \rangle, \quad \forall\, \mathbf{V} \in \mathcal{V}$$

In the sequel, we shall use the notation $\nabla J(\Omega) \stackrel{\text{def}}{=} \mathbf{G}(\Omega)$.

In the case of smooth domain, the gradient is only supported on the boundary $\Gamma$ and depends linearly on the normal vector field $\mathbf{n}$. This result, called the structure theorem[6], is recalled as follows,

**Shape derivative structure theorem:** *Let $J(.)$ be a differentiable shape functional at every shape $\Omega$ of class $\mathcal{C}^{k+1}$ for $k \geq 0$ with shape gradient $G(\Omega) \in \mathcal{D}(D; \mathbb{R}^3)'$. In this case, the shape gradient has the following representation,*

$$G(\Omega) = {}^*\gamma_\Gamma(g\,\mathbf{n})$$

*where $g(\Gamma) \in \mathcal{D}^{-k}(\Gamma)$ stands for a scalar distribution and ${}^*\gamma_\Gamma$ stands for the adjoint trace operator[7].*

This result is easier to understand when $g(\Gamma)$ is integrable over $\Gamma$, that is to say, $g \in L^1(\Gamma)$. Indeed in this case, it means that the directional shape derivative can always be written as follows,

$$dJ(\Omega; \mathbf{V}) = \int_\Gamma g\,\mathbf{V} \cdot \mathbf{n}\, d\Gamma$$

**Basic shape derivative calculus**

In the previous paragraph, we have introduced the notion of Eulerian derivative for shape functionals. This notion can be extended to functions defined on Banach or Hilbert spaces built on smooth domains $\Omega$. Hence, a function

---

[5]e.g.,

$$\mathcal{V} \stackrel{\text{def}}{=} \{\mathbf{V} \in \mathcal{C}^0(0, \varepsilon; \mathcal{C}^1(D; \mathbb{R}^3)), \operatorname{div} \mathbf{V} = 0, \quad \text{in } D, \langle \mathbf{V}, \mathbf{n} \rangle = 0 \quad \text{on } \partial D\}$$

[6]We refer the reader to [51] Theorem 3.5 for the case of non-smooth domains.

[7]i.e.,

$$\langle {}^*\gamma_\Gamma(g\,\mathbf{n}), \mathbf{V} \rangle_{\mathcal{D}(D;\mathbb{R}^3)',\mathcal{D}(D;\mathbb{R}^3)} = \langle g, \mathbf{V} \cdot \mathbf{n} \rangle_{\mathcal{D}'(\Gamma),\mathcal{D}(\Gamma)}$$

$y \in H(\Omega)$ admits a material derivative at $\Omega$ in the direction $\mathbf{V} \in \mathcal{V}$ if the following limit

$$\dot{y}(\Omega; \mathbf{V}) \overset{\text{def}}{=} \lim_{s \to 0} \frac{1}{s} \left[ y(\Omega_s(\mathbf{V})) \circ \mathbf{T}_s(\mathbf{V}) - y(\Omega) \right]$$

admits a limit in the Hilbert space[8] $H(\Omega)$.

Endowed with the following definition, it is possible to derive the Eulerian shape derivative of the following functionals,

$$J(\Omega) = \int_\Omega y(\Omega) \, d\Omega$$

If $y(\Omega)$ is weakly shape differentiable in $L^1(\Omega)$, then the functional $J(\Omega)$ is shape differentiable and its directional derivate writes,

$$dJ(\Omega; \mathbf{V}) = \int_\Omega \left[ \dot{y}(\Omega; \mathbf{V}) + y(\Omega) \operatorname{div} \mathbf{V}(0) \right] d\Omega$$

In order to apply the structure theorem, it is useful to define the notion of shape derivative for functions. Hence, if $y \in H(\Omega)$ admits a material derivative $\dot{y}(\Omega; \mathbf{V}) \in H(\Omega)$ and $\nabla y \cdot \mathbf{V}(0) \in H(\Omega)$ for all $\mathbf{V} \in \mathcal{V}$, we define the shape derivative as

$$y'(\Omega; \mathbf{V}) = \dot{y}(\Omega; \mathbf{V}) - \nabla y(\Omega) \cdot \mathbf{V}(0)$$

In this case, the Eulerian shape derivative of $J(\Omega)$ takes the following form,

$$dJ(\Omega; \mathbf{V}) = \int_\Omega \left[ y'(\Omega; \mathbf{V}) + \operatorname{div}(y(\Omega) \mathbf{V}(0)) \right] d\Omega$$

If $\Omega$ is class $\mathcal{C}^k$ with $k \geq 1$, then using the Stokes formula, we get

$$dJ(\Omega; \mathbf{V}) = \int_\Omega y'(\Omega; \mathbf{V}) \, d\Omega + \int_\Gamma y(\Omega) \mathbf{V} \cdot \mathbf{n} \, d\Gamma$$

**REMARK 1.1**    In the case where $y(\Omega) = Y|_\Omega$, where $Y \in H(D)$ with $\Omega \subset D$, its shape derivative is zero since $\dot{y}(\Omega; \mathbf{V}) = \nabla Y \cdot \mathbf{V}$. Hence,

$$dJ(\Omega; \mathbf{V}) = \int_\Gamma y(\Omega) \mathbf{V} \cdot \mathbf{n} \, d\Gamma$$

This is a simple illustration of the structure theorem.                    ▯

In the case of functionals involving integration over the boundary $\Gamma$, we need to introduce the notion of material derivative on $\Gamma$.

---

[8]e.g., $W^{m,p}(\Omega)$ or $L^2(0, \tau; W^{m,p}(\Omega))$.

Let $z \in W(\Gamma)$ where $W(\Gamma)$ is an Hilbert space of functions[9] defined over $\Gamma$. It is said that it admits a material derivative in the direction $\mathbf{V} \in \mathcal{V}$, if the following limit,

$$\dot{z}(\Gamma; \mathbf{V}) \stackrel{\text{def}}{=} \lim_{s \to 0} \frac{1}{s} \left[ z(\Gamma_s(\mathbf{V})) \circ \mathbf{T}_s(\mathbf{V}) - z(\Gamma) \right]$$

admits a limit in the Hilbert space $W(\Gamma)$.

As a consequence of this definition, it is possible to derive the Eulerian shape derivative of the following functional,

$$J(\Gamma) = \int_{\Gamma} z(\Gamma) \, d\Gamma$$

If $z(\Gamma)$ is weakly shape differentiable in $L^1(\Gamma)$, then the functional $J(\Gamma)$ is shape differentiable and its directional derivate writes

$$dJ(\Gamma; \mathbf{V}) = \int_{\Gamma} \left[ \dot{z}(\Gamma; \mathbf{V}) + z(\Gamma) \operatorname{div}_{\Gamma} \mathbf{V}(0) \right] d\Gamma$$

where

$$\operatorname{div}_{\Gamma} \mathbf{V} \stackrel{\text{def}}{=} \gamma_{\Gamma} \left[ \operatorname{div} \mathbf{V} - (\mathbf{D}\mathbf{V} \cdot \mathbf{n}) \cdot \mathbf{n} \right]$$

stands for the tangential divergence.

As in the previous case, it is also possible to introduce the notion of shape derivative for $z(\Gamma)$. Let $\Omega$ be of class $\mathcal{C}^k$ with $k \geq 2$. If $z \in W(\gamma)$ admits a material derivative $\dot{z}(\Gamma; \mathbf{V}) \in W(\Gamma)$ and $\nabla_{\Gamma} y \cdot \mathbf{V}(0) \in W(\Gamma)$ for all $\mathbf{V} \in \mathcal{V}$, we define the shape derivative as

$$z'(\Gamma; \mathbf{V}) = \dot{z}(\Gamma; \mathbf{V}) - \nabla_{\Gamma} z(\Gamma) \cdot \mathbf{V}(0)$$

where

$$\nabla_{\Gamma} z = \nabla Z|_{\Gamma} - (\nabla Z \cdot \mathbf{n}) \mathbf{n}$$

stands for the tangential gradient and $Z$ is any smooth extension of $z$ inside $\Omega$.

Using the above definition, it is possible to transform the expression of the differential as follows,

$$dJ(\Gamma; \mathbf{V}) = \int_{\Gamma} \left[ z'(\Gamma; \mathbf{V}) + H z(\Gamma) \mathbf{V}(0) \cdot \mathbf{n} \right] d\Gamma$$

where $H$ stands for the mean curvature of $\Gamma$.

**REMARK 1.2**  In the case where $z(\Gamma) = y(\Omega)|_{\Gamma}$, the Eulerian derivative takes the following form,

$$dJ(\Gamma; \mathbf{V}) = \int_{\Gamma} \left[ y'(\Omega; \mathbf{V}) + (\nabla y(\Omega) \cdot \mathbf{n} + H z(\Gamma) \mathbf{V}(0) \cdot \mathbf{n} \right] d\Gamma$$

⬚

---

[9] e.g., $W^{m,p}(\Gamma)$.

**Application to shape design**

Thanks to the framework introduced previously, it is possible to build a complete sensitivity analysis of shape functionals. Coming back to our optimal shape problem, we can state the following:
the shape gradient for the tracking functional

$$J(\Omega) = \int_{(0,\tau) \times \Omega} (\mathbf{u} - \mathbf{u}_d)^2 \, dx \, dt$$

is given by

$$\nabla J(\Omega) = {}^*\gamma_\Gamma \big[ \sigma(\mathbf{v}, \pi) \cdot \mathbf{n} \big] \tag{1.21}$$

where $(\mathbf{u}, p)$ is a solution of system (1.20) associated to the shape $\Omega$ and $(\mathbf{v}, \pi)$ is solution of the adjoint system,

$$\begin{cases} -\partial_t \mathbf{v} - D\,\mathbf{v} \cdot \mathbf{u} + {}^*D\,\mathbf{u} \cdot \mathbf{v} - \nu \, \Delta \mathbf{v} + \nabla \pi = -(\mathbf{u} - \mathbf{u}_d), & (0,\tau) \times \Omega \\ \operatorname{div} \mathbf{v} = 0, & (0,\tau) \times \Omega \\ \mathbf{v} = 0, & (0,\tau) \times \partial D \\ \mathbf{v} = 0, & (0,\tau) \times \Gamma \\ \mathbf{v}(t = \tau) = 0, & \Omega \end{cases} \tag{1.22}$$

**The associated shape differential equation**

Now as in the previous section, we can choose to solve the first-order optimality equation (1.21) using a continuous gradient-based method. That means that we write

$$J(\Omega_r(\mathbf{V})) - J(\Omega_0) = \int_0^r \langle \nabla J(\Omega_s(\mathbf{V})), \mathbf{V}(s) \rangle \, ds$$

Then solving the equation

$$\nabla J(\mathbf{V}(s)) + \mathbb{A}^{-1}(s) \cdot \mathbf{V}(s) = 0, \; s \in (0, +\infty) \tag{1.23}$$

leads to a decrease of the functional $J(\Omega_s(\mathbf{V}))$. The equation (1.23) is referred to as the shape differential equation and some of its properties are studied in chapter 4 of this book. Notably, we study its solvability in the case of smooth shape functionals. We also prove some results concerning the asymptotic behaviour of the solution of this equation, which hold essentially when the shape gradient has some continuity properties for an *ad-hoc* shape topology[10].

**The level-set framework**

In chapter 4, we also relate the shape differential equation to the Hamilton-Jacobi equation involved in the level-set setting. The level-set setting consists

---

[10]The Hausdorff-complementary topology.

in parametrizing the perturbed domain $\Omega_s$ as the positiveness set of a scalar function $\Phi : (0, \varepsilon) \times \bar{D} \to \mathbb{R}$,

$$\Omega_s = \Omega_s(\Phi) \overset{\text{def}}{=} \{\mathbf{x} \in D, \quad \Phi(s, \mathbf{x}) > 0\}$$

and its boundary is the zero-level set,

$$\Gamma_s = \Gamma_s(\Phi) \overset{\text{def}}{=} \{\mathbf{x} \in D, \quad \Phi(s, \mathbf{x}) = 0\}$$

This parametrization and the one introduced in the *speed method* can be linked thanks to the following identity,

$$\mathbf{V}(s) = -\partial_s \Phi(s) \frac{\nabla \Phi(s)}{\|\nabla \Phi(s)\|^2}$$

Both frameworks are equivalent if $\Phi(s)$ belongs to set of functions without steps, which means that $\|\nabla \Phi(s)\|$ is different from zero almost everywhere in $D$. We show how to build without step functions and we study the shape differential equation in this setting.

### 1.4.3 Dynamical shape design of a solid inside a fluid flow

We return to our model problem, but this time we consider that the shape of the solid is moving. Our goal is to control this motion in order to optimize some objective functionals. Our goal is to build gradient-based methods in order to find the optimal shape dynamic, i.e., we would like to solve the following problem,

$$\min_{Q \in \mathcal{E}} J(Q) \tag{1.24}$$

where $Q \in \mathcal{E}$ is a smooth evolution set, which means

$$Q \overset{\text{def}}{=} \bigcup_{t \in (0, \tau)} \{t\} \times \Omega_t$$

where $\Omega_t$ is a smooth domain of $\mathbb{R}^3$ with boundary $\Gamma_t$. The set,

$$\Sigma \overset{\text{def}}{=} \bigcup_{t \in (0, \tau)} \{t\} \times \Gamma_t$$

stands for the non-cylindrical lateral boundary. We call the set $Q$ a tube.

### The $\mathbb{R}^{d+1}$-approach

The optimal control of moving domain is a problem which is relevant of classical (non-linear) control theory as well as of classical shape theory. In fact on the pure theoretical level, the dynamical shape control theory can

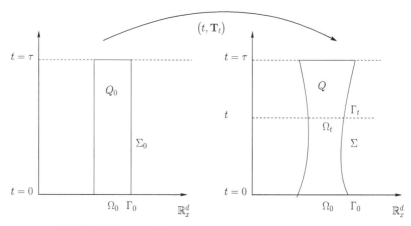

**FIGURE 1.3:**   Non-cylindrical space-time domain

be viewed as an application of the shape optimization theory for space-time manifolds.

Indeed the dynamical shape control consists in finding the optimal evolution of a spatial domain $\Omega_t$ in $\mathbb{R}^d$. Let us consider the mapping

$$\mathcal{S} : t \in \mathbb{R} \longrightarrow \Omega_t \in \mathcal{P}(\mathbb{R}^d)$$

where $\mathcal{P}(\mathbb{R}^d)$ stands for the set of parts inside $\mathbb{R}^d$.

Usually we would like to minimize some cost functional,

$$j(\mathcal{S}) = \int_0^\tau J(t, \mathcal{S}(t)) \, dt$$

Obviously, it is equivalent to the problem of finding the optimal tube

$$Q = \bigcup_{0 < t < \tau} \{t\} \times \Omega_t \in \mathbb{R}^{d+1}.$$

In fact the tube $Q$ is the graph in $\mathbb{R} \times \mathcal{P}(\mathbb{R}^d) \subset \mathbb{R}^{d+1}$ of the shape mapping $\mathcal{S}$. As for usual mappings defined from $\mathbb{R}$ in some space $E$, the graph $G \subset \mathbb{R} \times E$ and of course any subset $G$ is not a graph. Now under simple conditions on that set $G$, it becomes a graph. In the same way, any subset $Q \in \mathbb{R} \times \mathcal{P}(\mathbb{R}^d)$ will not be a tube. Intuitively we would say that we require some *causality* in the evolution of the set $\Omega_t$.

When the boundary of the set $\Omega_t$ is smooth enough [11], the idea is to avoid the normal field $\nu$ to the lateral boundary $\Sigma$ of the tube to be strictly vertical. To handle non-smooth situations, we adopt an Eulerian viewpoint that associates

---

[11]say there exists a tangent space.

to each tube $Q$ the non-empty closed convex set of speed vector fields $\mathbf{V}$ which transport (in a weak sense) the characteristic function of the moving domain. Wen we consider the tube $Q$ as a subset of $\mathbb{R}^{d+1}$, the control problem becomes a usual shape optimization problem ( as far as no real time consideration enters). The sensitivity analysis is then classically performed by considering *horizontal* vector fields

$$\tilde{\mathbf{Z}}(s,t,x) = (0, \mathbf{Z}(s,t,x)) \in \mathbb{R}^{d+1}$$

where $s$ is the perturbation parameter of the tube.

Then the $d+1$ dimensional shape optimization analysis fully applies and the so called *Shape differential Equation* furnishes descent direction, i.e., it furnishes the existence of a vector field $\mathbf{Z}^*$ such that, for some $\alpha > 0$,

$$J(Q_s) \le J(Q) - \alpha \int_0^s \|\mathbf{Z}^*(\sigma)\|^2 \, d\sigma$$

$\forall s > 0$.

We show that the existence of that field $\mathbf{Z}^*$ induces the existence of a usual vector field $\mathbf{V}(t,x) \in \mathbb{R}^d$ which builds that tube, i.e., $\Omega_t = \mathbf{T}(V)(\Omega_0)$.

## The $\mathbb{R}^d$-approach

In order to carry out the sensitivity analysis of functionals depending on the tube $Q$ , we assume that the domains are the images of the domain $\Omega_0 \overset{\text{def}}{=} \Omega_{t=0}$ through a given family of smooth maps $\mathbf{T}_t : \bar{D} \to \bar{D}$ , i.e.,

$$\Omega_t = \mathbf{T}_t(\Omega_0), \Gamma_t = \mathbf{T}_t(\Gamma_0)$$

Two major class of such mappings are given by :

- the Lagrangian parametrization,

$$\mathbf{T}_t = \boldsymbol{\theta}(t,.)$$

  where $\boldsymbol{\theta} : (0,\tau) \times \bar{D} \to \bar{D}$. In this case, the minimization problem (1.24) can be transformed as

$$\min_{\boldsymbol{\theta} \in \Theta} J(Q(\boldsymbol{\theta})) \tag{1.25}$$

- the Eulerian parametrization, where the transformation is the flow associated to a given velocity field $\mathbf{V}(t,\mathbf{x})$,

$$\begin{cases} \partial_t \mathbf{T}_t(\mathbf{x}) = \mathbf{V}(t, \mathbf{T}_t(\mathbf{x})), & (t,\mathbf{x}) \in (0,\tau) \times D \\ \mathbf{T}_{t=0}(\mathbf{x}) = \mathbf{x}, & \mathbf{x} \in D \end{cases}$$

  In this case, the minimization problem (1.24) can be transformed as

$$\min_{\mathbf{V} \in \mathcal{V}} J(Q(\mathbf{V})) \tag{1.26}$$

**Existence of tubes**

In the smooth case, the existence of tubes follows the Cauchy-Lipschitz theory on differential equations [147, 51]. In the non-smooth case, the Lipschitz regularity of the velocities $\mathbf{V}$ can be weakened using the equations satisfied by the characteristic functions $\xi(t, \mathbf{x})$ associated to the domain $\Omega_t(\mathbf{V})$,

$$\begin{cases} \partial_t \xi + \nabla \xi \cdot \mathbf{V} = 0, & (0, \tau) \times D \\ \xi_{t=0} = \chi_\Omega, & D \end{cases} \tag{1.27}$$

We shall consider velocity fields such that $\mathbf{V} \in L^1(0, \tau; L^2(D; \mathbb{R}^d))$ and the divergence positive part $(\operatorname{div} \mathbf{V})^+ \in L^1(0, \tau; L^\infty(D))$. In this case, using a Galerkin approximation and some energy estimates, we are able to derive an existence result of solutions with initial data given in $H^{-1/2}(D)$. For the time being, no uniqueness result has been obtained for this smoothness level.

Actually, when the field $\mathbf{V}$ and its divergence are simply $L^1$ functions, the notion of weak solutions associated to the convection problems (1.27) does not make sense. In this case, the correct modeling tool for shape evolution is to introduce the product space of elements $(\xi = \xi^2, \mathbf{V})$ equipped with a parabolic BV like topology for which the constraint (1.27) defines a closed subset $\mathcal{T}_\Omega$ which contains the weak closure of smooth elements

$$\mathcal{T}_\Omega \overset{\text{def}}{=} \left\{ (\chi_\Omega \circ \mathbf{T}_t^{-1}(\mathbf{V}), \mathbf{V}) \mid \mathbf{V} \in \mathcal{U}_{ad} \right\}$$

This approach consists in handling characteristic functions $\xi = \xi^2$ which belongs to $L^1(0, \tau; \mathrm{BV}(D))$ together with vector fields $\mathbf{V} \in L^2(0, \tau; L^2(D, \mathbb{R}^d))$ solution of problem (1.27). For a given element $(\xi, \mathbf{V}) \in \mathcal{T}_\Omega$, we consider the set of fields $\mathbf{W}$ such that $(\xi, \mathbf{W}) \in \mathcal{T}_\Omega$. It forms a closed convex set, noted $\mathcal{V}_\xi$. Hence, we can define the unique minimal norm energy element $\mathbf{V}_\xi$ in the convex set $\mathcal{V}_\xi$. For a given tube $\xi$, the element $\mathbf{V}_\xi$ is the unique (with minimal norm) vector field associated to $\xi$ via the convection equation (1.27).

We choose to adopt a different point of view inspired by the optimization problems framework. Indeed, our final goal is to apply the weak set evolution setting to the control problem arising in various fields such as free boundary problems or image processing. The usual situation can be described as follows. Let us consider a given smooth enough functional $J(\xi, \mathbf{V})$. We would like to solve the following optimization problem

$$\inf_{(\xi, \mathbf{V}) \in \mathcal{T}_\Omega} J(\xi, \mathbf{V}) \tag{1.28}$$

The space $\mathcal{U}_{ad}$ is a space of smooth velocities.

In most situations, such a problem does not admit solutions and we need to add some regularization terms to ensure its solvability. Consequently, we shall introduce different penalization terms which furnish compactness properties of the minimizing sequences inside an ad-hoc weak topology involving bounded variation constraints. Then, the new problem writes

$$\inf_{(\xi, \mathbf{V}) \in \mathcal{T}_\Omega} J(\xi, \mathbf{V}) + F(\xi, \mathbf{V}) \tag{1.29}$$

The penalization term $F(\xi, \mathbf{V})$ can be chosen using several approaches:

- We can first consider the time-space perimeter of the lateral boundary $\Sigma$ of the tube, developed in [155]. This approach easily draws part of the variational properties associated to the bounded variation functions space framework. In particular, it uses the compactness properties of tube family with bounded perimeters in $\mathbb{R}^{d+1}$. Nevertheless, this method leads to heavy variational analysis developments.

- We can rather consider the time integral of the spatial perimeter of the moving domain which builds the tube, as introduced in [157]. We shall extend these results to the case of vector fields living in $L^2((0, \tau) \times D; \mathbb{R}^d)$. In this case, only existence results for solutions of the convection equation can be handled and the uniqueness property is lost.

## Tube derivative

In this paragraph, we are interested in differentiability properties of integrals defined over moving domains,

$$J(Q(\mathbf{V})) = \int_{Q(\mathbf{V})} f(\mathbf{V}) \, dx \, dt$$

The transverse map $\mathcal{T}_\rho^t$ associated to two vector fields $(\mathbf{V}, \mathbf{W}) \in \mathcal{U}$ is defined as follows,

$$\mathcal{T}_\rho^t : \overline{\Omega_t} \longrightarrow \overline{\Omega_t^\rho} \overset{\text{def}}{=} \overline{\Omega_t(\mathbf{V} + \rho \mathbf{W})}$$
$$x \longmapsto T_t(\mathbf{V} + \rho \mathbf{W}) \circ T_t(\mathbf{V})^{-1}$$

**REMARK 1.3** The transverse map allows us to perform sensitivity analysis on functions defined on the unperturbed domain $\Omega_t(\mathbf{V})$. ◻

The following result states that the transverse map $\mathcal{T}_\rho^t$ can be considered as a dynamical flow with respect to the perturbation variable $\rho$. The transverse map $\mathcal{T}_\rho^t$ is the flow of a transverse field $\mathcal{Z}_\rho^t$ defined as follows:

$$\mathcal{Z}_\rho^t \overset{\text{def}}{=} \mathcal{Z}^t(\rho, .) = \left( \frac{\partial \mathcal{T}_\rho^t}{\partial \rho} \right) \circ (\mathcal{T}_\rho^t)^{-1} \tag{1.30}$$

i.e., is the solution of the following dynamical system :

$$T_t^\rho(\mathcal{Z}_\rho^t) : \overline{\Omega_t} \longrightarrow \overline{\Omega_t^\rho}$$
$$x \longmapsto x(\rho, x) \equiv T_t^\rho(\mathcal{Z}_\rho^t)(x)$$

with

$$\frac{dx(\rho)}{d\rho} = \mathcal{Z}^t(\rho, x(\rho)), \rho \geq 0$$
$$x(\rho = 0) = x, \qquad \text{in } \Omega_t(\mathbf{V}) \tag{1.31}$$

Since, we will mainly consider derivatives of perturbed functions at point $\rho = 0$, we set $\mathbf{Z}(t) \stackrel{\text{def}}{=} \mathcal{Z}^t_{\rho=0}$. A fundamental result lies in the fact that $\mathbf{Z}$ can be obtained as the solution of a linear time dynamical system depending on the vector fields $(\mathbf{V}, \mathbf{W}) \in \mathcal{U}$.

The vector field $\mathbf{Z}$ is the unique solution of the following Cauchy problem,

$$\begin{cases} \partial_t \mathbf{Z} + [\mathbf{Z}, \mathbf{V}] = \mathbf{W}, & (0, \tau) \times D \\ \mathbf{Z}_{t=0} = 0, & D \end{cases} \qquad (1.32)$$

where $[\mathbf{Z}, \mathbf{V}] \stackrel{\text{def}}{=} \mathrm{D}\, Z \cdot \mathbf{V} - \mathrm{D}\, \mathbf{V} \cdot \mathbf{Z}$ stands for the Lie bracket of the pair $(\mathbf{Z}, \mathbf{V})$. The derivative with respect to $\rho$ at point $\rho = 0$ of the following composed function,

$$f^\rho : [0, \rho_0] \to H(\Omega_t(\mathbf{V}))$$
$$\rho \mapsto f(\mathbf{V} + \rho \mathbf{W}) \circ T^t_\rho$$

$\dot{f}(\mathbf{V}; \mathbf{W})$ is called the eulerian material derivative of $f(\mathbf{V})$ at point $\mathbf{V} \in U$ in the direction $\mathbf{W} \in \mathcal{U}$. We shall use the notation,

$$\dot{f}(\mathbf{V}) \cdot \mathbf{W} = \dot{f}(\mathbf{V}; \mathbf{W}) \stackrel{\text{def}}{=} \frac{d}{d\rho} f^\rho \Big|_{\rho=0}$$

With the above definition, we can state the differentiability properties of non-cylindrical integrals with respect to their moving support.

For a bounded measurable domain $\Omega_0$ with boundary $\Gamma_0$, let us assume that for any direction $\mathbf{W} \in U$ the following hypothesis holds,

i) $f(\mathbf{V})$ admits an eulerian material derivative $\dot{f}(\mathbf{V}) \cdot \mathbf{W}$

then $J(\mathbf{V})$ is Gâteaux differentiable at point $\mathbf{V} \in \mathcal{U}$ and its derivative is given by the following expression,

$$J'(\mathbf{V}) \cdot \mathbf{W} = \int_{\Omega_t(\mathbf{V})} \left[ \dot{f}(\mathbf{V}) \cdot \mathbf{W} + f(\mathbf{V}) \operatorname{div} \mathbf{Z} \right] d\Omega \qquad (1.33)$$

Futhermore, if

ii) $f(\mathbf{V})$ admits an Eulerian shape derivative given by the following expression,

$$f'(\mathbf{V}) \cdot \mathbf{W} = \dot{f}(\mathbf{V}) \cdot \mathbf{W} - \nabla f(\mathbf{V}) \cdot \mathbf{Z} \qquad (1.34)$$

then

$$J'(\mathbf{V}) \cdot \mathbf{W} = \int_{\Omega_t(\mathbf{V})} [f'(\mathbf{V}) \cdot \mathbf{W} + \operatorname{div}(f(\mathbf{V}) \mathbf{Z})] \, d\Omega \qquad (1.35)$$

Furthermore, if $\Omega_0$ is an open domain with a Lipschitzian boundary $\Gamma_0$, then

$$J'(\mathbf{V}) \cdot \mathbf{W} = \int_{\Omega_t(\mathbf{V})} f'(\mathbf{V}) \cdot \mathbf{W} d\Omega + \int_{\Gamma_t(\mathbf{V})} f(\mathbf{V}) \mathbf{Z} \cdot \mathbf{n} \, d\Gamma \qquad (1.36)$$

**REMARK 1.4**    The last identity will be of great interest while trying to prove a gradient structure result for general non-cylindrical functionals.    ▯

It is possible to define the solution of the adjoint transverse system. For $\mathbf{F} \in L^2(0, \tau; (H^1(D))^d)$, there exists a unique field

$$\mathbf{\Lambda} \in \mathcal{C}^0([0, \tau]; (L^2(D))^d)$$

solution of the backward dynamical system,

$$\begin{cases} -\partial_t \mathbf{\Lambda} - D\,\mathbf{\Lambda} \cdot \mathbf{V} - {}^*D\,\mathbf{V} \cdot \mathbf{\Lambda} - (\operatorname{div} \mathbf{V}) \mathbf{\Lambda} = \mathbf{F}, \ (0, \tau) \\ \mathbf{\Lambda}(\tau) = 0 \end{cases} \tag{1.37}$$

**REMARK 1.5**    The field $\mathbf{\Lambda}$ is the dual variable associated to the transverse field $\mathbf{Z}$ and is a solution of the adjoint problem associated to the transverse dynamical system.    ▯

Actually, we shall deal with specific right-hand sides $\mathbf{F}$ of the form $\mathbf{F}(t) = {}^*\gamma_{\Gamma_t(\mathbf{V})}(f(t)\,\mathbf{n})$. In this case, the adjoint field $\mathbf{\Lambda}$ is supported on the moving boundary $\Gamma_t(\mathbf{V})$ and has the following structure,

$$\mathbf{\Lambda} = {}^*\gamma_{\Sigma(\mathbf{V})}(\lambda\,\mathbf{n}) \tag{1.38}$$

where $\lambda$ is the unique solution of the following boundary dynamical system,

$$\begin{cases} -\partial_t \lambda - \nabla_\Gamma \lambda \cdot \mathbf{V} - (\operatorname{div} \mathbf{V}) \lambda = f, \ \Sigma(\mathbf{V}) \\ \lambda(\tau) = 0, \qquad\qquad\qquad\qquad \Gamma_\tau(\mathbf{V}) \end{cases} \tag{1.39}$$

$p$ is the canonical projection on $\Gamma_t(\mathbf{V})$ and $\chi_{\Omega_t(\mathbf{V})}$ is the characteristic function of $\Omega_t(\mathbf{V})$ inside $D$.

Then in chapter 2 , we will establish the following adjoint identity,

$$\int_0^\tau \int_{\Gamma_t(\mathbf{V})} E\,\mathbf{Z} \cdot \mathbf{n} = -\int_0^\tau \int_{\Gamma_t(\mathbf{V})} \lambda\,\mathbf{W} \cdot \mathbf{n} \tag{1.40}$$

where $\lambda$ is the unique of problem (1.39) with $f = E$.

### Eulerian dynamical shape control

We come back to our design problem.  We would like to minimize the following tracking functional with respect to $\mathbf{V}$,

$$j(\mathbf{V}) = \int_{Q(\mathbf{V})} |\mathbf{u} - \mathbf{u}_d|^2 \tag{1.41}$$

where $(\mathbf{u}, p)$ is a solution of the following system,

$$\begin{cases} \partial_t \mathbf{u} + D\,\mathbf{u} \cdot \mathbf{u} - \nu\,\Delta \mathbf{u} + \nabla p = 0, \ Q(\mathbf{V}) \\ \operatorname{div} \mathbf{u} = 0, \qquad\qquad\qquad\quad Q(\mathbf{V}) \\ \mathbf{u} = \mathbf{u}_\infty, \qquad\qquad\qquad\quad (0, \tau) \times \partial D \\ \mathbf{u} = \mathbf{V}, \qquad\qquad\qquad\qquad \Sigma(\mathbf{V}) \\ \mathbf{u}(t = 0) = \mathbf{u}_0, \qquad\qquad \Omega_0 \end{cases} \tag{1.42}$$

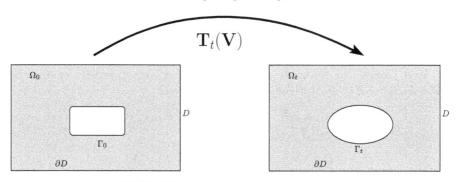

**FIGURE 1.4**: Shape optimization

Using the Eulerian shape analysis framework rapidly sketched in the previous section, we will prove in chapter 5 that this functional is differentiable and its gradient has the following form,

$$\nabla j(\mathbf{V}) = {}^*\gamma_{\Sigma(\mathbf{V})}\big[ - \lambda\,\mathbf{n} - \sigma(\mathbf{v}, \pi) \cdot \mathbf{n}\big] \tag{1.43}$$

where $(\mathbf{v}, \pi)$ stands for the adjoint fluid state solution of the following system,

$$\begin{cases} -\partial_t \mathbf{v} - \mathrm{D}\,\mathbf{v}\cdot\mathbf{u} + {}^*\mathrm{D}\,\mathbf{u}\cdot\mathbf{v} - \nu\,\Delta\mathbf{v} + \nabla\pi = -(\mathbf{u} - \mathbf{u}_d), & Q(\mathbf{V}) \\ \mathrm{div}(\mathbf{v}) = 0, & Q(\mathbf{V}) \\ \mathbf{v} = 0, & (0,\tau) \times \partial D \\ \mathbf{v} = 0, & \Sigma(\mathbf{V}) \\ \mathbf{v}(\tau) = 0, & \Omega_\tau(\mathbf{V}) \end{cases} \tag{1.44}$$

and $\lambda$ is the adjoint transverse boundary field, solution of the tangential dynamical system,

$$\begin{cases} -\partial_t\lambda - \nabla\lambda\cdot\mathbf{V} - (\mathrm{div}\,\mathbf{V})\,\lambda = f, & \Sigma(\mathbf{V}) \\ \lambda(\tau) = 0, & \Gamma_\tau(\mathbf{V}) \end{cases} \tag{1.45}$$

with

$$f = [-(\sigma(\mathbf{v}, \pi) \cdot \mathbf{n})] \cdot (\mathrm{D}\,\mathbf{V}\cdot\mathbf{n} - \mathrm{D}\,\mathbf{u}\cdot\mathbf{n}) + \frac{1}{2}|\mathbf{u} - \mathbf{u}_d|^2 \tag{1.46}$$

This result can be obtained using several techniques. One of them is the use of a min-max formulation involving a Lagrangian functional coupled with a function space embedding, particulary suited for non-homogeneous Dirichlet boundary problems. Let us give here the main steps of this proof. We rewrite the minimization problem as a min-max problem,

$$j(\mathbf{V}) = \min_{(\psi, r) \in X \times P} \max_{(\phi, q) \in Y \times Q} \mathcal{L}_{\mathbf{V}}(\psi, r; \phi, q) \tag{1.47}$$

with

$$\mathcal{L}_{\mathbf{V}}(\psi, r; \phi, q) = J_{\mathbf{V}}(\psi, r) - e_{\mathbf{V}}(\psi, r; \phi, q) \tag{1.48}$$

and

$$e_{\mathbf{V}}(\psi, r; \boldsymbol{\phi}, q) = \int_{Q(\mathbf{V})} [\partial_t \psi + D\,\psi \cdot \psi - \nu \Delta \psi + \nabla r] \cdot \boldsymbol{\phi} - \int_{Q(V)} q \operatorname{div} \psi$$
$$- \int_{\Sigma(\mathbf{V})} (\psi - \mathbf{V}) \cdot \sigma(\boldsymbol{\phi}, q) \cdot \mathbf{n}$$

stands for the weak fluid state operator. The state and multiplier variables are defined on the hold-all domain $D$, i.e.,

$$X = Y \stackrel{\text{def}}{=} H^1(0, \tau; H^2(D)), \; P = Q \stackrel{\text{def}}{=} H^1(0, \tau; H^1(D))$$

The saddle points $(\mathbf{u}, p; \mathbf{v}, \pi)$ are solutions of the first-order optimality conditions with respect to the multipliers $(\boldsymbol{\phi}, q)$ and the state variables $(\psi, r)$ which leads respectively to the primal system (1.20) and to the adjoint system (1.44).

The crucial point concerns the derivation with respect to the design variable $\mathbf{V} \in \mathcal{U}_{ad}$. We consider a perturbation vector field $W \in \mathcal{U}_{ad}$ with an increment parameter $\rho \geq 0$. Since the state and multiplier variables are defined in the hold-all domain $D$, the pertubed Lagrangian $\mathcal{L}^\rho$ only involves perturbed supports. Using the min-max derivation principle [51] recalled in chapter 5, then

$$\langle j'(\mathbf{V}), \mathbf{W} \rangle \stackrel{\text{def}}{=} \frac{d}{d\rho} j(\mathbf{V} + \rho \mathbf{W}) \Big|_{\rho=0} = \frac{d}{d\rho} \mathcal{L}^\rho \Big|_{\rho=0} (\mathbf{u}, p; \mathbf{v}, \pi) \quad (1.49)$$

We can state

$$\langle j'(\mathbf{V}), \mathbf{W} \rangle = \int_{\Sigma(\mathbf{V})} \left[ f\,\mathbf{Z} \cdot \mathbf{n} + \left( -\sigma(\mathbf{v}, \pi) \cdot \mathbf{n} \right) \cdot \mathbf{W} \right]$$

with $f$ given by equation (1.46) and where $\mathbf{Z}$ stands for the transverse vector field introduced previously. We finally use the transverse adjoint identity,

$$\int_{\Sigma(\mathbf{V})} f\,\mathbf{Z} \cdot \mathbf{n} = - \int_{\Sigma(\mathbf{V})} \lambda \mathbf{W} \cdot \mathbf{n}, \quad \forall \mathbf{W} \in \mathcal{U}_{ad}$$

where $\lambda$ is solution of equation (1.45) which corresponds to the adjoint system associated to the transverse dynamical system satisfied by $\mathbf{Z}$. This adjoint variable is only supported by the moving boundary $\Gamma_t(\mathbf{V})$ over $(0, \tau)$.

## Lagrangian dynamical shape control

In chapter 6, we use a Lagrangian parametrization of the shape motion. In this case, we prove that for $\mathbf{V} \stackrel{\text{def}}{=} \partial_t \theta \circ \theta^{-1} \in \mathcal{U}_{ad}$, the functional $j(\theta)$ possesses a gradient $\nabla j(\theta)$ which is supported on the moving boundary $\Sigma(\theta)$ and can be represented by the following expression,

$$\nabla j(\theta) = {}^*\gamma_{\Sigma(\theta)} \left[ \partial_t \mathbf{E} + (\operatorname{div}_\Gamma \mathbf{V}) \mathbf{E} + D\,\mathbf{E} \cdot \mathbf{V} + \mathbf{E} \cdot D\,\mathbf{u} + \frac{1}{2} |\mathbf{u} - \mathbf{u}_d|^2 \mathbf{n} \right] \quad (1.50)$$

where $\mathbf{E} = \sigma(\mathbf{v}, \pi) \cdot \mathbf{n}$ and $(\mathbf{v}, \pi)$ stands for the adjoint fluid state solution of the following system,

$$\begin{cases} -\partial_t \mathbf{v} - D\mathbf{v} \cdot \mathbf{u} + {}^*D\mathbf{u} \cdot \mathbf{v} - \nu\,\Delta\mathbf{v} + \nabla\pi = -(\mathbf{u} - \mathbf{u}_d), & Q(\theta) \\ \mathrm{div}(\mathbf{v}) = 0, & Q(\theta) \\ \mathbf{v} = 0, & (0, \tau) \times \partial D \quad (1.51) \\ \mathbf{v} = 0, & \Sigma(\theta) \\ \mathbf{v}(\tau) = 0, & \Omega_\tau \end{cases}$$

### 1.4.4 Control of an elastic solid inside a fluid flow

**A simple mechanical model**

We consider our model problem where now the solid is supposed to be a two dimensional elastically supported rigid motion. For the sake of simplicity, we only consider one degree of freedom for the structural motion : the vertical displacement $d(t)\,\mathbf{e}_2$ where $\mathbf{e}_2$ is the element of Cartesian basis $(\mathbf{e}_1, \mathbf{e}_2)$ in $\mathbb{R}^2$. As previously, we use an Eulerian parametrization, where the transformation is the flow associated to a given velocity field $\mathbf{V}(t, \mathbf{x})$,

$$\begin{cases} \partial_t \mathbf{T}_t(\mathbf{x}) = \mathbf{V}(t, \mathbf{T}_t(\mathbf{x})), & (t, \mathbf{x}) \in (0, \tau) \times D \\ \mathbf{T}_{t=0}(\mathbf{x}) = \mathbf{x}, & \mathbf{x} \in D \end{cases}$$

Hence, we have

$$\overline{\Omega_t^f} = \mathbf{T}_t(\overline{\Omega_0^f}), \; \overline{\Omega_t^s} = \mathbf{T}_t(\overline{\Omega_0^s})$$

Since we only consider one degree of freedom motion, we write

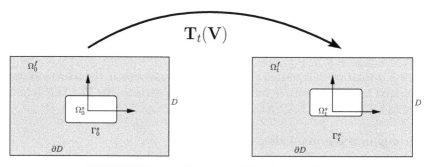

**FIGURE 1.5:** Rigid solid inside a fluid flow

$$\overline{\Omega_t^s} = \overline{\Omega_0^s} + d(t)\,\mathbf{e}_2$$

We set $\Sigma^s \equiv \bigcup\limits_{0 < t < \tau} (\{t\} \times \Gamma^s_t)$, $Q^f \equiv \bigcup\limits_{0 < t < \tau} (\{t\} \times \Omega^f_t)$ and $\Sigma^f \equiv (0, \tau) \times \partial D$

In our simple case, we can give an example of an appropriate flow vector field :

$$\begin{cases} \mathbf{V}(t, \mathbf{x}) = \dot{d}(t)\mathbf{e}_2, & \mathbf{x} \in \overline{\Omega^s_t} \\ \mathbf{V}(t, \mathbf{x}) = \mathrm{Ext}(\dot{d}(t)\mathbf{e}_2), & \mathbf{x} \in \Omega^f_t \\ \mathbf{V}(t, \mathbf{x}) \cdot \mathbf{n} = 0, & \mathbf{x} \in \partial D \end{cases} \tag{1.52}$$

where Ext is an arbitrary extension operator from $\Gamma^s_0$ into $\Omega^f_0$. The map $\mathbf{T}_t$ is usually referred to as the Arbitrary Euler-Lagrange map.

The solid is described by the evolution of its displacement and its velocity and the couple $(d, \dot{d})$ is the solution of the following ordinary second order differential equation :

$$\begin{cases} m\ddot{d} + kd = F_f, \\ \left[d, \dot{d}\right](t = 0) = [d_0, d_1] \end{cases} \tag{1.53}$$

where $(m, k)$ stand for the structural mass and stiffness. $F_f$ is the projection of the fluid loads on $\Gamma^s_t$ along the direction of motion $\mathbf{e}_2$.

The fluid is assumed to be a viscous incompressible Newtonian fluid. Its evolution is described by its velocity $\mathbf{u}$ and its pressure $p$. The couple $(\mathbf{u}, p)$ satisfies the classical Navier-Stokes equations written in non-conservative form

$$\begin{cases} \partial_t \mathbf{u} + \mathrm{D}\,\mathbf{u} \cdot \mathbf{u} - \nu\,\Delta\mathbf{u} + \nabla p = 0, & Q^f(\mathbf{V}) \\ \mathrm{div}(\mathbf{u}) = 0, & Q^f(\mathbf{V}) \\ \mathbf{u} = \mathbf{u}_\infty, & \Sigma^f \\ \mathbf{u}(t = 0) = \mathbf{u}_0, & \Omega^f_0 \end{cases} \tag{1.54}$$

where $\nu$ stands for the kinematic viscosity and $\mathbf{u}_\infty$ is the farfield velocity field. Hence, the projected fluid loads $F_f$ have the following expression :

$$F_f = -\left(\int_{\Gamma^s_t} \sigma(\mathbf{u}, p) \cdot \mathbf{n}\right) \cdot \mathbf{e}_2 \tag{1.55}$$

where $\sigma(\mathbf{u}, p) = -p\mathrm{I} + \nu\,(\mathrm{D}\,\mathbf{u} + {}^*\mathrm{D}\,\mathbf{u})$ stands for the fluid stress tensor inside $\Omega^f_t$.

We complete the whole system with kinematic continuity conditions at the fluid-structure interface $\Gamma^s_t$ :

$$\mathbf{u} = \mathbf{V} \stackrel{\mathrm{def}}{=} \dot{d}\,\mathbf{e}_2, \text{ on } \Sigma(\mathbf{V}) \tag{1.56}$$

To summarize, we get the following coupled system :

$$\begin{cases} \partial_t \mathbf{u} + \mathrm{D}\,\mathbf{u}\cdot\mathbf{u} - \nu\,\Delta\mathbf{u} + \nabla p = 0, & Q^f(\mathbf{V}) \\ \mathrm{div}(\mathbf{u}) = 0, & Q^f(\mathbf{V}) \\ \mathbf{u} = \mathbf{u}_\infty, & \Sigma^f \\ \mathbf{u} = \dot{d}\,\mathbf{e}_2, & \Sigma^s(\mathbf{V}) \\ m\,\ddot{d} + k\,d = -\left( \int_{\Gamma_t^s} \sigma(\mathbf{u}, p)\cdot\mathbf{n} \right)\cdot\mathbf{e}_2, & (0,\tau) \\ \left[\mathbf{u}, d, \dot{d}\right](t=0) = [\mathbf{u}_0, d_0, d_1], & \Omega_0^f \times \mathbb{R}^2 \end{cases} \tag{1.57}$$

**Optimization problem**

Here we choose the control variable as the far-field Dirichlet boundary condition on $\Sigma^f$. The goal is to minimize a functional depending on the state variables of the coupled fluid-structure problem,

$$\min_{\mathbf{u}_\infty \in \mathcal{U}_c} j(\mathbf{u}_\infty) \tag{1.58}$$

where $j(\mathbf{u}_\infty) = J_{\mathbf{u}_\infty}\left([\mathbf{u}, p, d](\mathbf{u}_\infty)\right)$ with $[\mathbf{u}, p, d](\mathbf{u}_\infty)$ is a weak solution of problem (1.57) and $J_{\mathbf{u}_\infty}$ is a real functional of the following form,

$$J_{\mathbf{u}_\infty}\left([\mathbf{u}, p, d](\mathbf{u}_\infty)\right) = \frac{1}{2}\int_0^\tau |d - d_g|^2\,\mathrm{d}t \tag{1.59}$$

This functional is a tracking functional. That means that we would like to control the motion of the solid thanks to the far-field boundary conditions[12]. The purpose is to illustrate how the moving shape analysis presented in the previous section can be applied to the analysis of optimization problems for various coupled systems involving a moving interface.

**Sensitivity analysis**

Using the Eulerian sensitivity analysis introduced in the last section, we prove in chapter 7 that the tracking functional is differentiable and its gradient takes the following form,

$$\nabla j(\mathbf{u}_\infty) = {}^*\gamma_{\Sigma^f}\left[\sigma(\mathbf{v}, \pi)\cdot\mathbf{n}\right] \tag{1.60}$$

with $(\mathbf{v}, \pi, b)$ solutions of the following adjoint system,

$$\begin{cases} -\partial_t\mathbf{v} - \mathrm{D}\,\mathbf{v}\cdot\mathbf{u} + {}^*\mathrm{D}\,\mathbf{u}\cdot\mathbf{v} - \nu\,\Delta\mathbf{v} + \nabla\pi = 0, & Q^f(\mathbf{V}) \\ \mathrm{div}(\mathbf{v}) = 0, & Q^f(\mathbf{V}) \\ \mathbf{v} = 0, & \Sigma^f \\ \mathbf{v} = b\,\mathbf{e}_2, & \Sigma^s(\mathbf{V}) \\ \mathbf{v}(\tau) = 0, & \Omega_\tau^f(\mathbf{V}) \end{cases} \tag{1.61}$$

---

[12] It can also be interpreted as an inverse problem.

$$
\begin{cases}
m\ddot{b} + k\,b = (d - d_g) + \partial_t \left( \int_{\Gamma_t^s(\mathbf{V})} \sigma(\mathbf{v}, \pi) \cdot \mathbf{n} \right) \cdot \mathbf{e}_2 \\[2ex]
+ \int_{\Gamma_t^s(\mathbf{V})} \left[ |\dot{d}|^2 \left( \mathbf{D}\mathbf{v} \cdot \mathbf{e}_2 \right) \cdot \mathbf{e}_2 - \nu \left( \mathbf{D}\mathbf{v} \cdot \mathbf{n} \right) \cdot \left( \mathbf{D}\mathbf{u} \cdot \mathbf{n} \right) \right] \cdot \mathbf{n} \ (0, \tau) \qquad (1.62) \\[2ex]
\left[ b, \dot{b} \right] (\tau) = [0, 0]
\end{cases}
$$

The proof of this result is based on the Min-Max derivation principle. The minimization problem can be put into a Min-Max formulation involving the Lagrangian functional defined as follows,

$$
\mathcal{L}_{\mathbf{u}_\infty} (\mathbf{u}, p, d_1, d_2; \mathbf{v}, \pi, b_1, b_2) \overset{\text{def}}{=} \\
J_{\mathbf{u}_\infty} (\mathbf{u}, p, d_1, d_2) - \langle e_{\mathbf{u}_\infty} (\mathbf{u}, p, d_1, d_2), (\mathbf{v}, \pi, b_1, b_2) \rangle
$$

where $e_{\mathbf{u}_\infty} (\mathbf{u}, p, d_1, d_2)$ stands for the global weak state operator and where the displacement $d_1$ and the velocity $d_2$ are considered as independent variables. Then derivating first-order optimality conditions with respect to the state variables $(\mathbf{u}, p, d_1, d_2)$ leads to the fluid and solid adjoint systems (1.61)-(1.62), where the adjoint state $b_1$ has been eliminated to the benefit of $b \overset{\text{def}}{=} b_2$.

The keystone is the sensitivity of the Navier-Stokes system with respect to pertubations of the solid velocity $\mathbf{V} = d_2\,\mathbf{e}_2$ as described in the last section. It involves the transverse adjoint state $\lambda$ as an intermediate sensitivity multiplier that can be eventually eliminated. Hence the expression of the cost function gradient in equation (1.60) is obtained using a bypass through the Min-Max problem.

## A general coupled fluid-structure model

In chapter 8, we consider a more general model where the solid is supposed to be a 3D non-linear elastic solid. For the sake of simpleness, the reader is referred to this chapter for more details since it requires a heavier analysis.

# Chapter 2

# An introductory example : the inverse Stefan problem

In this chapter, we shall introduce various concepts that will be further developed in the core of this lecture note. These concepts will be illustrated on a model problem arising in free boundary systems. More precisely, we consider the identification of a moving boundary that represents the isothermal interface between a solid phase and a liquid phase, from measurements on a fixed part of the solid boundary. This problem is referred in the literature as the inverse Stefan problem [61, 144]. We make use of the transverse derivative concepts introduced in [154, 155].

## 2.1  The mechanical and mathematical settings

We consider a fluid phase located at time $t \geq 0$ in a domain $\Omega^f(t) \subset D$ with boundary $\Gamma^f(t)$ where $D \subset \mathbb{R}^d$ is a hold-all domain. It is surrounded by a solid phase $\Omega^s(t) \stackrel{\text{def}}{=} D \setminus \Omega^f(t)$ with boundary $\partial\Omega^s(t) = \Gamma^s \cup \Gamma^f(t)$, as described in Figure (2.1) where $\Gamma^s \stackrel{\text{def}}{=} \partial D$. The motion of the interface $\Gamma^f(t)$ is characterized by its scalar Eulerian normal velocity $v^f(\mathbf{x}, t)$, $\mathbf{x} \in \Gamma^f(t)$. We consider the velocity field $\mathbf{V}(t, \mathbf{x}) : (0, \tau) \times D \to \mathbb{R}^d$ satisfying the following properties,

$$
\begin{cases}
\mathbf{V}(t, \mathbf{x}) \cdot \mathbf{n}^s(\mathbf{x}) = 0, & (t, \mathbf{x}) \in \Sigma^s \stackrel{\text{def}}{=} (0, \tau) \times \Gamma^s \\[2mm]
\mathbf{V}(t, \mathbf{x}) \cdot \mathbf{n}^f(t, \mathbf{x}) = v^f, & (t, \mathbf{x}) \in \Sigma^f \stackrel{\text{def}}{=} \bigcup_{t \in (0, \tau)} \{t\} \times \Gamma^f(t)
\end{cases}
\tag{2.1}
$$

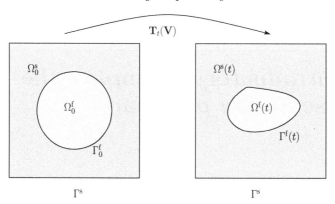

**FIGURE 2.1:** Melting front propagation

We consider an admissible space[1] $\mathcal{U}_{ad}$ for the velocity field. The map $\mathbf{V} \in \mathcal{U}_{ad}$ can be viewed as a non-autonomous velocity field $\{\mathbf{V}(t) : 0 \le t \le \tau\}$ :

$$\mathbf{V}(t) : D \to \mathbb{R}^d$$
$$\mathbf{x} \mapsto \mathbf{V}(t)(\mathbf{x}) \stackrel{\text{def}}{=} \mathbf{V}(t, \mathbf{x})$$

Let us define the field $\mathbf{x}_{\mathbf{V}}(t; \mathbf{X}) : [0, \tau] \to \mathbb{R}^d$ solution of the following Cauchy problem,

$$\begin{cases} \dfrac{d\mathbf{x}}{dt}(t) & = \mathbf{V}(t, \mathbf{x}(t)), \, t \in [0, \tau] \\ \mathbf{x}(t = 0) = \mathbf{X}, & \mathbf{X} \in D. \end{cases} \tag{2.2}$$

We build the corresponding Lagrangian flow transformation,

$$\mathbf{T}_t(\mathbf{V}) : D \to \mathbb{R}^d$$
$$\mathbf{X} \mapsto \mathbf{T}_t(\mathbf{V})(\mathbf{X}) \stackrel{\text{def}}{=} \mathbf{x}_{\mathbf{V}}(t; \mathbf{X})$$

---

[1]

$$\mathcal{U}_{ad} \stackrel{\text{def}}{=} \left\{ \mathbf{V} \in \mathcal{C}^0\left([0, \tau]; \mathcal{V}_0^k(D)\right), \quad \exists c > 0, \quad \mathscr{L}\mathrm{ip}_k(\mathbf{V}(t)) \le c, \, \forall t \in [0, \tau] \right\}$$

with $k \ge 0$ and where

$$\mathcal{V}_0^k(D) \stackrel{\text{def}}{=} \left\{ \mathbf{V} \in \left(\mathcal{C}^k(\bar{D})\right)^d, \mathbf{V} \cdot \mathbf{n}_{\partial D} = 0, \text{ on } D \right\}$$

with $\mathscr{L}\mathrm{ip}_k(\mathbf{V}) \stackrel{\text{def}}{=} \displaystyle\sum_{|\alpha|=k} \mathscr{L}\mathrm{ip}(\mathrm{D}^\alpha \mathbf{V})$ for $k \ge 1$ and $\mathscr{L}\mathrm{ip}(\mathbf{V}) \stackrel{\text{def}}{=} \displaystyle\sup_{\mathbf{y} \ne \mathbf{x}} \dfrac{|\mathbf{V}(\mathbf{y}) - \mathbf{V}(\mathbf{x})|}{|\mathbf{y} - \mathbf{x}|}$.

It is shown in [51, 60] that the conditions (2.1) imply that $\mathbf{T}_t(\mathbf{V})$ is a diffeo-morphism[2] that maps $\overline{D}$ into $\overline{D}$. In particular

$$\Omega^{\mathrm{f}}(t) = \mathbf{T}_t(\mathbf{V})(\Omega_0^{\mathrm{f}}), \ \Omega^{\mathrm{s}}(t) = \mathbf{T}_t(\mathbf{V})(\Omega_0^{\mathrm{s}})$$
$$\Gamma^{\mathrm{f}}(t) = \mathbf{T}_t(\mathbf{V})(\Gamma_0^{\mathrm{f}}), \quad \Gamma^{\mathrm{s}} = \mathbf{T}_t(\mathbf{V})(\Gamma^{\mathrm{s}})$$

In the sequel, we shall consider the velocity field $\mathbf{V}$ as the main variable characterizing the evolution of the melting front and define $\mathbf{v}^{\mathrm{f}} \overset{\text{def}}{=} \mathbf{V} \cdot \mathbf{n}^{\mathrm{f}}$ on $\Sigma^{\mathrm{f}}$.

For a given evolution of the boundary $\Gamma^{\mathrm{f}}(t)$, we consider the solution $y$ of the heat equation inside the solid phase. We impose a given heat flux $f(t,.)$ on the fixed solid boundary $\Gamma^{\mathrm{s}}$ and it is assumed that the temperature is equal to a fixed phase transition temperature $y_{\mathrm{f}}$ on the moving melting interface $\Gamma^{\mathrm{f}}(t)$ for $t \in (0, \tau)$. This leads to the following non-cylindrical system,

$$\begin{cases} \partial_t\, y - \Delta y = 0, \, Q^{\mathrm{s}} \equiv \bigcup_{0<t<\tau} (\{t\} \times \Omega^{\mathrm{s}}(t)), \\ \partial_{\mathbf{n}} y = f, \qquad \Sigma^{\mathrm{s}}, \\ y = y_{\mathrm{f}}, \qquad \Sigma^{\mathrm{f}}, \\ y(t=0) = y_0, \ \Omega_0^{\mathrm{s}} \end{cases} \qquad (2.3)$$

The space-time domain $Q^{\mathrm{s}}$ is called a tube.

For a smooth velocity $\mathbf{V} \in \mathcal{U}_{ad}$ and the data $(f, y_0) \in L^2\big((0, \tau); L^2(\Gamma^{\mathrm{s}})\big) \times L^2(\Omega_0^{\mathrm{s}})$, there exists a unique solution [59]

$$y \in L^2\big((0, \tau); H^1_{y^{\mathrm{f}}, \Gamma^{\mathrm{f}}(t)}(\Omega^{\mathrm{s}}(t))\big) \cap L^\infty\big((0, \tau); L^2(\Omega^{\mathrm{s}}(t))\big) \qquad (2.4)$$

where

$$H^1_{\varphi, \Gamma}(D) \overset{\text{def}}{=} \{\phi \in H^1(D), \quad \phi = \varphi \text{ on } \Gamma\}$$

---

## 2.2   The inverse problem setting

For a given evolution of the melting interface, we consider the solution of the heat equation (2.3) and we consider its trace on the fixed boundary $\Sigma^{\mathrm{s}}$.

---

2

$$\mathbf{T}(\mathbf{V}) \in \mathcal{C}^1\big([0, \tau]; (\mathcal{C}^k(\overline{D}))^d\big) \cap \mathcal{C}^0\big([0, \tau]; (W^{k+1, \infty}(D))^d\big)$$
$$\mathbf{V} \circ \mathbf{T}(\mathbf{V}) \in L^\infty\big((0, \tau); (W^{k+1, \infty}(D))^d\big)$$
$$\mathbf{T}^{-1}(\mathbf{V}) \in \mathcal{C}^1\big([0, \tau]; (\mathcal{C}^k(\overline{D}))^d\big)$$

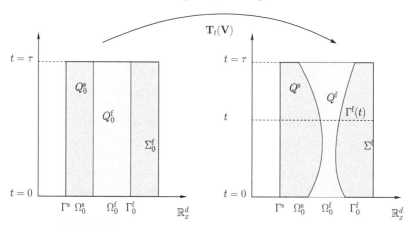

**FIGURE 2.2**:   Non-cylindrical space-time domain

On the mathematical viewpoint, we introduce the observation space

$$\mathcal{O} \overset{\text{def}}{=} L^2\big((0,\tau); L^2(\Gamma^{\text{s}})\big)$$

and the observation operator,

$$\mathcal{O} : \mathcal{U}_{ad} \to \mathcal{O}$$
$$\mathbf{V} \mapsto \mathcal{O}(\mathbf{V}) \overset{\text{def}}{=} \gamma_{\Sigma^{\text{s}}}(y(\mathbf{V})) \tag{2.5}$$

where $y(\mathbf{V})$ stands for the solution of equation (2.3) and $\gamma_{\Sigma^{\text{s}}}$ is the zero order trace operator on $\Sigma^{\text{s}}$.

The inverse Stefan problem consists in recovering the evolution of the melting front $\Gamma^{\text{f}}(t)$ from the knowledge of the temperature on the fixed solid boundary $\Gamma^{\text{s}}$. This means that for a given temperature $y_{\text{d}} \in L^2\big((0,\tau); L^2(\Gamma^{\text{s}})\big)$, we look for $\mathbf{V} \in \mathcal{U}_{ad}$ such that

$$\mathcal{O}(\mathbf{V}) = y_{\text{d}}, \quad \text{in } \mathcal{O} \tag{2.6}$$

It is a non-linear ill-posed inverse problem that can be solved using a least-square minimization problem regularized thanks to a Tikhonov zero order term. Hence we look for the solution $\mathbf{V}$ of the following optimization problem,

$$\min_{\mathbf{V} \in \mathcal{U}_{ad}} \frac{1}{2}\|\mathcal{O}(\mathbf{V}) - y_d\|_{\mathcal{O}}^2 + \frac{\alpha}{2}\|\mathbf{V}\|_{\mathcal{U}_{ad}}^2 \tag{2.7}$$

with $\alpha > 0$.

In the sequel, we will prove the following result,

**THEOREM 2.1**
*For $\mathbf{V} \in \mathcal{U}_{ad}$, $y_{\text{d}} \in \mathcal{O}$, we consider the functional*

$$j(\mathbf{V}) = \frac{1}{2}\int_0^\tau \int_{\Gamma^{\text{s}}} |y(\mathbf{V}) - y_{\text{d}}|^2 \, \mathrm{d}s \, \mathrm{d}t \tag{2.8}$$

*where $y(\mathbf{V})$ is the solution of the following system*

$$\begin{cases} \partial_t y - \Delta y = f, \, Q(\mathbf{V}), \\ \partial_\mathbf{n} y = 0, \quad\quad \Sigma^s, \\ y = y_f, \quad\quad \Sigma(\mathbf{V}), \\ y(t = 0) = y_0, \, \Omega_0 \end{cases} \quad (2.9)$$

*$j(\mathbf{V})$ is differentiable with respect to $\mathbf{V}$ and its gradient has the following structure*

$$\nabla j(\mathbf{V}) = -{}^*\gamma_{\Sigma(\mathbf{V})}(\lambda\mathbf{n}), \quad in\ \mathcal{U}_{ad}^* \quad (2.10)$$

*where the transverse adjoint state $\lambda \in \mathcal{C}^0\big([0,\tau]; H^1(\Gamma_t(\mathbf{V}))\big)$ is the solution of the backward dynamical system,*

$$\begin{cases} -\partial_t \lambda - \nabla_\Gamma \lambda \cdot \mathbf{V} - (\operatorname{div}\mathbf{V})\lambda = \partial_\mathbf{n} y \, \partial_\mathbf{n} \varphi, \, \Sigma(\mathbf{V}) \\ \lambda(t = \tau) = 0, \quad\quad\quad\quad\quad\quad\quad\quad\quad \Gamma_\tau(\mathbf{V}) \end{cases} \quad (2.11)$$

*and where the adjoint state $\varphi$ is the solution of the following backward system,*

$$\begin{cases} \partial_t \varphi + \Delta\varphi = 0, \quad Q(\mathbf{V}) \\ \partial_\mathbf{n}\varphi = y(\mathbf{V}) - y_d, \, \Sigma^s \\ \varphi = 0, \quad\quad\quad\quad \Sigma(\mathbf{V}) \\ \varphi(t = \tau) = 0, \quad\quad \Omega_\tau(\mathbf{V}) \end{cases} \quad (2.12)$$

## 2.3 The eulerian derivative and the transverse field

A possible choice in order to solve the above minimization problem is to use a gradient based method such as the conjugate gradient method. Hence, we need to evaluate the gradient with respect to $\mathbf{V}$ of the functional

$$j(\mathbf{V}) \overset{\text{def}}{=} \frac{1}{2}\|\mathscr{O}(\mathbf{V}) - y_d\|_{\mathscr{O}}^2 \quad (2.13)$$

where, for the sake of simpleness, we have dropped the regularizing term $\frac{\alpha}{2}\|\mathbf{V}\|_{\mathcal{U}_{ad}}^2$. Let us choose a perturbation direction $\mathbf{W} \in \mathcal{U}_{ad}$. We would like to compute the directional derivative of $j$,

$$[D_\mathbf{V}[j](\mathbf{V})] \cdot \mathbf{W} \overset{\text{def}}{=} \lim_{\rho \to 0} \frac{1}{\rho}\big(j(\mathbf{V} + \rho\mathbf{W}) - j(\mathbf{V})\big) \quad (2.14)$$

Then the goal is to evaluate the directional derivative of the element $y(\mathbf{V})$ which is a solution of the moving heat equation (2.3). In order to do so, we write the associated variational formulation satisfied $y(\mathbf{V})$,

$$\int_0^\tau \int_{\Omega_t(\mathbf{V})} [\partial_t y(\mathbf{V})\, \phi(\mathbf{V}) + \nabla y(\mathbf{V}) \cdot \nabla\phi(\mathbf{V})]\, \mathrm{dx}\, \mathrm{dt} = 0, \quad (2.15)$$
$$\forall \phi(\mathbf{V}) \in L^2\big((0,\tau); H^1_{0,\Gamma_t(\mathbf{V})}(\Omega_t(\mathbf{V}))\big)$$

where we have set without loss of generality, $(f, y_d, y_0) = (0, 0, 0)$ together with $\Omega_t(\mathbf{V}) \stackrel{\text{def}}{=} \Omega^s(t)$ and $\Gamma_t(\mathbf{V}) \stackrel{\text{def}}{=} \Gamma^f(t)$.

Looking at equation (2.15), it is clear that we need to establish how to differentiate the generic term

$$J(\mathbf{V}) = \int_0^\tau \int_{\Omega_t(\mathbf{V})} f(\mathbf{V}) \, d\mathbf{x} \, dt$$

with respect to $\mathbf{V}$.

To this end, we introduce the perturbated moving domain

$$\Omega_t(\mathbf{V} + \rho\,\mathbf{W}) \stackrel{\text{def}}{=} T_t(\mathbf{V} + \rho\,\mathbf{W})(\Omega_0)$$

This family generates a perturbed tube

$$Q(\mathbf{V} + \rho\,\mathbf{W}) \stackrel{\text{def}}{=} \bigcup_{0 \leq t \leq \tau} \left( \{t\} \times \Omega_t(\mathbf{V} + \rho\,\mathbf{W}) \right)$$

as described in Figure (2.3). Since the function $f(\mathbf{V})$ is defined on the non-

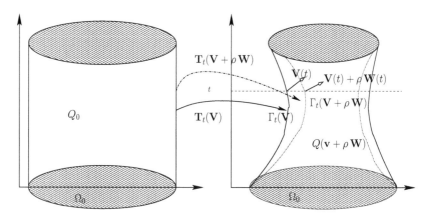

**FIGURE 2.3**:   Perturbed tube

cylindrical reference tube $Q(\mathbf{V})$, it is natural to introduce the transformation between $Q(\mathbf{V})$ and $Q(\mathbf{V} + \rho\,\mathbf{W})$. A canonical choice is furnished by

$$\mathcal{T}^t(\rho; \mathbf{x}) : \Omega_t(\mathbf{V}) \rightarrow \Omega_t(\mathbf{V} + \rho\,\mathbf{W})$$
$$\mathbf{x} \mapsto \mathcal{T}^t(\rho; \mathbf{x}) \stackrel{\text{def}}{=} \left[ \mathbf{T}_t(\mathbf{V} + \rho\,\mathbf{W}) \right] \circ \mathbf{T}_t(\mathbf{V})^{-1} \left] (\mathbf{x}) \right.$$

Hence, the perturbated functional can be written as follows,

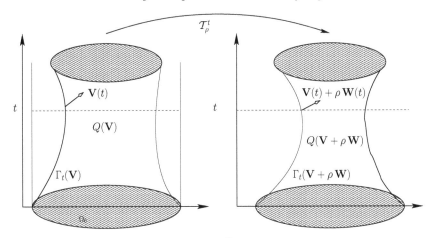

**FIGURE 2.4**: Transverse map

$$J(\mathbf{V} + \rho\,\mathbf{W}) = \int_0^\tau \int_{\Omega_t(\mathbf{V}+\rho\,\mathbf{W})} f(\mathbf{V} + \rho\,\mathbf{W})\,\mathrm{d}\mathbf{x}\,\mathrm{d}t$$

$$= \int_0^\tau \int_{T_\rho^t\left(\Omega_t(\mathbf{V})\right)} f(\mathbf{V} + \rho\,\mathbf{W})\,\mathrm{d}\mathbf{x}\,\mathrm{d}t$$

$$= \int_0^\tau \int_{\Omega_t(\mathbf{V})} \left(\det \mathrm{D}\,T_\rho^t\right) f(\mathbf{V} + \rho\,\mathbf{W}) \circ T_\rho^t\,\mathrm{d}\mathbf{x}\,\mathrm{d}t$$

where we have performed a transport into the moving reference domain $\Omega_t(\mathbf{V})$. Now we shall need to differentiate the terms inside the integral with respect to $\rho$ at point $\rho = 0$. The easiest way to do so is to connect this problem to the classical shape derivative calculus handled inside the speed method framework [147, 135]. This means that we need to identify a transverse velocity field that may generate the transverse map $T^t(\rho; \mathbf{x})$ as the solution of a dynamical system with respect to the parameter $\rho \in [0, \rho_0]$. Actually, it can be proven that $\mathbf{T}(\mathbf{V} + \rho\,\mathbf{W})$ is continuously differentiable[3] with respect to $\rho$ and that the transverse map $T_\rho^t$ can be considered as the flow with respect to $\rho$ of the transverse vector field

$$\mathcal{Z}(\rho; (t, \mathbf{x})) \stackrel{\mathrm{def}}{=} \left[\partial_\rho T^t(\rho)\right] \circ T^t(\rho)^{-1}(\mathbf{x})$$
$$= \left[\partial_\rho \mathbf{T}(\mathbf{V} + \rho\,\mathbf{W})\right] \circ \mathbf{T}(\mathbf{V} + \rho\,\mathbf{W})^{-1}(\mathbf{x})$$

Indeed the map $\mathcal{Z} \in \mathcal{Z}_{ad}$ can be viewed as a non-autonomous velocity field :

$$\mathcal{Z}(\rho) : (0, \tau) \times D \to \mathbb{R}^d$$
$$(t, \mathbf{x}) \mapsto \mathcal{Z}(\rho)(t, \mathbf{x}) \stackrel{\mathrm{def}}{=} \mathcal{Z}(\rho; (t, \mathbf{x}))$$

---

[3]in $\mathcal{Z}_{ad} \stackrel{\mathrm{def}}{=} \mathcal{C}^0\left([0, \tau]; (\mathcal{C}^{k-1}(\bar{D}))^d\right)$.

with $0 \leq \rho \leq \rho_0$.

We define the field $\mathcal{X}_{\mathcal{Z}}^t(\rho; \mathbf{x}) : [0, \rho_0] \to \mathbb{R}^d$ solution of the following Cauchy problem,

$$\begin{cases} \dfrac{d\mathcal{X}}{d\rho}(\rho) &= \mathcal{Z}(\rho; (t, \mathcal{X}(\rho))), \rho \in [0, \tau] \\ \mathcal{X}(\rho = 0) = \mathbf{x}, &\quad \mathbf{x} \in \Omega_t(\mathbf{V}) \end{cases} \tag{2.16}$$

We build the corresponding Lagrangian flow transformation,

$$\mathbf{T}_\rho(\mathcal{Z}) : (0, \tau) \times D \to \mathbb{R}^d$$

$$(t, \mathbf{x}) \mapsto \mathbf{T}_\rho(\mathcal{Z})(t, \mathbf{x}) \overset{\text{def}}{=} \mathcal{X}_{\mathcal{Z}}(\rho; (t, \mathbf{x}))$$

and we identify the transverse map $\mathcal{T}(\rho)$ with $\mathbf{T}_\rho(\mathcal{Z})$. This allows writing the perturbated functional as follows,

$$J(\mathbf{V} + \rho\,\mathbf{W}) = \int_0^\tau \int_{\Omega_t(\mathbf{V})} \left( \det \mathrm{D}\,\mathbf{T}_\rho(\mathcal{Z}_\rho^t) \right) f(\mathbf{V} + \rho\,\mathbf{W}) \circ \mathbf{T}_\rho(\mathcal{Z}_\rho^t)\, d\mathbf{x}\, dt$$

Compared with the classical speed method, the parameter $\rho$ plays the role of the peuso-time $t$, the field $\mathcal{Z}^t(\rho)$ plays the role of the speed $\mathbf{V}(t)$ and the transverse field $\mathbf{Z}(t) \overset{\text{def}}{=} \mathcal{Z}^t(\rho = 0)$ has the same role as $\mathbf{V}(0)$. Hence, the non-cylindrical shape derivative calculus reduces to apply the classical shape derivative calculus with the following identification : $\mathbf{V}(0) \leftrightarrow \mathbf{Z}(t)$. Following this idea, the following identity holds

**LEMMA 2.1**

$$\partial_\rho \big[ \det \mathrm{D}\,\mathbf{T}_\rho(\mathcal{Z}_\rho^t) \big]\big|_{\rho=0} = \operatorname{div} \mathbf{Z}(t) \tag{2.17}$$

In the classical shape derivative calculus, the notion of material derivative has been introduced. Here we define its non-cylindrical counterpart. Let $\mathcal{Y}(\mathbf{V})$ be a given functional space defined on $Q(\mathbf{V})$. An example is furnished by $\mathcal{Y}(\mathbf{V}) = L^2\big((0, \tau); H^1_{0, \Gamma_t(\mathbf{V})}(\Omega_t(\mathbf{V}))\big)$.

**DEFINITION 2.1**  *We say that a function $f(\mathbf{V}) \in \mathcal{Y}(\mathbf{V})$ for $\mathbf{V} \in \mathcal{U}_{ad}$ admits an Eulerian material derivative $\dot{f}(\mathbf{V}; \mathbf{W})$ in the direction $\mathbf{W} \in \mathcal{U}_{ad}$ if the following limit*

$$\dot{f}(\mathbf{V}; \mathbf{W}) \overset{\text{def}}{=} \lim_{\rho \to 0} \frac{f(\mathbf{V} + \rho\,\mathbf{W}) \circ \mathcal{T}_\rho^t(\mathbf{V}) - f(\mathbf{V})}{\rho} \tag{2.18}$$

*exists in $\mathcal{Y}(\mathbf{V})$.*

Using this definition, we easily establish the following result,

**LEMMA 2.2**
*Let $\mathbf{V} \in \mathcal{U}_{ad}$ and $f(\mathbf{V}) \in \mathcal{Y}(\mathbf{V})$ such that it admits an Eulerian material derivative $\dot{f}(\mathbf{V}; \mathbf{W})$ in the direction $\mathbf{W} \in \mathcal{U}_{ad}$. Then $J(\mathbf{V})$ is differentiable with respect to $\mathbf{V}$ and its derivative is furnished by the following expression,*

$$\mathrm{D_V}\left[J(\mathbf{V})\right] \cdot \mathbf{W} = \int_0^\tau \int_{\Omega_t(\mathbf{V})} \left[(\mathrm{div}\,\mathbf{Z})\,f(\mathbf{V}) + \dot{f}(\mathbf{V}; \mathbf{W})\right] \mathrm{d}\mathbf{x}\,\mathrm{d}t \qquad (2.19)$$

It remains to precise a specific rule to compute the transverse $\mathbf{Z}$ starting from the knowledge of $\mathbf{V}, \mathbf{W} \in \mathcal{U}_{ad}$. This is realized thanks to the following result,

**THEOREM 2.2 [154, 59]**
*The transverse field $\mathbf{Z}$ is the unique vector in $\mathcal{Z}_{ad}$ such that $\partial_t\,\mathbf{Z} + \mathrm{D}\,\mathbf{Z} \cdot \mathbf{V} \in \mathcal{Z}_{ad}$ and is the solution of the following Cauchy problem,*

$$\begin{cases} \partial_t\,\mathbf{Z} + \mathrm{D}\,\mathbf{Z} \cdot \mathbf{V} - \mathrm{D}\,\mathbf{V} \cdot \mathbf{Z} = \mathbf{W}, & (0, \tau) \times D \\ \mathbf{Z}(t = 0, .) = 0, & D \end{cases} \qquad (2.20)$$

**REMARK 2.1**  In the sequel, we shall use the notation

$$[\mathbf{Z}, \mathbf{V}] \overset{\mathrm{def}}{=} \mathrm{D}\,\mathbf{Z} \cdot \mathbf{V} - \mathrm{D}\,\mathbf{V} \cdot \mathbf{Z}$$

for the Lie bracket associated to $(\mathbf{Z}, \mathbf{V}) \in \mathcal{Z}_{ad} \times \mathcal{U}_{ad}$.  ⬛

**PROOF**  It consists of 4 steps :

1. We first prove the differentiability of the flow $\mathbf{T}(\mathbf{V} + \rho\,\mathbf{W})$ with respect to $\rho$,

**LEMMA 2.3**
*Let us consider $(\mathbf{V}, \mathbf{W}) \in \mathcal{U}_{ad} \times \mathcal{U}_{ad}$, then there exists $\rho_0 > 0$ such that the application*

$$]0, \rho_0[ \longrightarrow \mathcal{C}^0\left([0, \tau]; \left(\mathcal{C}^{k-1}(\bar{D})\right)^d\right)$$
$$\rho \mapsto \mathbf{T}(\mathbf{V} + \rho\,\mathbf{W})$$

*is continuously differentiable and its derivative satisfies for any $t \in [0, \tau]$,*

$$\partial_\rho\left[\mathbf{T}_t(\mathbf{V} + \rho\,\mathbf{W})\right] =$$
$$\int_0^t \mathrm{D}(\mathbf{V} + \rho\,\mathbf{W})(\mu, \mathbf{T}_\mu(\mathbf{V} + \rho\,\mathbf{W})) \cdot \partial_\rho\left[\mathbf{T}_\mu(\mathbf{V} + \rho\,\mathbf{W})\right] d\mu$$
$$+ \int_0^t \mathbf{W}\left(\mu, \mathbf{T}_\mu(\mathbf{V} + \rho\,\mathbf{W})\right) d\mu.$$

2. Let us set $\mathcal{S}(\rho) \stackrel{\text{def}}{=} \partial_\rho[\mathbf{T}(\mathbf{V} + \rho\,\mathbf{W})]$, then the transverse field writes

$$\mathcal{Z}(\rho) = \mathcal{S}(\rho) \circ \mathbf{T}(\mathbf{V} + \rho\,\mathbf{W})^{-1}$$

At $\rho = 0$, we set

$$\mathbf{S} = \mathbf{Z} \circ \mathbf{T}(\mathbf{V})$$

We can state the following result

### LEMMA 2.4

*The function $\mathbf{S}$ is the unique vector field, in $\mathcal{Z}_{ad} \stackrel{\text{def}}{=} \mathcal{C}^1([0, \tau]; (\mathcal{C}^{k-1}(\bar{D}))^d)$, satisfying*

$$\mathbf{S}(t) = \int_0^t \mathbf{W}(\mu, \mathbf{T}_\mu(\mathbf{V}))d\mu + \int_0^t \mathrm{D}\,\mathbf{V}(\mu, \mathbf{T}_\mu(\mathbf{V})) \cdot \mathbf{S}(\mu)\,d\mu. \quad (2.21)$$

3. Then we prove the equivalence between equation (2.20) and equation (2.21),

### LEMMA 2.5

(i) *Let us consider $\mathbf{Z} \in \mathcal{Z}_{ad}$ such that*

    (a) *the following regularity holds*

$$\partial_t \mathbf{Z} + \mathrm{D}\,\mathbf{Z} \cdot \mathbf{V} \in \mathcal{C}^0([0, \tau]; (\mathcal{C}^{k-1}(\bar{D}))^d)$$

    (b) *and it satisfies equation (2.20)*

    *then $\mathbf{Z} \circ \mathbf{T}(\mathbf{V})$ belongs to $\mathcal{Z}_{ad}$ and satisfies equation (2.21).*

(ii) *Let us consider $\mathbf{S} \in \mathcal{Z}_{ad}$ solution of equation (2.21), then $\mathbf{S} \circ \mathbf{T}(\mathbf{V})^{-1} \in \mathcal{Z}_{ad}$ is such that*

    (a) *the following regularity holds*

$$\partial_t\left[\mathbf{S} \circ \mathbf{T}(\mathbf{V})^{-1}\right] + \mathrm{D}\left[\mathbf{S} \circ \mathbf{T}(\mathbf{V})^{-1}\right] \cdot \mathbf{V} \in \mathcal{C}^0([0, \tau]; (\mathcal{C}^{k-1}(\bar{D}))^d)$$

    (b) *and it satisfies equation (2.20).*

4. Since we already have proven the unique solvability of equation (2.21), we use the above lemma to conclude.

□

## 2.4 The Eulerian material derivative of the state

With the previous framework, we are ready to differentiate the variational formulation (2.15) with respect to $\mathbf{V}$. We set $y_\rho = y(\mathbf{V} + \rho\,\mathbf{W})$, $\mathcal{F}_\rho = \mathrm{D}\,\mathcal{T}_\rho$ and $\mathcal{J}_\rho = \det(\mathcal{F}_\rho)$. The perturbated formulation writes

$$\int_0^\tau \int_{\Omega_t(\mathbf{V})} \mathcal{J}_\rho \left\{ \left[ (\partial_t\, y_\rho) \circ \mathcal{T}_\rho \right] \left[ \phi_\rho \circ \mathcal{T}_\rho \right] + (\nabla y_\rho) \circ \mathcal{T}_\rho \cdot (\nabla \phi_\rho) \circ \mathcal{T}_\rho \right\} \mathrm{d}\mathbf{x}\,\mathrm{d}t = 0,$$
$$\forall\, \phi_\rho \in \Theta(\rho)$$

where

$$\Theta(\rho) \overset{\mathrm{def}}{=} \left\{ \phi \circ (\mathcal{T}_\rho)^{-1} \mid \quad \phi \in L^2\big((0,\tau); H^1_{0,\Gamma_t(\mathbf{V})}(\Omega_t(\mathbf{V})) \big) \right\}$$

The following identities are easy to check,

$$(\nabla y_\rho) \circ \mathcal{T}_\rho = \nabla(y_\rho \circ \mathcal{T}_\rho) \mathcal{F}_\rho^{-1}$$
$$(\partial_t\, y_\rho) \circ \mathcal{T}_\rho = \partial_t\, (y_\rho \circ \mathcal{T}_\rho) - \nabla(y_\rho \circ \mathcal{T}_\rho) \mathcal{F}_\rho^{-1} \partial_t\, \mathcal{T}_\rho \qquad (2.22)$$
$$\partial_t\, \mathcal{T}_\rho = -\mathcal{F}_\rho \cdot \mathbf{V} + (\mathbf{V} + \rho\,\mathbf{W}) \circ \mathcal{T}_\rho$$

We set $y^\rho \overset{\mathrm{def}}{=} y_\rho \circ \mathcal{T}_\rho$, then we get

$$\int_0^\tau \int_{\Omega_t(\mathbf{V})} \mathcal{J}_\rho \left[ \left( \partial_t\, y^\rho - \nabla y^\rho \mathcal{F}_\rho^{-1} \partial_t\, \mathcal{T}_\rho \right) \phi + \nabla y^\rho \mathcal{F}_\rho \cdot \nabla\phi \mathcal{F}_\rho \right] \mathrm{d}\mathbf{x}\,\mathrm{d}t = 0,$$
$$\forall\, \phi \in \Theta$$

where $\Theta \overset{\mathrm{def}}{=} L^2\big((0,\tau); H^1_{0,\Gamma_t(\mathbf{V})}(\Omega_t(\mathbf{V})) \big)$.
The following identities are easy to check,

$$\partial_\rho \mathcal{J}_\rho |_{\rho=0} = \mathrm{div}\,\mathbf{Z}$$
$$\partial_\rho \mathcal{F}_\rho |_{\rho=0} = \mathrm{D}\,\mathbf{Z}$$
$$\partial_\rho [\mathcal{F}_\rho^{-1}] |_{\rho=0} = -\,\mathrm{D}\,\mathbf{Z} \qquad (2.23)$$
$$\partial_\rho [\partial_t\, \mathcal{T}_\rho] |_{\rho=0} = \mathbf{W} - [\mathbf{Z}, \mathbf{V}]$$

Recalling that $\mathcal{T}_\rho |_{\rho=0} = \mathrm{I}$ and $\dot{y} \overset{\mathrm{def}}{=} \partial_\rho y^\rho |_{\rho=0}$, the derivative of the variational formulation writes

$$\int_0^\tau \int_{\Omega_t(\mathbf{V})} \Big\{ \mathrm{div}\,\mathbf{Z}\, [\partial_t\, y\, \phi + \nabla y \cdot \nabla\phi] + [\partial_t\, \dot{y} - \nabla y \cdot (\mathbf{W} - [\mathbf{Z}, \mathbf{V}])]\, \phi$$
$$+ \nabla\dot{y} \cdot \nabla\phi - \nabla y\, \mathrm{D}\,\mathbf{Z} \cdot \nabla\phi - \nabla y \cdot \nabla\phi\, \mathrm{D}\,\mathbf{Z} \Big\} \mathrm{d}\mathbf{x}\,\mathrm{d}t = 0,$$
$$\forall\, \phi \in \Theta$$

Using the transverse dynamical system, we get

$$\int_0^\tau \int_{\Omega_t(\mathbf{V})} [\partial_t \dot{y}\,\phi + \nabla \dot{y} \cdot \nabla \phi]\,d\mathbf{x}\,dt + \int_0^\tau \int_{\Omega_t(\mathbf{V})} [\mathrm{div}\,\mathbf{Z}\,\partial_t\,y - \nabla y \cdot (\partial_t\,\mathbf{Z})]\,\phi\,d\mathbf{x}\,dt$$

$$- \int_0^\tau \int_{\Omega_t(\mathbf{V})} \nabla y\big(-(\mathrm{div}\,\mathbf{Z})\,\mathrm{I} + \mathrm{D}\,\mathbf{Z} + {}^*\mathrm{D}\,\mathbf{Z}\big) \cdot \nabla \phi\,d\mathbf{x}\,dt = 0,$$

$$\forall \phi \in \Theta$$

Performing integration by parts, we shall obtain the strong formulation satisfied by the material Eulerian derivative $\dot{y}$,

$$\begin{cases} \partial_t \dot{y} - \Delta \dot{y} = -(\mathrm{div}\,\mathbf{Z})\partial_t\,y + \nabla y \cdot (\partial_t\,\mathbf{Z}) \\ \qquad\qquad + \mathrm{div}\left[\nabla y\big((\mathrm{div}\,\mathbf{Z})\,\mathrm{I} - (\mathrm{D}\,\mathbf{Z} + {}^*\mathrm{D}\,\mathbf{Z})\big)\right] & Q(\mathbf{V}) \\ \partial_\mathbf{n} \dot{y} \quad = \nabla y\big(-(\mathrm{div}\,\mathbf{Z})\,\mathrm{I} + \mathrm{D}\,\mathbf{Z} + {}^*\mathrm{D}\,\mathbf{Z}\big) \cdot n, & \Sigma^s \qquad (2.24) \\ \dot{y} \qquad = 0, & \Sigma(\mathbf{V}) \\ \dot{y}(t=0) \ = 0, & \Omega_0 \end{cases}$$

together with

$$\begin{cases} \partial_t\,\mathbf{Z} + \mathrm{D}\,\mathbf{Z} \cdot \mathbf{V} - \mathrm{D}\,\mathbf{V} \cdot \mathbf{Z} = \mathbf{W}, \ (0,\tau) \times D \\ \mathbf{Z}(t=0,.) = 0, & D \end{cases} \qquad (2.25)$$

Finally using the fact that $\Gamma^s$ does not move, the directional derivative of $j(\mathbf{V})$ is given by the following expression

$$[\mathrm{D}_\mathbf{V}[j](\mathbf{V})] \cdot \mathbf{W} = \int_0^\tau \int_{\Gamma^s} (y(\mathbf{V}) - y_d)\,\dot{y}(\mathbf{V};\mathbf{W})\,d\mathbf{x}\,dt \qquad (2.26)$$

---

## 2.5   The Eulerian partial derivative of the state

Now, we would like to identify the gradient $\nabla j(\mathbf{V})$. To this end, we need to introduce several adjoint problems. Before proceeding, it may be convenient to define the notion of Eulerian partial derivative.

**DEFINITION 2.2**   *We say that a function $f(\mathbf{V}) \in \mathcal{Y}(\mathbf{V})$ for $\mathbf{V} \in \mathcal{U}_{ad}$ admits an Eulerian partial derivative $f'(\mathbf{V};\mathbf{W})$ in the direction $\mathbf{W} \in \mathcal{U}_{ad}$ if it admits an Eulerian material derivative $\dot{f}(\mathbf{V};\mathbf{W})$ and*

$$f'(\mathbf{V};\mathbf{W}) \overset{\text{def}}{=} \dot{f}(\mathbf{V};\mathbf{W}) - \nabla f \cdot \mathbf{Z} \qquad (2.27)$$

*exists in $\mathcal{Y}(\mathbf{V})$ and depends linearly on $\mathbf{W}$, where $\mathbf{Z}$ is the transverse field solution of the transverse dynamical system (2.20).*

Going back to the derivation of the generic functional $J(\mathbf{V})$, we are able to quote the following result,

**LEMMA 2.6**
*Let $\mathbf{V} \in \mathcal{U}_{ad}$ and $f(\mathbf{V}) \in \mathcal{Y}(\mathbf{V})$ such that it admits an Eulerian partial derivative $f'(\mathbf{V}; \mathbf{W})$ in the direction $\mathbf{W} \in \mathcal{U}_{ad}$. Then $J(\mathbf{V})$ is differentiable with respect to $\mathbf{V}$ and its derivative is furnished by the following expression,*

$$D_{\mathbf{V}}\left[J(\mathbf{V})\right] \cdot \mathbf{W} = \int_0^\tau \int_{\Omega_t(\mathbf{V})} [f'(\mathbf{V}; \mathbf{W}) + \operatorname{div}(f\,\mathbf{Z})]\, d\mathbf{x}\, dt \qquad (2.28)$$

*Furthermore if the boundary $\Gamma_t(\mathbf{V})$ is Lipschitz, then*

$$D_{\mathbf{V}}\left[J(\mathbf{V})\right] \cdot \mathbf{W} = \int_0^\tau \int_{\Omega_t(\mathbf{V})} f'(\mathbf{V}; \mathbf{W})\, d\mathbf{x}\, dt + \int_0^\tau \int_{\Gamma_t(\mathbf{V})} f(\mathbf{V})\langle \mathbf{Z}, \mathbf{n}\rangle\, ds\, dt$$
$$(2.29)$$

The main interest in the use of the Eulerian partial derivative is that it commutes with time and space derivatives. In order to obtain, the system satisfied by $y'$, we consider the variational formulation (2.15) with $\phi \in L^2((0, \tau) \times D)$,

$$\int_0^\tau \int_{\Omega_t(\mathbf{V})} [-y(\mathbf{V})\,\partial_t\,\phi + \nabla y(\mathbf{V}) \cdot \nabla\phi]\, d\mathbf{x}\, dt = 0,$$
$$\forall \phi \in L^2((0, \tau) \times D)$$

Differentiation with respect to $\mathbf{V}$ leads to

$$\int_0^\tau \int_{\Omega_t(\mathbf{V})} [-y'\,\partial_t\,\phi + \nabla y' \cdot \nabla\phi]\, d\mathbf{x}\, dt$$
$$+ \int_0^\tau \int_{\Gamma^s \cup \Gamma_t(\mathbf{V})} [-y(\mathbf{V})\,\partial_t\,\phi + \nabla y(\mathbf{V}) \cdot \nabla\phi]\,\langle \mathbf{Z}, \mathbf{n}\rangle$$
$$\forall \phi \in L^2((0, \tau) \times D)$$

Integration by parts inside the last equation with $\phi \in \mathcal{D}(Q(\mathbf{V}))$ leads to

$$\partial_t\, y' - \Delta y' = 0,\ Q(\mathbf{V})$$
$$y'(t = 0) = 0,\ \ \Omega_0$$

Using that $\langle \mathbf{Z}, \mathbf{n}\rangle = 0$ on $\Gamma^s$, we get

$$\partial_{\mathbf{n}} y' = 0,\ \Gamma^s$$

The Dirichlet boundary condition,

$$y = y_f,\ \Gamma_t(\mathbf{V})$$

differentiates as follows,

$$y' = -\partial_{\mathbf{n}} y \langle \mathbf{Z}, \mathbf{n} \rangle, \ \Gamma_t(\mathbf{V})$$

Hence, the Eulerian partial derivative of $y(\mathbf{V})$ is the solution of the following linearized system,

$$\begin{cases} \partial_t \, y' - \Delta y' = 0, & Q(\mathbf{V}) \\ \partial_{\mathbf{n}} y' = 0, & \Sigma^{\mathrm{s}} \\ y' = -\partial_{\mathbf{n}} y \langle \mathbf{Z}, \mathbf{n} \rangle, & \Sigma(\mathbf{V}) \\ y'(t = 0) = 0, & \Omega_0 \end{cases} \tag{2.30}$$

Finally, the derivative of $j(\mathbf{V})$ writes

$$[D_{\mathbf{V}}[j](\mathbf{V})] \cdot \mathbf{W} = \int_0^\tau \int_{\Gamma^{\mathrm{s}}} \big(y(\mathbf{V}) - y_{\mathrm{d}}\big) \, y'(\mathbf{V}; \mathbf{W}) \, \mathrm{d}s \, \mathrm{d}t \tag{2.31}$$

---

## 2.6   The adjoint state and the adjoint transverse field

In order to identify the gradient $\nabla j(\mathbf{V})$, we introduce the following adjoint problem,

$$\begin{cases} \partial_t \, \varphi + \Delta \varphi = 0, & Q(\mathbf{V}) \\ \partial_{\mathbf{n}} \varphi = y(\mathbf{V}) - y_{\mathrm{d}}, & \Sigma^{\mathrm{s}} \\ \varphi = 0, & \Sigma(\mathbf{V}) \\ \varphi(t = \tau) = 0, & \Omega_\tau(\mathbf{V}) \end{cases} \tag{2.32}$$

Using several integrations by parts, it turns that the derivative of $j(\mathbf{V})$ can be written

$$[D_{\mathbf{V}}[j](\mathbf{V})] \cdot \mathbf{W} = \int_0^\tau \int_{\Gamma_t(\mathbf{V})} \partial_{\mathbf{n}} \, y \, \partial_{\mathbf{n}} \, \varphi \, \langle \mathbf{Z}, \mathbf{n} \rangle \, \mathrm{d}s \, \mathrm{d}t \tag{2.33}$$

This expression does not allow to identify $\nabla j(\mathbf{V})$ since it does not explicitly involve the vector field $\mathbf{W}$.

Let us consider a distributed term involving the transverse field $\mathbf{Z}$,

$$\int_{0,\tau} \int_D \mathbf{F} \cdot \mathbf{Z} \, \mathrm{d}x \, \mathrm{d}t$$

with $F \in L^2(0, \tau; H(D))$ where $H(D)$ is an *ad-hoc*[4] Hilbert space of functions defined in the hold-all domain $D$. Using several integrations by parts the

---

[4]stable by multiplication of functions in the space $\mathcal{C}^{k-1}(\bar{D})$.

following identity holds,

$$\int_0^\tau \int_D [\partial_t \mathbf{Z} + D\mathbf{Z} \cdot \mathbf{V} - D\mathbf{V} \cdot \mathbf{Z}] \cdot \mathbf{\Lambda} = \int_0^\tau \int_D \mathbf{W} \cdot \mathbf{\Lambda}$$

$$\int_0^\tau \int_D [-\partial_t \mathbf{\Lambda} - D\mathbf{\Lambda} \cdot \mathbf{V} - {}^*D\mathbf{V} \cdot \mathbf{\Lambda} - (\operatorname{div} \mathbf{V}) \mathbf{\Lambda}] \cdot \mathbf{Z} = \int_0^\tau \int_D \mathbf{W} \cdot \mathbf{\Lambda}$$

for any $\mathbf{\Lambda} \in \mathcal{D}((0, \tau) \times D)$. Hence, we introduce the adjoint transverse field $\mathbf{\Lambda}$ as the unique solution in an *ad-hoc* space[5] of the following adjoint transverse dynamical problem,

$$\begin{cases} -\partial_t \mathbf{\Lambda} - D\mathbf{\Lambda} \cdot \mathbf{V} - {}^*D\mathbf{V} \cdot \mathbf{\Lambda} - (\operatorname{div} \mathbf{V}) \mathbf{\Lambda} = \mathbf{F}, & (0, \tau) \times D \\ \mathbf{\Lambda}(t = \tau) = 0, & D \end{cases} \tag{2.34}$$

Hence the following adjoint identity holds,

$$\int_0^\tau \int_D \mathbf{F} \cdot \mathbf{Z} \, dx \, dt = \int_0^\tau \int_D \mathbf{\Lambda} \cdot \mathbf{W} \, dx \, dt$$

In the case where $F = {}^*\gamma_{\Gamma_t(\mathbf{V})}(f \, \mathbf{n})$, where $f \in L^2(\Sigma(\mathbf{V}))$, the distributed term turns out to be supported by the lateral boundary,

$$\int_0^\tau \int_D {}^*\gamma_{\Gamma_t(\mathbf{V})} f \, \mathbf{n} \cdot \mathbf{Z} \, dx \, dt = \int_0^\tau \int_{\Gamma_t(\mathbf{V})} f \, \mathbf{Z} \cdot \mathbf{n}$$

Then in this case, the adjoint field is also supported by $\Sigma(\mathbf{V})$ and it can be expressed as

$$\mathbf{\Lambda} = -{}^*\gamma_{\Gamma_t(\mathbf{V})}(\lambda \, \mathbf{n})$$

where $\lambda$ is the unique solution[6] of the following adjoint tangential transverse problem

$$\begin{cases} -\partial_t \lambda - \nabla_\Gamma \lambda \cdot \mathbf{V} - (\operatorname{div} \mathbf{V}) \lambda = f, & \Sigma(\mathbf{V}) \\ \lambda(t = \tau) = 0, & \Gamma_\tau(\mathbf{V}) \end{cases} \tag{2.35}$$

This means that we have

$$\int_0^\tau \int_{\Gamma_t(\mathbf{V})} f \, \mathbf{Z} \cdot \mathbf{n} \, ds \, dt = -\int_0^\tau \int_{\Gamma_t(\mathbf{V})} \lambda \, \mathbf{n} \cdot \mathbf{W} \, ds \, dt \tag{2.36}$$

This identity allows to identify the gradient of $j(\mathbf{V})$ with respect to $\mathbf{V}$ by setting $f = \partial_{\mathbf{n}} y \, \partial_{\mathbf{n}} \varphi$.

---

[5] $\Lambda_{ad} \stackrel{\text{def}}{=} \{\mathbf{\Lambda} \in \mathcal{C}^0([0, \tau]; H(D)), \quad \partial_t \mathbf{\Lambda} + D\mathbf{\Lambda} \cdot \mathbf{V} \in L^2(0, \tau; H(D))\}.$

[6] in $\lambda_{ad} \stackrel{\text{def}}{=} \{\lambda \in \mathcal{C}^0([0, \tau]; L^2(\Gamma_t(\mathbf{V}))), \quad \partial_t \lambda + \nabla_\Gamma \lambda \cdot \mathbf{V} \in L^2(0, \tau; H(D))\}.$

# Chapter 3

# Weak evolution of sets and tube derivatives

*As a first reading, it is kindly advised for someone new to this field, to be focused on Section 4 and 5. The Sections 2 and 3 furnish the necessary mathematical framework that justify the different calculus tools described later on. These tools are usually built when the dynamic of the domains are smooth enough, which means that the velocity field is Lipschitz. However in most cases such as the tube deformations or other intrinsic objects, this regularity is not necessary. This is why we will concentrate on the weak evolution of sets.*

## 3.1  Introduction

The weak convection of measurable sets has been introduced in [157], [155] to deal with the shape differential equation and its applications. We enlarge this approach to variational problems related to the tubes evolution ( this terminology is borrowed from [8] where it is defined thanks to mutational concepts). We focus on compactness results using different kinds of boundedness assumptions. This leads to existence results connected to the "parabolic version " of the compact inclusion property from the bounded variation functions space into the integrable functions space, known as the Helly's compactness theorem. One of the boundedness hypotheses is based on the use of the "Density Perimeter" properties [22], [21].

The so-called "Speed Method" has been developed in relation with the shape optimization of systems governed by Partial Differential Equations (PDE) in e.g., [147], [145], [146], [150]. In the strong version, we considered the flow mapping $T_t(\mathbf{V})$ of a smooth vector field $\mathbf{V} \in \mathcal{C}^0([0, \infty[, \mathcal{C}^k(\mathbb{R}^d, \mathbb{R}^d) \cap L^\infty(\mathbb{R}^d, \mathbb{R}^d))$. Then for any set $\Omega_0 \subset \mathbb{R}^d$, the transported set $\Omega_t(\mathbf{V}) \stackrel{\text{def}}{=} T_t(\mathbf{V})(\Omega_0)$ is defined at all times. Hence the shape differential equation has been studied in [145],[150],[79] for shape functional governed by several linear classical boundary value problems (with use of the extractor estimate for the shape gradient [24]), and in [155] for the case of a non-linear viscous flow.

In order to weaken the above regularity requirement, we can notice that the characteristic function of the evolution domain, $\xi_t \overset{\text{def}}{=} \xi_{\Omega_t(\mathbf{V})}$, can be written as $\xi_t = \xi_0 \circ [\mathbf{T}_t(\mathbf{V})]^{-1}$ and is the solution of the following convection equation,

$$\partial_t \xi + \nabla \xi \cdot \mathbf{V} = 0, \ \xi(t = 0) = \xi_{\Omega_0}$$

This problem admits solutions when the vector field satisfies $\mathbf{V} \in L^2$, div $\mathbf{V} \in L^2$ and some growth assumption on the positive part $(\text{div }\mathbf{V})^+$. The incompressible situation was already introduced in [24], [155]. We shall add several results about the continuity and the compactness of the solutions.

---

## 3.2 Weak convection of characteristic functions

### 3.2.1 The convection equation

Let $\mathbf{V}$ be a smooth vector field in $\mathbb{R}^d$ and $\mathbf{T}_t(\mathbf{V})$ its associated flow mapping. For any measurable set $\Omega \subset \mathbb{R}^d$, we consider the perturbed set $\Omega_t(\mathbf{V}) \overset{\text{def}}{=} \mathbf{T}_t(\mathbf{V})(\Omega)$. Its characteristic function satisfies the following strong transport property,

$$\xi_{\Omega_t} = \xi_\Omega \circ [\mathbf{T}_t(\mathbf{V})]^{-1} \tag{3.1}$$

and the function $\xi(t, x) \overset{\text{def}}{=} \xi_{\Omega_t}(x)$ satifies, in a weak sense, the following convection equation,

$$\begin{cases} \partial_t \xi + \nabla \xi \cdot \mathbf{V} = 0, \ (0, \tau) \times D \\ \xi(t = 0) = \xi_\Omega, \quad D \end{cases} \tag{3.2}$$

**REMARK 3.1** Notice that the term $\nabla \xi.\mathbf{V} = \text{div}(\xi \mathbf{V}) - \xi \text{ div } \mathbf{V}$ does make sense as a distribution, whenever $\xi \in L^\infty((0, \tau) \times D)$, $\mathbf{V} \in L^1((0, \tau) \times D; \mathbb{R}^d)$ and div $\mathbf{V} \in L^1((0, \tau) \times D)$. ∎

When $\mathbf{V}$ is a smooth vector field, e.g., $\mathbf{V} \in L^1(0, \tau; W^{1,\infty}(D, \mathbb{R}^d))$ where $D$ stands for a bounded hold-all domain[1] containing the perturbed sets for any time $t \geq 0$, the associated flow mapping is well defined for time $t \in [0, \tau]$ and the identities (3.2) - (3.1) are equivalent.

Many questions arise concerning the convection equation (3.2) when the vector field $\mathbf{V}$ is non-smooth. Nevertheless, to our mind, this is the correct

---

[1]With the following boundary condition associated to the vector field $\mathbf{V}$,

$$\langle \mathbf{V}, \mathbf{n} \rangle = 0, \ \text{ on } \partial D$$

or any viability condition in the case of a non-smooth boundary [51].

approach for many shape problems such as shape evolution, shape identification or optimization, free boundary problems, fluid-structure interaction problems, computer vision or problems where the topology of the set $\Omega_t$ may be changing. Actually, the problem (3.2) is equivalent to the following one:

$$\int_0^\tau \int_D \xi \left( \partial_t \phi + \langle V, \nabla \phi \rangle + \phi \operatorname{div} V \right) dx\, dt + \int_\Omega \phi(0, x)\, dx = 0 \tag{3.3}$$

$$\forall \phi \in C^\infty([0, \tau] \times D), \quad \phi(\tau, .) = 0$$

We shall consider velocity fields such that $V \in L^1(0, \tau; L^2(D; \mathbb{R}^d))$ and the divergence positive part[2] $(\operatorname{div} V)^+ \in L^1(0, \tau; L^\infty(D))$. In this case, using a Galerkin approximation and some energy estimate, we are able to derive an existence result of solutions with initial data given in $H^{-1/2}(D)$. For the time being, any uniqueness result has been obtained for this smoothness level, but there exists an estimate (3.5) for such variational weak solutions.

Actually, when the field $V$ and its divergence are simply $L^1$ functions, the notion of weak solutions associated to the convection problems (3.2)-(3.3) does not make sense. In this case, the correct modeling tool for shape evolution is to introduce the product space of elements $(\xi = \xi^2, V)$ equipped with a parabolic BV like topology for which the constraint (3.3) defines a closed subset $\mathcal{T}_\Omega$ which contains the weak closure of smooth elements

$$\{(\xi_\Omega \circ [T_t(V)]^{-1}, V), \quad V \in C^\infty([0, \tau] \times \bar{D}; \mathbb{R}^d)\}$$

This approach consists in handling characteristic functions $\xi = \xi^2$ which belongs to $L^1(0, \tau; BV(D))$ together with vector fields $V \in L^2(0, \tau; L^2(D, \mathbb{R}^d))$ solution of problem (3.3). For a given element $(\xi, V) \in \mathcal{T}_\Omega$, we consider the set of fields $W$ such that $(\xi, W) \in \mathcal{T}_\Omega$. It forms a closed convex set, noted $\mathcal{V}_\xi$. Hence, we can define the unique minimal norm energy element $V_\xi$ in the convex set $\mathcal{V}_\xi$. For a given tube[3] $\xi$, the element $V_\xi$ is the unique (with minimal norm) vector field associated to $\xi$ via the convection equation (3.2).

### 3.2.2 The Galerkin approximation

Let us consider the following evolution problem,

$$\begin{cases} \partial_t u + \nabla u \cdot V = f, & (0, \tau) \times D \\ u(t = 0) = u_0, & D \end{cases} \tag{3.4}$$

We shall state the following existence result,

---

[2] To deal with the dual evolution problem (3.14), we shall assume

$$(\operatorname{div} V)^- \in L^1(0, \tau; L^\infty(D)).$$

[3] We say that $\xi \in \mathcal{T}_\Omega$ is a tube, i.e., a measurable non-cylindrical subset in $(0, \tau) \times D$, defined up to a zero measure set with bottom $\Omega$.

**PROPOSITION 3.1**
*Let $\mathbf{V} \in L^1(0, \tau; L^2(D, \mathbb{R}^3))$ with*

$$\text{div }\mathbf{V} \in L^1(0, \tau; L^2(D, \mathbb{R}^3)),$$
$$\|(\text{div }\mathbf{V}(t))^+\|_{L^\infty(D)} \in L^1(0, \tau),$$
$$\langle \mathbf{V}, \mathbf{n} \rangle = 0, \ \ in \ L^1(0, \tau; H^{-\frac{1}{2}}(\partial D))$$

*and $f \in L^1(0, \tau; L^2(D))$, $u_0 \in L^2(D)$. Then, equation (3.4) admits at least one weak solution $u \in L^\infty(0, \tau; L^2(D)) \cap C^0([0, \tau]; W^{-1,1}(D))$ together with*

$$\partial_t u \in L^1(0, \tau; W^{-1,1}(D))$$

*Moreover, there exists a constant $M > 0$ such that the following estimate holds true,*

$$\|u\|_{L^\infty(0,\tau;L^2(D))} \leq M \left[ \|u_0\|_{L^2(D)} + \|f\|_{L^1(0,\tau;L^2(D))} \right] \Big[ 1 +$$

$$+ \left( \int_0^\tau \left[ (\|(\text{div }\mathbf{V}(s))^+\|_{L^\infty(D,\mathbb{R}^3)} + \|f(s)\|_{L^2(D,\mathbb{R}^3)}) \int_s^\tau (\|(\text{div }\mathbf{V}(\sigma))^+\|_{L^\infty(D)} + \right.\right.$$

$$\|f(\sigma)\|_{L^2(D)}) d\sigma \Big] \ ds \Bigg) \Bigg] \quad (3.5)$$

**REMARK 3.2**
$$\text{div }\mathbf{V} = (\text{div }\mathbf{V})^+ - (\text{div }\mathbf{V})^-.$$

◻

**PROOF**   Let us consider a dense family $\{e_k\}_{k \geq 1}$ in $L^2(D)$ with $e_k \in C_c^\infty(D)$, $\forall k \geq 1$. Consider the element

$$u^m(t, x) = \sum_{i=1}^m u_i^m(t) \, e_i(x)$$

solution of the following linear ordinary differential system:

$$\int_D [\partial_t u^m(t, x) + \langle \mathbf{V}(t, x), \nabla u^m(t, x) \rangle] \, e_j(x) \, dx = \int_D f(t, x) \, e_j(x) \, dx,$$

$$1 \leq j \leq m \ (3.6)$$

Using the vector notation $(U^m)_{1 \leq i \leq m} \overset{\text{def}}{=} (u_i^m)_{1 \leq i \leq m}$, the last system is equivalent to the following one,

$$\partial_t U^m(t) + M^{-1} \cdot A(t) \cdot U^m(t) = F(t) \quad (3.7)$$

where

$$[M_{i,j}]_{1 \leq i,j \leq m} = \int_D e_i(x)\, e_j(x)\, dx$$

$$[A_{i,j}]_{1 \leq i,j \leq m}(t) = \int_D \langle \mathbf{V}(t,x), \nabla e_i(x) \rangle e_j(x)\, dx$$

Classically, this linear differential system admits a global solution when $\mathbf{V} \in L^1(0, \tau; L^2(D, \mathbb{R}^3))$. We shall derive classical energy estimates, by multiplying the system (3.6) by $u^m(t)$ and by using the following identity,

$$\int_D \langle \mathbf{V}(t,x), \nabla u^m(t,x) \rangle u^m(t,x)\, dx = -\frac{1}{2} \int_D |u^m(t,x)|^2 \operatorname{div} \mathbf{V}(t,x)\, dx \quad (3.8)$$

Then we get

$$\frac{1}{2} \int_D \partial_t |u^m(t)|^2 \leq \frac{1}{2} \int_D |u^m(t)|^2 (\operatorname{div} \mathbf{V}(t))^+ + \int_D f(t)\, u^m(t) \quad (3.9)$$

Integrating over $(0, \tau)$, we obtain

$$\|u^m(\tau)\|^2_{L^2(D)} \leq \|u^m(0)\|^2_{L^2(D)} + \int_0^\tau \int_D |u^m(s,x)|^2 (\operatorname{div} \mathbf{V}(s,x))^+ \, dx\, ds$$

$$+ 2 \int_0^\tau \int_D f(s,x)\, u^m(s,x)\, dx\, ds \quad (3.10)$$

Now we set

$$\psi(s) = \|(\operatorname{div} \mathbf{V}(s,.))^+\|_{L^\infty(D)}$$

We have

$$\|u^m(\tau)\|^2_{L^2(D)} \leq \|u^m(0)\|^2_{L^2(D)} + \int_0^\tau \psi(s) \|u^m(s)\|^2_{L^2(D)}\, ds + 2 \int_0^\tau \int_D f(s)\, u(s)$$

Using Cauchy-Schwartz inequality, we have

$$2 \int_D f(s)\, u(s)\, dx \leq 2 \|f(s)\|_{L^2(D)} \|u^m(s)\|_{L^2(D)}$$

And using Young inequality, we get

$$\|u^m(s)\|_{L^2(D)} \leq \frac{1}{2} \left[ 1 + \|u^m\|^2_{L^2(D)} \right]$$

from which we deduce

$$\|u^m(\tau)\|^2_{L^2(D)} \leq \left[ \|u^m(0)\|^2_{L^2(D)} + \|f\|_{L^1(0,\tau;L^2(D))} \right]$$

$$+ \int_0^\tau \left[ \psi(s) + \|f(s)\|_{L^2(D)} \right] \|u^m(s)\|^2_{L^2(D)}\, ds$$

Using that $u_0 \in L^2(D)$, we get

$$\|u^m(\tau)\|^2_{L^2(D)} \leq M \left[ \|u_0\|^2_{L^2(D)} + \|f\|_{L^1(0,\tau;L^2(D))} \right]$$
$$+ \int_0^\tau \left[ \psi(s) + \|f(s)\|_{L^2(D)} \right] \|u^m(s)\|^2_{L^2(D)} \, ds$$

From Gronwall's inequality, we derive:

$$\|u^m(\tau)\|_{L^2(D)} \leq M \left[ \|u_0\|_{L^2(D)} + \|f\|_{L^1(0,\tau;L^2(D))} \right] \left[ 1+ \right.$$

$$\left. \int_0^\tau \left[ \psi(s) + \|f(s)\|_{L^2(D)} \right] \exp\left( \int_s^\tau \left[ \psi(\sigma) + \|f(\sigma)\|_{L^2(D)} \right] d\sigma \right) ds \right] \quad (3.11)$$

Hence $u^m$ remains bounded in $L^\infty(0,\tau;L^2(D))$ and there exists an element $u \in L^\infty(0,\tau;L^2(D))$ and a subsequence still denoted $u^m$ which converges to $u$ for the weak-* topology. In the limit case, the element $u$ satisfies itself to the previous estimate and it can be checked that $u$ is the solution of equation (3.2) in the distribution sense, i.e,

$$-\int_0^\tau \int_D u \left( \partial_t \phi + \operatorname{div}(\phi \mathbf{V}) \right) dx \, dt = \int_D \phi(0) \, u_0 \, dx + \int_0^\tau \int_D f \, \phi \, dx \, dt$$
$$\forall \phi \in H_0^1(0,\tau;L^2(D)) \cap L^2(0,\tau;H_0^1(D)), \quad \phi(\tau) = 0 \quad (3.12)$$

□

**REMARK 3.3**    Notice that the identity (3.8) can also be written in the following form

$$\int_D \operatorname{div}(\mathbf{V}(t,x) \, u^m(t,x)) \, u^m(t,x) \, dx = +\frac{1}{2} \int_D \langle u^m(t,x), u^m(t,x) \rangle \operatorname{div} \mathbf{V}(t,x) \, dx$$
$$(3.13)$$

Hence when $(\operatorname{div} \mathbf{V})^+$ is turned into $(\operatorname{div} \mathbf{V})^-$, the previous existence results apply for the following evolution problem,

$$\begin{cases} \partial_t \, p(t) + \operatorname{div}(p(t) \, \mathbf{V}(t)) = g, & (0,\tau) \times D \\ p(t=0) = p_0, & D \end{cases} \quad (3.14)$$

□

## 3.3    Tube evolution in the context of optimization problems

In the previous section, we have stated an existence result for weak solutions of the convection equation (3.2). Unfortunately, this result does not guarantee neither uniqueness nor the stability of characteristic functions. A possible strategy is to assume different regularity levels for the velocity field in order to recover uniqueness and stability properties for initial characteristic functions [5]. This makes the mathematical setting even more complex. Here, we chose to adopt a different point of view inspired by the optimization problems framework. Indeed, our final goal is to apply the weak set evolution setting to the control problem arising in various fields such as free boundary problems or image processing. The usual situation can be described as follows: let us consider a given smooth enough functional $J(\xi, \mathbf{V})$. We would like to solve the following optimization problem

$$\inf_{(\xi, \mathbf{V}) \in \mathcal{T}_\Omega^{\mathscr{L}\mathrm{ip}}} J(\xi, \mathbf{V}) \tag{3.15}$$

where

$$\mathcal{T}_\Omega^{\mathscr{L}\mathrm{ip}} \overset{\text{def}}{=} \left\{ (\chi_\Omega \circ \mathbf{T}_t^{-1}(\mathbf{V}), \mathbf{V}) \mid \mathbf{V} \in \mathcal{U}_{ad} \right\}$$

stands for the space of smooth admissible couples. The space $\mathcal{U}_{ad}$ is a space of smooth velocities[4].

In most situations, such a problem does not admit solutions and we need to add some regularization terms to ensure its solvability. Consequently, we shall introduce different penalization terms which furnish compactness properties of the minimizing sequences inside an ad-hoc weak topology involving bounded variation constraints. Then, the new problem writes

$$\inf_{(\xi, \mathbf{V}) \in \mathcal{T}_\Omega^{\mathscr{L}\mathrm{ip}}} J(\xi, \mathbf{V}) + F(\xi, \mathbf{V}) \tag{3.16}$$

The penalization term $F(\xi, \mathbf{V})$ can be chosen using several approaches:

- We can first consider the time-space perimeter of the lateral boundary $\Sigma$ of the tube, developed in [155]. This approach easily draws part of the variational properties associated to the bounded variation functions space framework. In particular, it uses the compactness properties of tubes family with bounded perimeters in $\mathbb{R}^{d+1}$. Nevertheless, this method leads to heavy variational analysis developments.

- We can rather consider the time integral of the spatial perimeter of the moving domain which builds the tube, as introduced in [157]. We

---

[4]
$$\mathcal{U}_{ad} \overset{\text{def}}{=} \left\{ \mathbf{V} \in \mathcal{C}^0([0, \tau]; \mathcal{V}_0^k(D)), \quad \exists c > 0, \quad \mathscr{L}\mathrm{ip}_k(\mathbf{V}(t)) \leq c, \forall t \in [0, \tau] \right\}$$

with $k \geq 0$ and where

$$\mathcal{V}_0^k(D) \overset{\text{def}}{=} \left\{ \mathbf{V} \in (\mathcal{C}^k(\bar{D}))^d, \mathbf{V} \cdot n_{\partial D} = 0, \text{ on } D \right\}$$

with $\mathscr{L}\mathrm{ip}_k(\mathbf{V}) \overset{\text{def}}{=} \sum_{|\alpha|=k} \mathscr{L}\mathrm{ip}(\mathrm{D}^\alpha \mathbf{V})$ for $k \geq 1$ and $\mathscr{L}\mathrm{ip}(\mathbf{V}) \overset{\text{def}}{=} \sup_{y \neq \mathbf{x}} \dfrac{|\mathbf{V}(y) - \mathbf{V}(\mathbf{x})|}{|y - \mathbf{x}|}$.

shall extend these results to the case of vector fields living in $L^2((0,\tau) \times D; \mathbb{R}^d)$. In this case, only existence results for solutions of the convection equation can be handled and the uniqueness property is lost. The main idea, here, is to consider the set of pairs $(\xi, \mathbf{V})$ solving the convection equation with a given initial set $\Omega$ inside $D$.

### 3.3.1 Penalization using the generalized perimeter's time integral

In this section, we shall consider the optimization problem (3.16) with a particular penalization term of the following type,

$$F(\xi, \mathbf{V}) = \alpha \int_0^\tau \|\xi(t)\|_{\mathrm{BV}(D)} \, dt + \beta \int_0^\tau \left[ \|\mathbf{V}(t)\|_{L^2(D;\mathbb{R}^d)}^2 + \|\operatorname{div} \mathbf{V}(t)\|_{L^2(D)}^2 \right] dt$$

(3.17)

This penalization term, involving the $L^1$ norm of the generalized perimeter in $D$ associated to the set characterized by $\xi$, will furnish the required compactness properties for the minimizing sequences of problem (3.16) in order to get an existence result in some ad-hoc functional spaces endowed with specific weak topologies.

**REMARK 3.4**   However, with the above penalization, we can not guarantee the generated solution set to be an open set. This drawback will be solved in the section 3.3.3.                                                          ⏹

To be more precise, given a measurable subset $\Omega \subset D$, we consider the following sets equipped with their respective weak topologies:

**DEFINITION 3.1**

$$\mathcal{A} = \left\{ \xi^2 = \xi, \quad \xi \in L^\infty(0, \tau; L^\infty(D)), \quad \nabla \xi \in L^1(0, \tau; M^1(D, \mathbb{R}^d)) \right\}$$

*where $M^1(D; \mathbb{R}^d)$ stands for the sets of Radon[5] measures on $D$ with values in $\mathbb{R}^d$.*

**DEFINITION 3.2**   *The sequence $\xi_n \to \xi$ in $\mathcal{A}$ if and only if the following conditions hold:*

$$\int_0^\tau \int_D (\xi_n - \xi) \, \phi \to 0, \quad \forall \phi \in L^1((0, \tau) \times D),$$

$$\int_0^\tau \int_D (\nabla \xi_n - \nabla \xi) \cdot g \to 0, \forall g \in L^1(0, \tau; \mathcal{C}_c^0(D, \mathbb{R}^d))$$

---

[5]This space can be defined as the dual space associated to the space $(\mathcal{C}_c^0(D; \mathbb{R}^d), \|.\|_{L^\infty})$. A useful property says that the bounded sets inside $L^1(D, \mathbb{R}^d)$ are relatively compact inside $M^1$ for the weak-* $\sigma(M^1, \mathcal{C}_c^0)$ topology.

**DEFINITION 3.3**

$$\mathcal{B} = \left\{ \mathbf{V} \in L^2(0, \tau; L^2(D, \mathbb{R}^d)), \quad \operatorname{div} \mathbf{V} \in L^2(0, \tau; L^2(D)) \right\}$$

Let us consider the following set:

$$\mathcal{T}_\Omega = \{(\xi, \mathbf{V}) \in \mathcal{A} \times \mathcal{B}, \text{ with } \partial_t \xi + \nabla \xi \cdot \mathbf{V} = 0, \quad \xi(t = 0) = \chi_\Omega\} \qquad (3.18)$$

**THEOREM 3.1**
*The set $\mathcal{T}_\Omega$ is closed in $\mathcal{A} \times \mathcal{B}$.*

**PROOF** Let $(\xi_n, \mathbf{V}_n)$ be a sequence inside $\mathcal{T}_\Omega$ converging towards $(\xi, \mathbf{V})$. From the Banach-Steinhaus theorem, we get the boundedness of the following $L^1(0, \tau; \mathrm{BV}(D))$ norm:

$$\int_0^\tau \|\nabla \xi_n\|_{M^1(D)} \, \mathrm{d}t \leq M$$

Using the next Proposition[6] 3.2, we deduce that there exists a subsequence still denoted $\xi_n$ which converges in $L^1$ for the strong topology. Then, the limit element satisfies $\xi^2 = \xi$. Finally, in the limit case, the weak formulation of problem (3.3),

$$\int_0^\tau \int_D \xi_n(\partial_t \phi + \langle \mathbf{V}_n, \nabla \phi \rangle - \phi \operatorname{div} \mathbf{V}_n) + \int_\Omega \phi(0, x) = 0,$$
$$\forall \phi \in C^\infty([0, \tau] \times D), \quad \phi(\tau, .) = 0$$

is satisfied by $(\xi, \mathbf{V})$ since $\mathbf{V}_n$ and $\operatorname{div} \mathbf{V}_n$ weakly converge in $L^2$. Hence, the limit pair $(\xi, \mathbf{V})$ turns out to be an element of $\mathcal{T}_\Omega$.
⬜

### 3.3.2 Parabolic version of Helly's compactness theorem

Let us define the space,

$$\mathcal{W} \stackrel{\text{def}}{=} \{ f \in L^1(0, \tau; \mathrm{BV}(D)), \quad \partial_t f \in L^2(0, \tau; H^{-2}(D)) \} \qquad (3.19)$$

The following compactness result holds,

---

[6]with
$$f_n = \xi_n \in L^1(0, \tau; \mathrm{BV}(D))$$
and
$$f'_n = \xi'_n = -\operatorname{div}(\xi_n \mathbf{V}_n) + \xi_n \operatorname{div} \mathbf{V}_n \in L^2(0, \tau; H^{-2}(D)).$$

**PROPOSITION 3.2**
*Uniformly bounded sets in $\mathcal{W}$ are sequentially relatively compact in*

$$L^1\big(0,\tau; L^1(D)\big)$$

**PROOF**   For $d \leq 3$ the following continuous injection holds,

$$L^1(D) \hookrightarrow H^{-2}(D).$$

Furthermore using Helly's theorem (Theorem 1.19 in [77] page 17) the following continous, compact injection holds,

$$\mathrm{BV}(D) \hookrightarrow_c L^1(D).$$

Then, we have the following embedding chain,

$$\mathrm{BV}(D) \hookrightarrow_c L^1(D) \hookrightarrow H^{-2}(D)$$

We adapt the proof of the compactness result (Theorem 5.1 page 58 in [97]) to the non-reflexive situation.

Let us consider $\{f_n\}_{n \geq 0}$ a uniformly bounded sequence in $\mathcal{W}$. Using the result proven in the next Lemma 3.1, we deduce for $f \in L^1\big(0,\tau; \mathrm{BV}(D)\big)$, $\forall n > 0$, there exists a constant $d_n > 0$ such that

$$\|f\|_{L^1\big(0,\tau; L^1(D)\big)} \leq n\|f\|_{L^1\big(0,\tau; \mathrm{BV}(D)\big)} + d_n\|f\|_{L^1\big(0,\tau; H^{-2}(D)\big)}$$

We apply this inequality with $f = f_{n,m} \stackrel{\mathrm{def}}{=} f_n - f_m$, for $m > n$.

Let us fix $\varepsilon > 0$, since there exists $M > 0$ such that

$$\|f_{n,m}\|_{L^1(0,\tau; \mathrm{BV}(D))} \leq M$$

We can choose $n$ such that $n\, M \leq \frac{\varepsilon}{2}$ and we get

$$\|f_{n,m}\|_{L^1(0,\tau; L^1(D))} \leq \frac{\varepsilon}{2} + d_n\|f_{n,m}\|_{L^1(0,\tau; H^{-2}(D))}$$

At this stage, if we establish the strong convergence towards zero of $f_{n,m}$ in the space $L^1(0,\tau; H^{-2}(D))$, then the proof is complete.

As $L^1(D) \subset H^{-2}(D)$ (for $d \leq 3$), using Lemma 1.2 page 7 in [97], we have

$$f_{n,m} \in W^{1,1}(0,\tau; H^{-2}(D)) \hookrightarrow \mathcal{C}^0([0,\tau], H^{-2}(D))$$

Then, we have the following uniform estimate

$$\|f_{n,m}(t_0))\|_{H^{-2}(D)} \leq M, \quad \forall t_0 \in [0,\tau]$$

Using the Lebesgue dominated convergence theorem, it will be sufficient to prove the pointwise convergence of $f_{n,m}(t_0)$ strongly to zero in $H^{-2}(D)$. To this end, we use a technique introduced by R.Temam as reported in [97] (footnote of the proof of Theorem 5.1 page 59, see also [139] page 273). It is easy to show that the following expression holds true,

$$f_{n,m}(t_0) = a_{n,m} + b_{n,m}$$

with

$$a_{n,m} = 1/s \int_{t_0}^{t_0+s} f_{n,m}(t)\, dt, \; b_{n,m} = -\frac{1}{s} \int_{t_0}^{t_0+s} (s-t)\, f'_{n,m}(t)\, dt$$

Let us choose $s$ such that

$$d_n \, \|b_{n,m}\|_{H^{-2}(D)} \leq \int_{t_0}^{t_0+s} \|f'_{n,m}(t)\|_{H^{-2}(D)} \, dt \leq$$

$$\leq d_n \, s^{1/2} \left( \int_{t_0}^{t_0+s} \|f'_{n,m}(t)\|^2_{H^{-2}(D)} \right)^{1/2}$$

$$\leq \frac{\varepsilon}{4}$$

Finally, we have the estimate

$$\|a_{n,m}\|_{BV(D)} \leq \frac{1}{s}\|f_{n,m}\|_{L^1(0,\tau;BV(D))} \leq \frac{M}{s}$$

Again using Helly's compactness result, we deduce that $a_n$ converges strongly in $H^{-2}(D)$ towards an element $a \in H^{-2}(D)$ . This limit is actually zero since by the same arguments we can show that $1/s \int_{t_0}^{t_0+s} f_n(t)\, dt$ converges in $H^{-2}(D)$ for the strong topology. Then it is a Cauchy sequence and we deduce that $a = 0$. Then, there exists $N \leq 0$, such that for $m > n \geq N$,

$$d_n \, \|a_{n,m}\|_{H^{-2}(D)} \leq \frac{\varepsilon}{4}$$

We conclude that

$$\forall \varepsilon > 0, \exists N \geq 0, \forall m > n \geq N,$$
$$\|f_{n,m}\|_{L^1(0,\tau;L^1(D))} \leq \frac{\varepsilon}{2} + \frac{\varepsilon}{4} + \frac{\varepsilon}{4} \leq \varepsilon$$

Then the sequence $\{f_n\}_{n\geq 0}$ is a Cauchy sequence in the Banach space $L^1(0, \tau; L^1(D))$. Then it is a convergent sequence for the strong topology.
□

### LEMMA 3.1

$$\forall \eta > 0, \text{there exists a constant } c_\eta \; \text{ with } \forall \phi \in BV(D),$$
$$\|\phi\|_{L^1(D)} \leq \eta \|\phi\|_{BV(D)} + c_\eta \|\phi\|_{H^{-2}(D)}$$

**PROOF**    Let us assume that the above estimate does not hold. Then, $\forall \eta > 0$, there exists $\phi_n \in BV(D)$ and $c_n \to \infty$ such that

$$\|\phi_n\|_{L^1(D)} \geq \eta \|\phi_n\|_{BV(D)} + c_n \|\phi_n\|_{H^{-2}(D)}$$

We introduce $\psi_n = \phi_n / \|\phi_n\|_{BV(D)}$, and we derive:

$$\|\psi_n\|_{L^1(D)} \geq \eta + c_n \|\psi_n\|_{H^{-2}(D)} \geq \eta$$

But also $\|\psi_n\|_{L^1(D)} \leq c \|\psi_n\|_{BV(D)} = c$, for some constant $c$. Then, we have

$$\|\psi_n\|_{H^{-2}(D)} \to 0$$

But as $\|\psi_n\|_{BV(D)} = 1$, there exists a subsequence strongly convergent in $L^1(D) \subset H^{-2}(D)$, which turns to be strongly convergent to zero. This is in contradiction with the fact that $\|\psi_n\|_{L^1(D))} \geq \eta$. $\qquad \Box$

### 3.3.3    Generation of clean open tubes

**Clean open tubes**

**DEFINITION 3.4**    *A clean open tube is a set $\tilde{Q}$ in $]0, \tau[\times D$ such that for a.e. $t > 0$, the set*

$$\tilde{\Omega}_t = \left\{ x \in D, \quad (t, x) \in \tilde{Q} \right\} \qquad (3.20)$$

*is an open set in $D$ satisfying for a.e. $t \in (0, \tau)$ the following cleanness property:*

$$\text{meas}(\partial \tilde{\Omega}_t) = 0, \ \tilde{\Omega}_t = \text{int}(\text{cl}(\tilde{\Omega}_t)). \qquad (3.21)$$

**REMARK 3.5**    If the set $\tilde{\Omega}_t$ is an open set in $D$ for a.e. $t \in (0, \tau)$, the set $\tilde{Q}$ is not necessarily an open subset in $]0, \tau[\times D$. Nethertheless when the field **V** is smooth, the tube $\bigcup_{0 < t < \tau} \{t\} \times T_t(\mathbf{V})(\Omega_0)$ is open (resp. open and clean open) when the initial set $\Omega_0$ is open (resp. open and clean open) in $D$. $\qquad \Box$

**REMARK 3.6**    Two tubes $Q, Q'$ are said to be equivalent if

$$\chi_Q = \chi_{Q'}, \ \text{in } L^2(0, \tau; L^2(D))$$

This means that $\Omega_t$ and $\Omega'_t$ are the same up to zero-measure sets. $\qquad \Box$

**LEMMA 3.2**
*Let $Q$ be a measurable set in $]0, \tau[\times D$. If there exists a clean open tube $\tilde{Q}$ such that*

$$\chi_Q = \chi_{\tilde{Q}}$$

*Then, this clean tube is unique.*

**PROOF**    Assume that there exists two equivalent clean tubes $\tilde{Q}$ and $\tilde{Q}'$. Then a.e. $t \in (0, \tau)$, we have $\tilde{\Omega}_t = \tilde{\Omega}'_t$ up to a measurable set $E_t$ with $\text{meas}(E_t) = 0$. As these two open sets satisfy the cleanness property (3.21), they must be equals. ▢

**DEFINITION 3.5**    *Let $\Omega_0$ be a clean open set in $D$ and $Q$ a clean open tube in $]0, \tau[\times D$ with $\xi \overset{\text{def}}{=} \chi_Q \in C^0([0, \tau]; H^{-1/2}(D))$. If there exists a divergence free field $\mathbf{V} \in \mathcal{B}$ such that:*

$$\partial_t \xi + \nabla \xi \cdot \mathbf{V} = 0, \; \xi(t = 0) = \chi_{\Omega_0} \tag{3.22}$$

*we say that $\mathbf{V}$ builds the tube and we note $Q = Q_V$.*

**DEFINITION 3.6**    *We define the set of fields that build the clean open tube $Q$,*

$$\mathcal{V}_Q = \{\mathbf{V} \in \mathcal{B}, \quad Q_V = Q\} \tag{3.23}$$

**PROPOSITION 3.3**
*If the set $\mathcal{V}_Q$ is non-empty, then the following minimization problem,*

$$e_D(Q_V) \overset{\text{def}}{=} \min_{\mathbf{V} \in \mathcal{V}_Q} \|\mathbf{V}\|_{L^2(0,\tau; L^2(D; \mathbb{R}^d))} \tag{3.24}$$

*admits a unique solution $\mathbf{V}_Q \in \mathcal{B} \cap L^2(0, \tau; L^2(D; \mathbb{R}^d))$.*

**PROOF**    It is easy to prove that if the set $\mathcal{V}_Q$ is non-empty, then it is closed and convex in $\mathcal{B}$. Hence, we use the projection theorem in the Hilbert space $L^2(0, \tau; L^2(D; \mathbb{R}^d))$. ▢

## Properties of the density perimeter

In this section, we introduce the notion of density perimeter [22], [21] for a closed set $A$ inside $D$. This notion has been introduced in order to have a control on the boundary measure of the limit set associated to a convergent family of open sets. This allow us to build a penalization term that will generate open set solutions to the optimization problem (3.16).

**DEFINITION 3.7**    *The $\varepsilon$-dilatation of the set $A$ is given by the following set,*

$$A^\varepsilon \overset{\text{def}}{=} \bigcup_{x \in A} B(x, \varepsilon) \tag{3.25}$$

**DEFINITION 3.8**    *Let $\gamma > 0$, the density perimeter of the set $A$ is given by the following expression,*

$$P_\gamma(A) = \sup_{\varepsilon \in (0,\gamma)} \left[ \frac{\text{meas}(A^\varepsilon)}{2\,\varepsilon} \right] \tag{3.26}$$

**Property 3.3.1 ([22])** *The mapping $\Omega \longrightarrow P_\gamma(\partial\Omega)$ is lower semi-continuous for the Hausdorff[7] complementary topology $H^c$.*

**Property 3.3.2 ([22])** *If the boundary set $\partial\Omega$ is such that*

$$P_\gamma(\partial\Omega) < \infty$$

*then* $\text{meas}(\partial\Omega) = 0$ *and* $\Omega \setminus \partial\Omega$ *is open in* $D$

**Property 3.3.3 ([22])** *Let us consider a family $\{\Omega_n\}_{n \geq 0}$ of open sets in $D$. Let us suppose that this family converges to some subset $\Omega \subset D$ for the $H^c$-topology and that there exists $M > 0$ such that $P_\gamma(\partial\Omega_n) \leq M$. Then the convergence holds for the char-topology[8].*

### Penalization using the *parabolic* generalized density perimeter

In this section, we shall describe a particular type of clean tubes construction for which the set $\mathcal{V}_Q$ is non-empty. Actually we shall combine the compactness properties of the generalized perimeter with the openness control properties of the density perimeter.

For any smooth free divergence vector field, $\mathbf{V} \in \mathcal{C}^0([0,\tau]; W_0^{1,\infty}(D, \mathbb{R}^d))$, we consider the following minimization problem,

$$\Theta_\gamma(\mathbf{V}, \Omega_0) = \min_{\mu \in \mathcal{M}_\gamma(\mathbf{V}, \Omega_0)} \|\dot{\mu}\|^2_{L^2(0,\tau)} \tag{3.27}$$

---

[7]The Hausdorff complementary topology is given by the metric,

$$d_{H^c}(\Omega_1, \Omega_2) = d_{H^d}(\Omega_1^c, \Omega_2^c)$$

where $\Omega_1, \Omega_2$ are open subsets of $D$ and $\Omega_1^c, \Omega_2^c$ their associated complementary sets in $D$. Furthermore $d_{H^d}$ stands for the Hausdorff metric defined by

$$d_{H^d}(F_1, F_2) = \max(\rho(F_1, F_2), \rho(F_2, F_1))$$

for any closed sets $F_1, F_2$ in $D$ and where

$$\rho(X, Y) = \sup_{x \in X} \inf_{y \in Y} \|x - y\|$$

[8]The char-topology is defined on the family of measurable subsets of $\mathbb{R}^d$ by the $L^2$-metric,

$$d_{char}(A_1, A_2) = \int_{\mathbb{R}^d} |\chi_{A_1} - \chi_{A_2}|$$

where

$$\mathcal{M}_\gamma(\mathbf{V},\Omega_0) = \left\{ \mu \in H^1(0,\tau), \quad \begin{array}{l} P_\gamma(\partial\Omega_t(\mathbf{V})) \leq \mu(t) \text{ a.e. } t \in (0,\tau), \\[2mm] \mu(t=0) \leq (1+\gamma)\,P_\gamma(\partial\Omega_0) \end{array} \right\} \tag{3.28}$$

When this set is empty, we choose to set $\Theta_\gamma(\mathbf{V},\Omega_0) = +\infty$.

**REMARK 3.7**    When the mapping

$$p : (0,\tau) \longrightarrow \mathbb{R}$$
$$t \mapsto P_\gamma(\partial\Omega_t(\mathbf{V}))$$

is an element of $H^1(0,\tau)$ , i.e., $p \in \mathcal{M}_\gamma(\mathbf{V},\Omega_0)$, we have

$$\Theta(\mathbf{V},\Omega_0) < \|\dot{p}\|^2_{L^2(0,\tau)}$$

as the minimizer will escape to possible variations of the function $p$.    ⬜

**PROPOSITION 3.4**
*For any $\mathbf{V} \in \mathcal{C}^0([0,\tau]; W_0^{1,\infty}(D,\mathbb{R}^d))$ with free divergence, the following inequality holds,*

$$P_\gamma(\partial\Omega_t(\mathbf{V})) \leq 2\,P_\gamma(\partial\Omega_0) + \sqrt{\tau}\,\Theta(\mathbf{V},\Omega_0)^{1/2} \tag{3.29}$$

**PROOF**    By definition,

$$P_\gamma(\partial\Omega_t(\mathbf{V})) \leq \mu(t) \leq 2\,P_\gamma(\partial\Omega_0) + \int_0^\tau \dot{\mu}(t)\,\mathrm{d}t$$

Then, using Cauchy-Schwartz inequality, we get

$$P_\gamma(\partial\Omega_t(\mathbf{V}_n)) \leq \mu(t) \leq 2\,P_\gamma(\partial\Omega_0) + \sqrt{\tau}\left(\int_0^\tau |\dot{\mu}(t)|^2\,\mathrm{d}t\right)^{1/2}$$

Minimizing over $\mu \in \mathcal{M}_\gamma(\mathbf{V},\Omega_0)$ furnishes the correct estimation.    ⬜

**PROPOSITION 3.5**
*Let $\mathbf{V}_n \in \mathcal{C}^0([0,\tau]; W_0^{1,\infty}(D,\mathbb{R}^d))$ with the following convergence property,*

$$\mathbf{V}_n \longrightarrow \mathbf{V} \text{ in } L^2(0,\tau; L^2(D,\mathbb{R}^d)) \tag{3.30}$$

*and the uniform boundedness,*

$$\exists M > 0, \ \Theta(\mathbf{V}_n,\Omega_0) \leq M \tag{3.31}$$

*Then*

$$\Theta(\mathbf{V}, \Omega_0) \le \liminf_{n\to\infty} \Theta(\mathbf{V}_n, \Omega_0) \tag{3.32}$$

**PROOF**    Let $\mu_n$ be the unique minimizer in $H^1(0,\tau)$ associated with $\Theta(\mathbf{V}_n, \Omega_0)$. Using the boundedness assumption and the estimation (3.29), we get

$$P_\gamma(\partial\Omega_t(\mathbf{V}_n)) \le \underbrace{2P_\gamma(\partial\Omega_0) + \sqrt{\tau}\, M^{1/2}}_{C}$$

We deduce that there exists a subsequence, still denoted $\mu_n$, which weakly converges to an element $\mu \in H^1(0,\tau)$. Furthermore, this convergence holds in $L^2(0,\tau)$ for the strong topology, and as a consequence a.e. $t \in (0.\tau)$. By definition, we have

$$P_\gamma(\partial\Omega_t(\mathbf{V}_n)) \le \mu_n \quad \text{a.e. } t \in (0, \tau)$$

Using the l-s-c property of the density perimeter (Theorem 1.9 page 7 in [77]), we have

$$P_\gamma(\partial\Omega_t(\mathbf{V})) \le \liminf_{n\to\infty} P_\gamma(\partial\Omega_t(\mathbf{V}_n)) \le \mu(t) \quad \text{a.e. } t \in (0, \tau)$$

But the square of the norm being also w-l-s-c, we have

$$\int_0^\tau |\dot\mu(t)|^2 \, dt \le \liminf_{n\to\infty} \int_0^\tau |\dot\mu_n(t)|^2 \, dt$$

This leads to

$$\Theta(\mathbf{V}, \Omega_0) \le \int_0^\tau |\dot\mu(t)|^2 \, dt \le \liminf_{n\to\infty} \Theta(\mathbf{V}_n, \Omega_0)$$

$\square$

### PROPOSITION 3.6
*Let $\mathbf{V}_n \in C^0([0,\tau]; W_0^{1,\infty}(D, \mathbb{R}^d))$ with convergence property (3.30) and bound estimate (3.31). We assume that $\Omega_0$ is an open subset in $D$ verifying*

$$\Omega_0 = \mathrm{int}(\mathrm{cl}(\Omega_0))$$

*Then, there exists a clean open tube*

$$\tilde{Q} = \bigcup_{0<t<\tau} \{t\} \times \tilde\Omega_t$$

*such that, for a.e. $t \in (0, \tau)$*

$$\chi_{\Omega_t(\mathbf{V}_n)} \to \chi_{\tilde\Omega_t}, \quad \text{in } L^2(D),$$
$$\Omega_t(\mathbf{V}_n) \to \tilde\Omega_t, \quad \text{in } H^c\text{-topology}$$

Moreover $\tilde{\Omega}_t$ *is the single open set such that cleanness condition* (3.21) *is satisfied and whose characteristic function* $\chi_{\tilde{\Omega}_t}$ *solves the convection problem:*

$$\partial_t \xi + \nabla \xi \cdot \mathbf{V}(t) = 0, \ \xi(t = 0) = \chi_{\Omega_0}$$

*This means that* $\tilde{Q}$ *is the unique clean tube equivalent to the limit tube* $Q_V$ *built by* $\mathbf{V}$.

**PROOF** We have $\chi_{Q_{V_n}} \to \xi_{Q_V}$ in $L^2((0,\tau) \times D)$. Then, for almost every $t \in (0,\tau)$, we have $\chi_{\Omega_t(\mathbf{V}_n)} \to \chi_{\Omega_t(\mathbf{V})}$ in $L^2(D)$. At each time $t \in (0,\tau)$, there exists a subsequence (depending on $t$) which converges, for the $H^c$-topology, to an open set $\omega_t$.

Meanwhile, we also have $\Omega_t(\mathbf{V}_n) \to \Omega_t(\mathbf{V})$ in the char-topology for a.e. $t \in (0,\tau)$. Then, $\chi_{\omega_t} = \chi_{\Omega_t(\mathbf{V})}$ for a.e. $t \in (0,\tau)$. From the boundedness of $P_\gamma(\Omega_t(\mathbf{V}_n))$ and using property (3.3.2), we derive that $\omega_t(\mathbf{V})$ is an open set in $D$ and meas$(\partial \omega_t(\mathbf{V})) = 0$.

Hence, we set $\tilde{\Omega}_t = \mathrm{cl}(\omega_t) - \partial \omega_t$. $\qquad\qquad\square$

**DEFINITION 3.9** *For smooth vector fields* $\mathbf{V}$, *we define the parabolic generalized density perimeter of the moving domain* $\Omega_t(\mathbf{V})$ *over the time period* $(0,\tau)$,

$$p_\gamma(\mathbf{V}, \Omega_0) = \int_0^\tau P_D(\Omega_t(\mathbf{V})) \, \mathrm{d}t + \gamma \, \Theta_\gamma(\mathbf{V}, \Omega_0) \tag{3.33}$$

*where* $P_D(\Omega_t(\mathbf{V}))$ *stands for the generalized perimeter[9] of* $\Omega_t(\mathbf{V})$.

**THEOREM 3.2**
Let $\mathbf{V}_n \in C^0([0,\tau]; W_0^{1,\infty}(D, \mathbb{R}^d))$ *which weakly converges in* $L^2(0,\tau; L^2(D, \mathbb{R}^d))$ *to* $\mathbf{V}$ *with the boundedness condition:*

$$p_\gamma(\mathbf{V}_n, \Omega_0) \leq M \tag{3.34}$$

*Then there exists a clean open tube* $\tilde{Q}_V$ *built by* $\mathbf{V}$ *with the following convergence: for a.e.* $t \in (0,\tau)$,

$$\xi_{\mathbf{V}_n}(t) \to \xi_{\mathbf{V}}(t), \quad \text{in } L^2(D),$$
$$\Omega_t(\mathbf{V}_n) \to \tilde{\Omega}_t, \quad \text{in } H^c\text{-topology}$$

---

[9]The generalized perimeter of a measurable set $A$ in $D$ is given by

$$P_D(A) = \|\nabla \chi_A\|_{M^1(D, \mathbb{R}^d)} = \sup_{\substack{\varphi \in C_c^\infty(D, \mathbb{R}^d), \\ \|\varphi\|_\infty \leq 1}} - \int_A \mathrm{div} \, \varphi \, \mathrm{d}x$$

*Furthermore,*

$$p_\gamma(\mathbf{V}) \leq \liminf_{n\to\infty} p_\gamma(\mathbf{V}_n)$$

Finally, if we define the following penalization term

$$F(\xi, \mathbf{V}) = \alpha\, p_\gamma(\mathbf{V}, \Omega_0) + \beta \int_0^\tau \left[ \|\mathbf{V}(t)\|^2_{L^2(D;\mathbb{R}^d)} + \|\operatorname{div}\mathbf{V}(t)\|^2_{L^2(D)} \right] \mathrm{d}t \quad (3.35)$$

we can obtain an existence result for the optimization problem (3.16) with the property that the associated tube solution is clean open if $\Omega_0$ is open.

---

## 3.4   Tube derivative concepts

In this section, we introduce alternative tools to handle the sensivity analysis of tube functionals. The Eulerian setting has been presented in Chapter 2 and it is based on the introduction of a velocity field $\mathbf{V} \in \mathcal{U}_{ad}$ which is smooth enough [10] for its Lagrangian flow to be defined in a classical manner[11]. Hence we have a parametrization of the moving set based on the couple $(\mathbf{V}, \mathbf{T}(\mathbf{V}))$. Introducing the transverse field which builds the flow mapping the reference moving domain into the perturbed one, allowed us to differentiate the following Eulerian integral

$$\int_0^\tau \int_{\Omega_t(\mathbf{V})} F(\mathbf{V})(\mathbf{x}, t)\, \mathrm{d}x\, \mathrm{d}t$$

with respect to $\mathbf{V}$.

A second choice consists in parametrizing the moving set thanks to the couple $(\mathbf{V}, \chi(\mathbf{V}))$ where $\chi(\mathbf{V})$ is the characteristic function of the moving set $\Omega_t(\mathbf{V})$ and is the solution of the transport equation (3.2). In this setting, it is possible to perform a sensitivity analysis of the following integral,

$$\int_0^\tau \int_D \chi(x, t)\, F(\chi)(\mathbf{x}, t)\, \mathrm{d}x\, \mathrm{d}t.$$

---

[10]
$$\mathcal{U}_{ad} \stackrel{\text{def}}{=} \left\{ \mathbf{V} \in \mathcal{C}^0\big([0,\tau]; V_0^k(D)\big), \quad \exists c > 0, \quad \mathscr{L}\mathrm{ip}_k(\mathbf{V}(t)) \leq c,\ \forall t \in [0,\tau] \right\}$$

with $k \geq 0$ and where

$$V_0^k(D) \stackrel{\text{def}}{=} \left\{ \mathbf{V} \in \big(\mathcal{C}^k(\bar{D})\big)^d, \mathbf{V} \cdot n_{\partial D} = 0,\ \text{on } D \right\}$$

with $\mathscr{L}\mathrm{ip}_k(\mathbf{V}) \stackrel{\text{def}}{=} \sum_{|\alpha|=k} \mathscr{L}\mathrm{ip}(\mathrm{D}^\alpha \mathbf{V})$ for $k \geq 1$ and $\mathscr{L}\mathrm{ip}(\mathbf{V}) \stackrel{\text{def}}{=} \sup_{y\neq\mathbf{x}} \dfrac{|\mathbf{V}(y) - \mathbf{V}(\mathbf{x})|}{|y - \mathbf{x}|}$.

[11]
$$\mathbf{T}(\mathbf{V}) \in \mathcal{C}^1\big([0,\tau]; (\mathcal{C}^k(\bar{D}))^d\big) \cap \mathcal{C}^0\big([0,\tau]; (W^{k+1,\infty}(D))^d\big)$$

A third choice consists in working directly on the tube inside the time-space Euclidian space $\mathbb{R}^{d+1} \overset{\text{def}}{=} \mathbb{R}_t \times \mathbb{R}_{\mathbf{x}}$. Hence it is possible to work in the classical Eulerian setting with extended velocity field $\tilde{\mathbf{V}} \overset{\text{def}}{=} *(0, \mathbf{V})$. This allows dealing with tube integrals of the following form,

$$\int_{Q(\tilde{\mathbf{V}})} F(\tilde{\mathbf{V}}) dQ + \int_{\Sigma(\tilde{\mathbf{V}})} f(\tilde{\mathbf{V}}) d\Sigma$$

where $\Sigma$ stands for the boundary of the tube $Q$. A motivation for this setting is the introduction of perimeter constraints inside optimization problems for tube evolution. As recalled in section 3.3, we can either define the time integral of the space perimeter,

$$\int_0^\tau \int_{\Omega(t)} d\Gamma \, dt$$

or the time-space perimeter,

$$\int_\Sigma d\Sigma$$

The boundedness of the latter characterizes the BV compactness of the tube $Q$ inside $\mathbb{R}^{d+1}$. Furthermore it has been proved in section 3.3 that the boundedness of the first perimeter furnishes the same compactness result. These perimeter constraints are associated in a variational way to the notion of surface tension which is highly required while studying free boundary problems as the one described in [153] concerning the hydrodynamical model of water waves.

### 3.4.1 Characteristics versus Eulerian flow setting

Let $(\xi, \mathbf{V}) \in \mathcal{T}_\Omega$ and a vector field $\mathbf{W}$ such that for all $s$, $|s| \leq s_1$ there exists $\xi^s$ with $(\xi^s, \mathbf{V} + s\mathbf{W}) \in \mathcal{T}_\Omega$. First, let us consider heuristically the term (if it exists)

$$\dot{\xi} = \partial_s \xi^s|_{s=0}$$

as a measure over $(0, \tau) \times D$. This measure should solve the evolution problem (3.4) with a measure term as right hand side:

$$\partial_t \dot{\xi} + \nabla \dot{\xi} \cdot \mathbf{V} = -\nabla \xi \cdot \mathbf{W}, \ \dot{\xi}(t = 0) = 0 \tag{3.36}$$

We consider a non-cylindrical functional depending on the vector field $\mathbf{V}$,

$$j(\mathbf{V}) = \int_0^\tau \int_{\Omega_t(\mathbf{V})} F_{\Omega_t(\mathbf{V})} + \int_0^\tau \int_{\Gamma_t(\mathbf{V})} f_{\Gamma_t(\mathbf{V})} + \int_{\Omega_\tau} g_{\Omega_\tau(\mathbf{V})} \tag{3.37}$$

We first consider the case where the functional only involves a distributed term defined on $Q(\mathbf{V})$, i.e;, $f_{\Gamma_t(\mathbf{V})} = 0$ and $g_{\Omega_\tau(\mathbf{V})} = 0$. We set $\xi \overset{\text{def}}{=} \chi_{\Omega_t(\mathbf{V})}$, and we can write

$$j(\xi, \mathbf{V}) = \int_0^\tau \int_D \xi \, F_\xi$$

where $F_{\Omega_t(\mathbf{V})} = F_\xi|_{\Omega_t(\mathbf{V})}$ and $F_\xi$ is defined in $(0,\tau) \times D$. We can define the derivative of the functional $j(\mathbf{V})$ with respect to $\mathbf{V}$ in the direction $\mathbf{W}$ as follows,

$$j'(\mathbf{V}, \mathbf{W}) \overset{\text{def}}{=} \partial_s j(\xi^s, \mathbf{V} + s\,\mathbf{W})|_{s=0} \tag{3.38}$$

Here, it leads to

$$j'(\mathbf{V}, \mathbf{W}) = \int_0^\tau \int_D \left[ \dot{\xi}\, F_\xi + \xi\, F' \right] \tag{3.39}$$

where $F' \overset{\text{def}}{=} \partial_s F(\xi^s)|_{s=0}$.

We introduce the following non-cylindrical adjoint state $\lambda$:

$$-\partial_t\, \lambda - \operatorname{div}(\lambda\,\mathbf{V}) = F_\xi, \ \lambda(\tau) = 0 \tag{3.40}$$

For the sake of simplicity, we assume that $F' = 0$, and we get

$$j'(\mathbf{V}, \mathbf{W}) = \int_0^\tau \int_D -\dot{\xi}\, [\partial_t\, \lambda + \operatorname{div}(\lambda\,\mathbf{V})]$$

Using integration by parts and equation (3.36), we have

$$j'(\mathbf{V}, \mathbf{W}) = \int_0^\tau \int_D \lambda \left[ \partial_t\, \dot{\xi} + \nabla\dot{\xi} \cdot \mathbf{V} \right] = -\int_0^\tau \int_D \lambda\, \nabla\xi \cdot \mathbf{W}$$

Integrating again by parts and using Stokes formula, we get

$$j'(\mathbf{V}, \mathbf{W}) = \int_0^\tau \int_D \xi\, \operatorname{div}(\lambda\,\mathbf{W}),$$

$$= \int_0^\tau \int_{\Gamma_t(\mathbf{V})} \lambda\, \langle \mathbf{W}, \mathbf{n}_t \rangle \tag{3.41}$$

In [154], [59], we have introduced the transverse field $\mathbf{Z}$ such that

$$\partial_s \left[ \int_0^\tau \int_{\Omega_t(\mathbf{V}+s\mathbf{W})} F \right]\Bigg|_{s=0} = \int_0^\tau \int_{\partial\Omega_t(\mathbf{V})} F\, \langle \mathbf{Z}, \mathbf{n}_t \rangle\, \mathrm{d}\Gamma_t\, \mathrm{d}t \tag{3.42}$$

We deduce that, as a measure on $]0, \tau[ \times D$, the element $\dot{\xi}$ satisfies the following identity,

$$\int_0^\tau \int_D \dot{\xi}\, F = \int_0^\tau \int_{\partial\Omega_t(\mathbf{V})} F\, \langle \mathbf{Z}, \mathbf{n}_t \rangle\, \mathrm{d}\Gamma_t\, \mathrm{d}t,$$

$$= \int_{\Sigma(\mathbf{V})} F\, \langle \mathbf{Z}, \mathbf{n}_t \rangle (1 + v^2)^{-1/2}\, \mathrm{d}\Sigma \tag{3.43}$$

where $v \overset{\text{def}}{=} \langle \mathbf{V}, \mathbf{n}_t \rangle$.

We introduce $\gamma_\Sigma \in \mathcal{L}(\mathcal{C}^0(]0, \tau[ \times D), \mathcal{C}^0(\Sigma))$, the trace operator on the lateral boundary $\Sigma$ of the tube $Q(\mathbf{V})$ and $\gamma_\Sigma^* \in \mathcal{L}(\mathcal{M}(\Sigma), \mathcal{M}(]0, \tau[\times D))$ its adjoint operator. Hence, we deduce

$$\dot{\xi} = \gamma_\Sigma^* \cdot \left( \frac{\langle \mathbf{Z}, \mathbf{n}_t \rangle}{\sqrt{(1 + v^2)}} \right) \tag{3.44}$$

Comparing equation (3.41) and (3.43), we derive the following result,

$$\int_0^\tau \int_{\Gamma_t(\mathbf{V})} F \langle \mathbf{Z}, \mathbf{n}_t \rangle = \int_0^\tau \int_{\Gamma_t(\mathbf{V})} \lambda \langle \mathbf{W}, \mathbf{n}_t \rangle \tag{3.45}$$

In the case where we keep the final time term $g_{\Omega_\tau}$, we set,

$$g_{\Omega_\tau} = g|_{\Omega_\tau}$$

where $g$ is defined on $D$. Introducing the adjoint problem,

$$-\partial_t \bar{\lambda} - \operatorname{div}(\bar{\lambda} \mathbf{V}) = F_\xi, \ \bar{\lambda}(\tau) = g \tag{3.46}$$

We have the following duality identity,

$$\int_{\Gamma_\tau(\mathbf{V})} g \langle \mathbf{Z}(\tau), \mathbf{n}_\tau \rangle \, \mathrm{d}\Gamma_\tau(\mathbf{V}) = \int_0^\tau \int_{\Gamma_t(\mathbf{V})} \bar{\lambda} \langle \mathbf{W}(t), \mathbf{n}_t \rangle \, \mathrm{d}\Gamma_t \, \mathrm{d}t \tag{3.47}$$

which leads to the functional derivative,

$$j'(\mathbf{V}, \mathbf{W}) = \int_0^\tau \int_{\Gamma_t(\mathbf{V})} \bar{\lambda} \langle \mathbf{W}, \mathbf{n}_t \rangle$$

### 3.4.2 Tangential calculus for tubes

In the last section, we have proven that the directional derivative of distributed non-cylindrical functionals involves the normal trace

$$z_t \overset{\text{def}}{=} \langle \mathbf{Z}, n_t \rangle, \ \text{ on } \Sigma(\mathbf{V}) \tag{3.48}$$

Let us recall the following characterization of $z_t$,

**LEMMA 3.3 [58],[59]**

$$z_t \circ \mathbf{T}_t(\mathbf{V}) =$$
$$\int_0^t \langle \mathbf{W}(\sigma), n_\sigma \rangle \circ \mathbf{T}_\sigma(\mathbf{V}) \exp \left( \int_\sigma^t \langle \mathrm{D} \, \mathbf{V}(\sigma) \cdot n_r, n_r \rangle \circ \mathbf{T}_r(\mathbf{V}) \, \mathrm{d}r \right) \mathrm{d}\sigma,$$

$$\text{on } \Gamma_0$$

**REMARK 3.8**  When the vector field $\mathbf{V}$ is chosen in the canonical form

$$\mathbf{V} = \mathbf{V} \circ p_t$$

the following identity holds,

$$\mathrm{D}\,\mathbf{V} \cdot \mathbf{n}_t = 0, \quad \text{on } \Gamma_t(\mathbf{V})$$

and we get

$$z_t \circ \mathbf{T}_t(\mathbf{V}) = \int_0^t \langle \mathbf{W}(\sigma), n_\sigma \rangle \circ \mathbf{T}_\sigma(\mathbf{V}) \, \mathrm{d}\sigma, \text{ on } \Gamma_0 \qquad (3.49)$$

☐

**DEFINITION 3.10**  *Let us consider the noncylindrical gradient operator,*

$$\langle \nabla_\Sigma \lambda, \nu_\tau \rangle = \partial_t \lambda + \nabla \lambda \cdot \mathbf{V} \qquad (3.50)$$

*where the vector* $\nu_\tau \overset{\text{def}}{=} (1, \mathbf{V}), \quad$ *on* $\Sigma$.

**REMARK 3.9**  The vector $\nu_\tau$ is tangent to the lateral surface $\Sigma$ since the outgoing normal field is given by

$$\nu \overset{\text{def}}{=} \frac{1}{\sqrt{1+v^2}}(-v, \mathbf{n}_t) \qquad (3.51)$$

with $v \overset{\text{def}}{=} \langle \mathbf{V}, n_t \rangle$.

☐

The duality identity (3.45) writes

$$\int_\Sigma F \frac{z_t}{\sqrt{1+v^2}} \, \mathrm{d}\Sigma = \int_\Sigma \lambda \frac{\langle \mathbf{W}, \mathbf{n}_t \rangle}{\sqrt{1+v^2}} \, \mathrm{d}\Sigma \qquad (3.52)$$

Using the operator $\nabla_\Sigma$ and the equation (3.40) satisfied by $\lambda$, we get

$$\int_\Sigma [\nabla_\Sigma \lambda \cdot \nu_\tau + \lambda \operatorname{div} \mathbf{V}] \frac{z_t}{\sqrt{1+v^2}} \, \mathrm{d}\Sigma = -\int_\Sigma \lambda \frac{\langle \mathbf{W}, \mathbf{n}_t \rangle}{\sqrt{1+v^2}} \, \mathrm{d}\Sigma$$

as

$$\lambda z_t = 0, \quad \text{on } \partial\Sigma$$

Performing an integration by parts on the manifold $\Sigma$, we get the following identity,

$$\int_\Sigma \left( -\lambda \operatorname{div}_\Sigma \left[ \frac{z_t}{\sqrt{1+v^2}} \nu_\tau \right] + \lambda \operatorname{div} \mathbf{V} \frac{z_t}{\sqrt{1+v^2}} \right) \mathrm{d}\Sigma$$

$$= -\int_\Sigma \lambda \frac{\langle \mathbf{W}, \mathbf{n}_t \rangle}{\sqrt{1+v^2}} \, \mathrm{d}\Sigma \qquad (3.53)$$

So we can prove the following proposition,

**PROPOSITION 3.7**

$$- \operatorname{div} \mathbf{V}(t) \, z_t + \sqrt{1 + v^2} \, \operatorname{div}_\Sigma \left[ \frac{z_t}{\sqrt{1 + v^2}} \, \nu_\tau \right] = \langle \mathbf{W}(t), \mathbf{n}_t \rangle \qquad (3.54)$$

**PROPOSITION 3.8**
*The normal transverse field $z_t$ satisfies the following identity,*

$$\partial_t e + \operatorname{div}(e \, \mathbf{V}) - (\operatorname{div} \mathbf{V}) \, e + \frac{1}{1 + v^2} \left[ v \left( \partial_{n_t} e + \langle \partial_t (e \, \mathbf{V}), n_t - v \partial_t e \rangle \right) \right]$$

$$- \frac{1}{1 + v^2} \langle \mathrm{D}(e \, \mathbf{V}) \cdot n_t, n_t \rangle = \frac{1}{\sqrt{1 + v^2}} \langle \mathbf{W}, n_t \rangle \quad (3.55)$$

*where $e = \dfrac{z_t}{\sqrt{1 + v^2}}$.*

**PROOF** We shall set

$$E = {}^*(E_1, E_2) = e \, {}^*(1, \mathbf{V})$$

Hence, using classical differential operators defined on $(0, \tau) \times D$

$$\operatorname{div}_\Sigma E = \operatorname{div}_Q \tilde{E} - \langle \mathrm{D}_Q \tilde{E} \cdot \nu, \nu \rangle \qquad (3.56)$$

where $\tilde{E}$ stands for an arbitrary extension of $E \overset{\text{def}}{=} {}^*(E_1, E_2)$ in $(0, \tau) \times D$ and

$$\operatorname{div}_Q E = \partial_t E_1 + \operatorname{div} E_2$$

$$\mathrm{D}_Q E = \begin{bmatrix} \partial_t E_1 \ \mathrm{D} \, E_1 \\ \partial_t E_2 \ \mathrm{D} \, E_2 \end{bmatrix}$$

Recalling that $\nu = \frac{1}{\sqrt{1+v^2}} \, {}^*(-\mathbf{v}, \mathbf{n}_t)$, we get

$$\operatorname{div}_\Sigma E = \operatorname{div}_Q {}^*(E_1, E_2) - \frac{1}{1 + v^2} \langle \mathrm{D}_Q \, {}^*(E_1, E_2) \cdot {}^*(-v, \mathbf{n}_t), {}^*(-v, \mathbf{n}_t) \rangle$$

$$= \partial_t E_1 + \operatorname{div} E_2 - \frac{1}{1 + v^2} \langle {}^*(-v \partial_t E_1 + \partial_{n_t} E_1, -v \partial_t E_2$$

$$+ \mathrm{D} \, E_2 \cdot n_t), {}^*(-v, \mathbf{n}_t) \rangle$$

$$= \partial_t E_1 + \operatorname{div} E_2 - \frac{1}{1 + v^2} \left[ v^2 \partial_t E_1 - v \partial_{n_t} E_1 - v \, \partial_t E_2 \cdot n_t \right.$$

$$+ \langle \mathrm{D} \, E_2 \cdot n_t, n_t \rangle \big]$$

$\square$

### 3.4.3 Classical shape analysis for tubes

Let us consider a tube $Q$ with lateral boundary $\Sigma$, and an horizontal perturbation leading to the perturbed tube $Q^s$ given in the following form:

- We consider an horizontal field $\tilde{\mathbf{Z}} \stackrel{\text{def}}{=} {}^*(0, \mathbf{Z})$ defined in $\mathbb{R}^{d+1} = \mathbb{R}_t \times \mathbb{R}_x^d$ and its associated flow mapping,

$$
\begin{aligned}
T_s(\tilde{\mathbf{Z}}) : \mathbb{R}_t \times \mathbb{R}_x^d &\longrightarrow \mathbb{R}_t \times \mathbb{R}_x^d \\
(t, x) &\mapsto (t, T_s(\mathbf{Z})(x))
\end{aligned}
$$

We designate by $\tilde{\mathbf{Z}}_\Sigma$ the tangential trace of the vector field $\tilde{\mathbf{Z}}$ on the lateral surface $\Sigma$. From the expression of the normal field $\nu$, we easily derive

$$
\tilde{\mathbf{Z}}_\Sigma = \frac{1}{1 + v^2} {}^*(v\, z_t,\, \mathbf{Z} - z_t\, \mathbf{n}_t) \tag{3.57}
$$

Here $\nu$ is to be understood as any extension of the normal field to a neighbourhood of the lateral boundary of the tube. For example, we can choose $n_t = \nabla b_{\Omega_t}$ where $b_{\Omega_t}$ stands for the oriented distance to the section of the tube at time $t$ and $\mathbf{V}$ can be understood as $\mathbf{V} \circ p_t$ where $p_t$ is the $\mathbb{R}^d$ projection on $\Gamma_t = \partial \Omega_t$.

### Mean curvature of the lateral boundary of tubes

**PROPOSITION 3.9**
*Assume that the field $\mathbf{V}$ verifies for each $t \in (0, \tau)$:*

$$
\mathbf{V}(t) = \mathbf{V}(t) \circ p_t
$$

*Then, on the boundary $\Gamma_t(\mathbf{V})$, we can define the time-space mean curvature of the lateral boundary $\Sigma$,*

$$
\mathcal{H}_\Sigma \stackrel{\text{def}}{=} \operatorname{div}_Q \nu
$$
$$
= -\frac{1}{(\sqrt{1 + v^2})^3} \left[ \langle \partial_t \mathbf{V}, \mathbf{n}_t \rangle - \langle [\nabla_{\Gamma_t} \langle \mathbf{V}, \mathbf{n}_t \rangle], \mathbf{V}|_{\Gamma_t} \rangle \right] + \frac{H_t}{\sqrt{1 + v^2}} \tag{3.58}
$$

**PROOF**  We choose $v = \langle \mathbf{V}, \nabla b_{\Omega_t}(\mathbf{V}) \rangle$. It is obvious that

$$
\partial_t \left( \frac{v}{\sqrt{1 + v^2}} \right) = \frac{1}{(\sqrt{1 + v^2})^3} \partial_t v
$$

But

$$
\partial_t v = \langle \partial_t \mathbf{V}, \nabla b \rangle + \langle \partial_t \nabla b, \mathbf{V} \rangle
$$

From the definition of the oriented distance, we got

$$\partial_t b_{\Omega_t(\mathbf{V})} = -\langle \mathbf{V}, \mathbf{n}_t \rangle \circ p_t \tag{3.59}$$

and

$$\partial_t \nabla b_{\Omega_t(\mathbf{V})} = -\left[ \nabla_{\Gamma_t} \langle \mathbf{V}, \mathbf{n}_t \rangle \right] \circ p_t$$

Then we get

$$\partial_t v = \langle \partial_t \mathbf{V}, \mathbf{n}_t \rangle - \langle \left[ \nabla_{\Gamma_t} \langle \mathbf{V}, \mathbf{n}_t \rangle \right], \mathbf{V}|_{\Gamma_t} \rangle$$

On the other hand,

$$\mathrm{div}\left( \frac{n_t}{\sqrt{1+v^2}} \right) = -\langle \nabla\left( \frac{1}{\sqrt{1+v^2}} \right), \mathbf{n}_t \rangle + \frac{1}{\sqrt{1+v^2}} \, \mathrm{div}\, \mathbf{n}_t$$

$$= -\frac{1}{(\sqrt{1+v^2})^3} \langle \varepsilon(\mathbf{V}) \cdot \mathbf{n}_t, \mathbf{n}_t \rangle + \frac{H_t}{\sqrt{1+v^2}}$$

where $\varepsilon(\mathbf{V}) = \frac{1}{2}(\mathrm{D}\,\mathbf{V} + {}^*\mathrm{D}\,\mathbf{V})$ is the deformation tensor.
We consider the situation in which the field $\mathbf{V}$ verifies the following property:

$$\mathbf{V} = \mathbf{V} \circ p_t \tag{3.60}$$

with

$$p_t = \mathrm{I} - b_{\Omega_t(\mathbf{V})} \, \nabla b_{\Omega_t(\mathbf{V})}$$

from which we deduce

$$\partial_t \, p_t = -\partial_t \, b_{\Omega_t(\mathbf{V})} \nabla b_{\Omega_t(\mathbf{V})} - b_{\Omega_t(\mathbf{V})} \nabla(\partial_t \, b_{\Omega_t(\mathbf{V})})$$

The restriction to the boundary $\Gamma_t$ leads to the distance $b_{\Omega_t(\mathbf{V})} = 0$, so the expression simplifies as follows,

$$\partial_t \, p_t|_{\Gamma_t} = \langle \mathbf{V}, \mathbf{n}_t \rangle \, \mathbf{n}_t$$

and

$$\mathrm{D}\,\mathbf{V} \cdot \mathbf{n}_t = 0, \quad \text{on } \Gamma_t(\mathbf{V})$$

☐

**DEFINITION 3.11**  *Consider $f(\Sigma)$ a function defined on the lateral boundary $\Sigma$. We call the Eulerian material derivative of $f(\Sigma)$ in the direction $\tilde{\mathbf{Z}}$ the following quantity,*

$$\dot{f}_\Sigma(\tilde{\mathbf{Z}}) \overset{\mathrm{def}}{=} \partial_s \left[ f(\Sigma_s) \circ \mathcal{T}_s(\tilde{\mathbf{Z}}) \right]\Big|_{s=0} \tag{3.61}$$

*and its associated partial Eulerian derivative is given by the following expression,*

$$f'_\Sigma(\tilde{\mathbf{Z}}) \overset{\mathrm{def}}{=} \dot{f}_\Sigma(\tilde{\mathbf{Z}}) - \langle \nabla_\Sigma f(\sigma), \tilde{\mathbf{Z}}_\Sigma \rangle \tag{3.62}$$

Using the expression of the tangential operator $\nabla_\Sigma$, we get

$$f'_\Sigma(\tilde{\mathbf{Z}}) = \dot{f}_\Sigma(\tilde{\mathbf{Z}}) - \frac{v\,z_t}{1+v^2}\,\partial_t\,f - \langle \nabla_\Gamma f, \mathbf{Z} - \frac{z_t}{1+v^2}\,\mathbf{n}_t\rangle$$

**REMARK 3.10**    When $f(\Sigma)$ is the restriction to the lateral boundary $\Sigma$ of a function $F$ defined over $\mathbb{R}^{d+1}$, we get

$$f'_\Sigma(\tilde{\mathbf{Z}}) = F'_Q(\tilde{\mathbf{Z}})|_\Sigma + \partial_\nu F\,\langle \tilde{\mathbf{Z}}, \nu\rangle$$
$$= \partial_\nu F\,\langle \tilde{\mathbf{Z}}, \nu\rangle$$

as $F'_Q(\tilde{\mathbf{Z}}) = 0$ since F does not depend on $\Sigma$.                          ▯

**Shape derivative on the lateral boundary of tubes**

Let us now consider a function $F \in C^1([0,\tau] \times \bar{D})$. First, we assume that $F$ is zero in the neighbourhood of $t = \tau$ so that the following derivative of the lateral boundary integral could be considered as the derivative of integral on the total boundary of the tube (as it will generate no term on the top $t = \tau$ of the tube). We set

$$\Sigma^s = \{(t, \mathbf{T}_t(\mathbf{V} + s\mathbf{W})(x)), \quad x \in \partial\Omega_0\}$$

*PROPOSITION 3.10*
*Assume the vector field* $\mathbf{V}$ *in the canonical form* $\mathbf{V}(t) = \mathbf{V}(t) \circ p_t$ *in a neighbourhood of the lateral boundary* $\Sigma$ *and set* $v = \langle \mathbf{V}(t), \mathbf{n}_t\rangle$ *on* $\Gamma_t$. *The following identity holds,*

$$\partial_s\left(\int_{\Sigma^s} F\,d\Sigma^s\right)\bigg|_{s=0} = \int_0^\tau \int_{\Gamma_t}\left(\frac{1}{\sqrt{1+v^2}}\,[-v\,\partial_t\,F + \partial_{n_t}F]\right.$$
$$+[-\frac{1}{(\sqrt{1+v^2})^3}\,[\langle \partial_t\mathbf{V}, \mathbf{n}_t\rangle - \langle \nabla_{\Gamma_t}\,v, \mathbf{V}|_{\Gamma_t}\rangle]$$
$$\left.+\frac{H_t}{\sqrt{1+v^2}}\right]\,F\bigg)\,\langle \mathbf{Z}, \mathbf{n}_t\rangle\,d\Gamma_t\,dt \quad (3.63)$$

**PROOF**    The classical shape calculus can be applied in $\mathbb{R}_t \times \mathbb{R}^d$,

$$\partial_s\left(\int_{\Sigma^s} F\,d\Sigma^s\right)\bigg|_{s=0} = \int_\Sigma (\partial_\nu F + \mathcal{H}_\Sigma F)\,\langle \tilde{\mathbf{Z}}, \nu\rangle\,d\Sigma$$

We have

$$\langle \tilde{\mathbf{Z}}, \nu\rangle = \frac{z_t}{\sqrt{1+v^2}}$$

and

$$\partial_\nu F = D_Q\,F \cdot \nu$$
$$= \frac{1}{\sqrt{1+v^2}}\,[-v\,\partial_t\,F + \partial_{n_t}F]$$

Using Proposition (3.9), we get

$$
\partial_s \left( \int_{\Sigma^s} F \, d\Sigma^s \right) \bigg|_{s=0} = \int_\Sigma \left( \frac{1}{\sqrt{1+v^2}} \left[ -v \, \partial_t F + \partial_{n_t} F \right] \right.
$$

$$
\left[ -\frac{1}{(\sqrt{1+v^2})^3} \left[ \langle \partial_t \mathbf{V}, \mathbf{n}_t \rangle - \langle \nabla_{\Gamma_t} v, \mathbf{V}|_{\Gamma_t} \rangle \right] \right.
$$

$$
\left. \left. + \frac{H_t}{\sqrt{1+v^2}} \right] F \right) \langle \mathbf{Z}, \mathbf{n}_t \rangle \frac{1}{\sqrt{1+v^2}} \, d\Sigma
$$

We conclude using $d\Sigma = \sqrt{1+v^2} \, d\Gamma_t \, dt$. $\qquad\qquad$ ☐

**REMARK 3.11** We can easily obtain the optimality condition for a minimal surface tube, by setting $F = 1$ in equation (3.63),

$$
\partial_s \left( \int_{\Sigma^s} d\Sigma^s \right) \bigg|_{s=0} = \int_0^\tau \int_{\Gamma_t} \left[ -\frac{1}{(\sqrt{1+v^2})^3} \left[ \langle \partial_t \mathbf{V}, \mathbf{n}_t \rangle - \langle \nabla_{\Gamma_t} v, \mathbf{V}|_{\Gamma_t} \rangle \right] \right.
$$

$$
\left. + \frac{H_t}{\sqrt{1+v^2}} \right] \langle \mathbf{Z}, \mathbf{n}_t \rangle \, d\Gamma_t \, dt
$$

We introduce the adjoint field $\lambda$ such that

$$
-\partial_t \lambda - \operatorname{div}(\lambda \, \mathbf{V}) = -\frac{1}{(\sqrt{1+v^2})^3} \left[ \langle \partial_t \mathbf{V}, \mathbf{n}_t \rangle - \langle \nabla_{\Gamma_t} v, \mathbf{V}|_{\Gamma_t} \rangle \right] + \frac{H_t}{\sqrt{1+v^2}}
$$

with $\lambda(t = \tau) = 0$.
Then, using the duality identity (3.45), we deduce that the optimality condition writes

$$
\int_\Sigma \lambda \langle \mathbf{W}, \mathbf{n}_t \rangle \, d\Sigma = 0, \quad \forall \mathbf{W}
$$

from which we deduce $\lambda = 0$ on $\Sigma$. This implies that the right-hand side of the backward equation is zero, i.e.,

$$
\langle \partial_t \mathbf{V}, \mathbf{n}_t \rangle - \langle \nabla_{\Gamma_t} v, \mathbf{V}|_{\Gamma_t} \rangle = (1+v^2) H_t
$$

$\qquad\qquad$ ☐

---

## 3.5 A first example : optimal trajectory problem

In this section, we investigate a problem related to optimal trajectories. Let $D$ be a compact domain in $\mathbb{R}^d$ with a smooth boundary $\partial D$. Let $a$ be a point inside $D$ and for any point $x \in \partial D$, we consider the family of $\mathcal{C}^1$ curves which join these two points in $D$. We set

$$\mathcal{C}_{a,x} = \{\mathcal{C}^1 \text{ curves } C_{a,x} \subset D, \text{ with extreme points } a \text{ and } x\} \qquad (3.64)$$

For any curve $C_{a,x} \in \mathcal{C}_{a,x}$, there exists an injective mapping $\gamma \in C^1([0,1];D)$ satisfying the following properties,

$$\begin{cases} C_{a,x} = \gamma([0,1]), \\ \gamma(0) = a, \qquad \gamma(1) = x \end{cases}$$

Let $g_C \in L^1(C_{a,x})$. Our objective is to minimize with respect to the curves $C_{a,x}$ the integral quantity,

$$J(C_{a,x}) = \int_{C_{a,x}} g_C \, \mathrm{d}C$$

Let $\mathbf{V}$ be a vector field in $L^1(0,1;\mathcal{C}^1(\bar{D},\mathbb{R}^d))$ with $\langle \mathbf{V}, \mathbf{n} \rangle = 0$ on the boundary $\partial D$, then we define the perturbed curve

$$C_{a,x}^V \overset{\text{def}}{=} \mathbf{T}_1(\mathbf{V})(C_{a,x}^0) \in \mathcal{C}_{a,x}$$

The claim is that from any element $C_{a,x}^0$ in $\mathcal{C}_{a,x}$ the element $C_{a,x}^V$ furnishes all the curves when $\mathbf{V}$ described the linear space $L^1((0,1);\mathcal{C}^1(\bar{D},\mathbb{R}^d))$, i.e.,

$$\mathcal{C}_{a,x} = \{C_{a,x}^V, \quad \mathbf{V} \in L^1(0,1;\mathcal{C}^1(\bar{D},\mathbb{R}^d)), \, \langle \mathbf{V}, \mathbf{n} \rangle = 0 \text{ on } \partial D\}$$

Hence, the minimization problem can be written in the following terms,

$$\min_{\mathbf{V} \in \mathcal{E}} j(\mathbf{V}) \qquad (3.65)$$

where

$$j(\mathbf{V}) = \int_{C_{a,x}^V} g_{C_{a,x}^V} \, \mathrm{d}C_{a,x}^V$$

with

$$\mathcal{E} = \{\mathbf{V} \in L^1(0,1;\mathcal{C}^1(\bar{D},\mathbb{R}^d)), \quad \langle \mathbf{V}, \mathbf{n} \rangle = 0 \text{ on } \partial D\}$$

### 3.5.1 Optimality conditions : case of planar parametric curves

In order to solve the above problem, we need to compute the following cost function derivative,

$$j'(\mathbf{V}, \mathbf{W}) = \partial_s j(\mathbf{V} + s\mathbf{W})|_{s=0}, \quad \forall (\mathbf{V}, \mathbf{W}) \in \mathcal{E} \times \mathcal{E}$$

First, we consider the simple case in which the density function $g_C$ is the restriction of a given function $G \in C^1(\bar{D})$, that is $g_C = G|_C$. In that case, the parameter $s$ of perturbations only occurs in the measure element on the curves. Let $\gamma$ be a parametrization of the reference curve $\mathcal{C}_{a,x}^0$. We have

$$j(\mathbf{V}+s\mathbf{W}) = \int_0^1 G \circ \mathbf{T}_1(\mathbf{V}+s\mathbf{W}) \circ \gamma(\sigma) \, \| \left[ \mathrm{D}\,\mathbf{T}_1(\mathbf{V}+s\mathbf{W}) \circ \gamma \cdot \gamma' \right](\sigma) \| \, d\sigma$$

We introduce

$$S_t(\mathbf{V};\mathbf{W}) = \partial_s \mathbf{T}_t(\mathbf{V}+s\mathbf{W})|_{s=0}$$

We have

$$\partial_s \| \left[ \mathrm{D}\,\mathbf{T}_1(\mathbf{V}+s\mathbf{W}) \circ \gamma \cdot \gamma' \right](\sigma) \|^2 \,|_{s=0}$$
$$= 2 \left\langle \mathrm{D}\,S_1(\mathbf{V};\mathbf{W}) \circ \gamma \cdot \gamma', \mathrm{D}\,\mathbf{T}_1(\mathbf{V}) \circ \gamma \cdot \gamma' \right\rangle$$

from which we deduce

$$\partial_s \, \| \left[ \mathrm{D}\,\mathbf{T}_1(\mathbf{V}+s\mathbf{W}) \circ \gamma \cdot \gamma' \right](\sigma) \||_{s=0} = \left\langle \mathrm{D}\,S_1(\mathbf{V};\mathbf{W}) \circ \gamma \cdot \gamma'(\sigma), \tau(\sigma) \right\rangle$$
$$= \frac{\left\langle \mathrm{D}\,S_1(\mathbf{V};\mathbf{W}) \cdot \tau, \tau_1 \circ \mathbf{T}_1(\mathbf{V}) \right\rangle \circ \gamma(\sigma)}{\|\gamma'\|}$$

where $\tau_1$ is the unitary tangential vector on the curve $\mathcal{C}_{a,x}^V$ while $\tau$ is the unitary tangential vector to the reference curve $\mathcal{C}_{a,x}^0$.
For $0 \le t \le 1$, $S_t$ is the solution of the following dynamical system [154, 59],

$$\partial_t S_t - \mathrm{D}\,\mathbf{V}(t) \circ \mathbf{T}_t(\mathbf{V}) \cdot S_t = \mathbf{W}(t) \circ \mathbf{T}_t(\mathbf{V}), \ S_t(t=0) = 0 \qquad (3.66)$$

This leads to the following derivative expression,

$$j'(\mathbf{V},\mathbf{W}) = \int_0^1 \left[ \langle (\nabla G) \circ \mathbf{T}_1(\mathbf{V})), S_1 \rangle (\gamma(\sigma)) \right.$$
$$\left. + G(\mathbf{T}_1(\mathbf{V})(\gamma(\sigma)) \langle \mathrm{D}\,S_1(\gamma(\sigma)) \cdot \tau(\gamma(\sigma)), \tau_1(\mathbf{T}_1(\gamma(\sigma))) \rangle \right] \|\gamma'(\sigma)\| \, d\sigma$$

Eliminating the parametrization $\gamma$, we get

$$j'(\mathbf{V},\mathbf{W}) = \int_{\mathcal{C}_{a,x}^0} \left[ \langle \nabla G(\mathbf{T}_1(\mathbf{V})), S_1 \rangle + G(\mathbf{T}_1(\mathbf{V})) \langle \mathrm{D}\,S_1 \cdot \tau, \tau_1(\mathbf{T}_1(\mathbf{V})) \rangle \right] dC$$
$$\qquad (3.67)$$

Obviously $j'(\mathbf{V},\mathbf{W})$ depends linearly on $\mathbf{W}$ through the term $S_1$. The Jacobian matrix $\mathrm{D}\,S_1$ is itself a solution of the following dynamical system,[12]

$$\begin{cases} \partial_t \, \mathrm{D}\,S_t - \left[ \mathrm{D}^2\,\mathbf{V} \circ \mathbf{T}_t(\mathbf{V}) \cdot \mathrm{D}\,\mathbf{T}_t(\mathbf{V}) \right] \cdot S_t - \mathrm{D}\,\mathbf{V} \circ \mathbf{T}_t(\mathbf{V}) \cdot \mathrm{D}\,S_t \\ \qquad = \mathrm{D}(\mathbf{W} \circ \mathbf{T}_t(\mathbf{V})), \\ \\ \mathrm{D}\,S(t=0) = 0 \end{cases} \qquad (3.68)$$

---

[12]We shall assume in this case $\mathbf{V} \in \mathcal{C}^0([0,1]; C^2(D, \mathbb{R}^d))$.

The couple $(S_t, DS_t)$ is the solution of the dynamical system (3.66), (3.68). We shall now introduce the backward adjoint dynamical system. Its solution $(\theta, A)$ allows to clarify the linear contribution of the field $\mathbf{W}$ in the expression of $j'(\mathbf{V}, \mathbf{W})$.

The vector $\theta$ and the matrix $A$ are defined on $[0, 1] \times \mathcal{C}_{a,x}^0$ with variables $(t, x = \gamma(\sigma))$ where the parameter[13] $\sigma \in [\sigma_0, \sigma_1]$.

Let $\theta$ solve the following backward dynamical system,

$$
\begin{cases}
-\partial_t \theta - {}^* \mathrm{D}\, \mathbf{V}(t) \circ \mathbf{T}_t(\mathbf{V}) \cdot \theta = 0, \\[2mm]
\theta(1, \gamma(\sigma)) = \nabla G(\mathbf{T}_1(\mathbf{V})(\gamma(\sigma)))
\end{cases}
\tag{3.69}
$$

and the matrix $A$ solves the following backward dynamical system,

$$
\begin{cases}
-\partial_t A - {}^* \left[ \mathrm{D}^2\, \mathbf{V} \circ \mathbf{T}_t(\mathbf{V}) \cdot \mathrm{D}\, \mathbf{T}_t(\mathbf{V}) \right] \cdot A - {}^* \mathrm{D}\, \mathbf{V} \circ \mathbf{T}_t(\mathbf{V}) \cdot A = 0, \\[2mm]
A(1, \gamma(\sigma)) = G(\mathbf{T}_1(\mathbf{V}))(\gamma(\sigma))\, \tau(\gamma(\sigma)) \cdot {}^* \tau_1(\mathbf{T}_1(\mathbf{V})(\gamma(\sigma)))
\end{cases}
\tag{3.70}
$$

In this setting, the derivative of the functional takes the following form,

$$
\begin{aligned}
j'(\mathbf{V}, \mathbf{W}) &= \int_0^1 \left( \langle \theta(1), S_1 \rangle + A(1) \cdot\cdot\, \mathrm{D}\, S_1 \right) \circ \gamma(\sigma)\, \|\gamma'(\sigma)\|\, \mathrm{d}\sigma \\
&= \int_{\mathcal{C}_{a,x}^0} \left( \langle \theta(1), S_1 \rangle + A(1) \cdot\cdot\, \mathrm{D}\, S_1 \right) \mathrm{d}C \\
&= \int_0^1 \mathrm{d}t \left[ \int_0^1 (\theta(t)\, \mathbf{W}(t) \circ \mathbf{T}_t(\mathbf{V}) \right. \\
&\qquad \left. + \mathrm{D}\, \theta(t) \cdot\cdot\, \mathrm{D}(\mathbf{W} \circ \mathbf{T}_t(\mathbf{V}))) \circ \gamma(\sigma)\, \mathrm{d}\sigma \right]
\end{aligned}
\tag{3.71}
$$

Then we get a backward calculus for the optimal field $\mathbf{V}$ along the reference trajectory $\mathcal{C}_{a,x}^0$ parametrized by $\gamma$. Actually the previous Eulerian approach, developed for planar curves, leads to a more explicit expression in the general setting of functional derivatives, as we shall see in the next section.

## 3.5.2 Optimality conditions : case of the general Eulerian setting

We consider now a calculus method which shall never refer to the reference curve $\mathcal{C}_{a,x}^0$ but only to the moving curves $\mathcal{C}_{a,x}^V$ and $\mathcal{C}_{a,x}^{V+sW}$.

Let us consider the following transverse mapping,

$$
\mathcal{T}_s^t = \mathbf{T}_t(\mathbf{V} + s\mathbf{W}) \circ \mathbf{T}_t(\mathbf{V})^{-1}
\tag{3.72}
$$

---

[13] Of course $\sigma$ could be chosen as the arc length of the reference curve $\mathcal{C}_{a,x}^0$. In this case, we would have $\|\gamma''(\sigma)\| = 1$.

This map sends the curve $C_{a,x}^V$ onto $C^s \overset{\text{def}}{=} C_{a,x}^{V+sW}$. Then we can write

$$j(\mathbf{V} + s\mathbf{W}) = \int_{C^s} g_{C^s}\, dC^s$$

Here $s$ is understood as the shape perturbation parameter in the classical setting. Hence, we consider the associated speed vector

$$\mathbf{Z}^t(s,x) = \left(\frac{\partial}{\partial s} T_s^t\right) \circ \left(T_s^t\right)^{-1} \tag{3.73}$$

Hence, the moving curve $C^s$ is obtained from the reference curve through the flow mapping associated to the vector field $\mathcal{Z}^t$. This flow evolves with respect to the parameter $s$ for a fixed time $t = 1$. We introduced the terminology transverse flow,

$$C^s = \mathbf{T}_s(\mathcal{Z}^t)(C_{a,x}^V)$$

Using classical differentiation results for boundary integrals[14], we obtain

$$j'(\mathbf{V}, \mathbf{W}) = \int_{C_{a,x}^V} \left[g'_{C;\mathbf{z}} + H\left\langle \mathbf{Z}(1), \mathbf{n}\right\rangle\right] dC_{a,x}^V \tag{3.74}$$

with

$$\mathbf{Z}(t,x) \overset{\text{def}}{=} \mathcal{Z}^t(0,x), \quad \mathbf{Z}(1) = \mathbf{Z}(1,.)$$

where $H$ stands for the curvature of $C_{a,x}^V$.

The term $g'_{C;\mathbf{z}}$ is the so-called boundary shape derivative of the function $g(C)$ on $C_{a,x}^V$ in the direction of the vector field $\mathbf{Z}(1)$. We recall here the very definition,

$$g'_{C;\mathbf{z}} = \partial_s\left[g_{C^s} \circ T_s^1\right]\big|_{s=0} - \left\langle \nabla_\tau g_{C_{a,x}^V}, \mathbf{Z}(1)\right\rangle \tag{3.75}$$

Here $\nabla_\tau g = \nabla G - \partial_n G\,\mathbf{n}$ stands for the tangential derivative of $g_C$ along the curve and is independent on the choice of the extension $G$ of $g_C$ outside of the curve. We shall now study two specific situations, where $g_C$ is either the trace of a distributed function defined over $(0,\tau) \times D$ or depends on the curvature of the curve $C_{a,x}^V$.

## Distributed density

In the very simple case where the function $g_C$ is the restriction to the curve $C$ of a smooth function: $g_C = G|_C$, we get

$$g'_{C;\mathbf{z}} = \partial_n G\left\langle \mathbf{Z}(1), \mathbf{n}\right\rangle \quad \text{on } C_{a,x}^V$$

---

[14]Here, we assume that the dimension is $d = 2$, so that the curves can be considered as a part of the boundary of a moving set.

which leads to the following derivative,

$$j'(\mathbf{V}, \mathbf{W}) = \int_{C_{a,x}^V} [\partial_n G + H] \langle \mathbf{Z}(1), \mathbf{n} \rangle \, dC_{a,x}^V \tag{3.76}$$

We recall that the transverse field $\mathbf{Z}$ solves the following dynamical system,

$$\begin{cases} \partial_t \mathbf{Z} + [\mathbf{Z}, \mathbf{V}] = \mathbf{W}, & (0,1) \times D \\ \mathbf{Z}(t=0) = 0 & D \end{cases} \tag{3.77}$$

with $[\mathbf{Z}, \mathbf{V}] \overset{\text{def}}{=} D\mathbf{Z} \cdot \mathbf{V} - D\mathbf{V} \cdot \mathbf{Z}$. We introduce the adjoint state $\Lambda$ solution of the following backward dynamical system,

$$\begin{cases} \Lambda(1) = \gamma_{C_{a,x}^V}^* \cdot [(\partial_n G + H) \, \mathbf{n}], & D \\ -\partial_t \Lambda - D\mathbf{V} \cdot \Lambda - {}^*D\Lambda \cdot \mathbf{V} + (\text{div } \mathbf{V}) \Lambda = 0, & (0,1) \times D \end{cases} \tag{3.78}$$

where $\gamma_{C_{a,x}^V} \in \mathcal{L}(\mathcal{C}^0(\bar{D}), C_{a,x}^V)$ is the trace operator on the curve $C_{a,x}^V$. This leads to the following formula,

$$j'(\mathbf{V}, \mathbf{W}) = \langle \Lambda(1), \mathbf{Z}(1) \rangle_{\mathcal{M}(D) \times \mathcal{C}^0(D)}$$

We may use the following integration by parts formula,

$$\int_0^1 \langle \Lambda, \partial_t \mathbf{Z} + [\mathbf{Z}, \mathbf{V}] \rangle_{\mathcal{M}(D) \times \mathcal{C}^0(D)} \, dt$$

$$= \int_0^1 \langle -\partial_t \Lambda - D\mathbf{V} \cdot \Lambda - {}^*D\Lambda \cdot \mathbf{V} + (\text{div } \mathbf{V}) \Lambda, \mathbf{Z} \rangle_{\mathcal{M}(D) \times \mathcal{C}^0(D)} \, dt$$

$$+ \langle \Lambda(1), \mathbf{Z}(1) \rangle_{\mathcal{M}(D) \times \mathcal{C}^0(D)} - \langle \Lambda(0), \mathbf{Z}(0) \rangle_{\mathcal{M}(D) \times \mathcal{C}^0(D)} \tag{3.79}$$

As $\mathbf{Z}(0) = 0$ and $\Lambda$ is the solution of the adjoint system (3.78), we get the following functional derivative,

$$j'(\mathbf{V}, \mathbf{W}) = \int_0^1 \langle \Lambda, \mathbf{W} \rangle_{\mathcal{M}(D) \times \mathcal{C}^0(D)} \, dt \tag{3.80}$$

Actually, it can be proven that the measure $\Lambda$ is of the following form,

$$\Lambda(t) = \gamma_{\mathbf{T}_t(\mathbf{V})(C_{a,x}^\circ)}^* (\tilde{\lambda} \, \mathbf{n}_t), \quad t \in (0,1)$$

We can also observe that directly from (3.76), using (3.47), we have

$$j'(\mathbf{V}, \mathbf{W}) = \int_0^1 \int_{C_{a,x}^V} \bar{\lambda} \langle \mathbf{W}(t), \mathbf{n}_t \rangle \, dC_{a,x}^V \, dt \tag{3.81}$$

where $\bar{\lambda}$ solves the backward problem,

$$\begin{cases} \bar{\lambda}(1) = \tilde{k}, & D \\ \partial_t \bar{\lambda} + \text{div}(\bar{\lambda} \, \mathbf{V}) = 0, & (0,1) \times D \end{cases} \tag{3.82}$$

with $\tilde{k}$ any extension of the function $k \stackrel{\text{def}}{=} \partial_n G + H$ to the domain $D$.
If div $\mathbf{V} = 0$, then the equation for $\lambda$ turns to be a backward convection of the ending term. We set $\tilde{\lambda}(t) = \bar{\lambda}(1-t)$ and $\tilde{\mathbf{V}}(t) = -\mathbf{V}(1-t)$, then we get

$$\begin{cases} \partial_t \tilde{\lambda} + \nabla \tilde{\lambda} \cdot \tilde{\mathbf{V}} = 0, & (0,1) \times D \\ \tilde{\lambda}(0) = \tilde{k}, & D \end{cases}$$

and we get

$$\tilde{\lambda}(t) = k \circ \mathbf{T}_t(\tilde{\mathbf{V}})^{-1}$$

Using the fact that $\mathbf{T}_t(\mathbf{V})^{-1} = \mathbf{T}_t(\mathbf{W}_t)$ with $\mathbf{W}_t(s,y) = -\mathbf{V}(t-s,y)$, it leads to

$$\bar{\lambda}(t) = \tilde{\lambda}(1-t) = k \circ \mathbf{T}_{1-t}(\mathbf{V})$$

Then, we get

$$j'(\mathbf{V}, \mathbf{W}) = \int_0^1 \int_{C_{a,x}^V} k \circ \mathbf{T}_{1-t}(\mathbf{V}) \langle \mathbf{W}(t), \mathbf{n}_t \rangle \, dC_{a,x}^V \, dt$$

and

$$\nabla j(\mathbf{V}) = {}^* \gamma_{C_{a,x}^V} \left( k \circ \mathbf{T}_{1-t}(\mathbf{V}) \, \mathbf{n}_t \right) \tag{3.83}$$

**Curvature dependent density**

Now, we assume that the density function $g$ depends on the curvature of the curves. Here, we choose

$$g(y) = \bar{g}(H(y), y), \quad y \in \mathscr{V}(C_{a,x}^V)$$

where $H(y) \stackrel{\text{def}}{=} \Delta b(x)$. Then, we get

$$g'_{C;\mathbf{Z}(1)} = \partial_H \bar{g}(H(y), y) \, H'_{C,\mathbf{Z}(1)}(y) + \langle \nabla_y \bar{g}(H(y), y), \mathbf{n}(y) \rangle \langle \mathbf{Z}(1), \mathbf{n} \rangle$$

Furthermore, we have

$$H'_{C,\mathbf{Z}(1)} = -\Delta_C(\langle \mathbf{Z}(1), \mathbf{n} \rangle)$$

This leads to

$$j'(\mathbf{V}, \mathbf{W}) = \int_{C_{a,x}^V} [-\partial_H \bar{g} \, \Delta_C(\langle \mathbf{Z}(1), \mathbf{n} \rangle) + (H \, g + \langle \nabla \bar{g}, \mathbf{n} \rangle) \langle \mathbf{Z}(1), \mathbf{n} \rangle] \, dC_{a,x}^V$$

Using tangential Stokes formula [53] on $C_{a,x}^V$, we have

$$j'(\mathbf{V}, \mathbf{W}) = \int_{C_{a,x}^V} [-\Delta_C(\partial_H \bar{g}) + H \, g + \langle \nabla \bar{g}, \mathbf{n} \rangle] \langle \mathbf{Z}(1), \mathbf{n} \rangle \, dC_{a,x}^V$$

$$+ \partial_\tau(\partial_H \bar{g}(x)) \langle \mathbf{Z}(1)(x), \mathbf{n}(x) \rangle - \partial_\tau(\partial_H \bar{g}(a)) \langle \mathbf{Z}(1)(a), \mathbf{n}(a) \rangle$$

$$- \partial_H \bar{g}(x) \partial_\tau \langle \mathbf{Z}(1)(x), \mathbf{n}(x) \rangle + \partial_H \bar{g}(a) \partial_\tau \langle \mathbf{Z}(1)(a), \mathbf{n}(a) \rangle \tag{3.84}$$

Since the extreme points $x$ and $a$ are fixed, their velocities are zero, i.e.,

$$\mathbf{V}(a) = \mathbf{V}(x) = \mathbf{W}(a) = \mathbf{W}(x) = 0$$

which implies $\mathbf{Z}(1)(a) = \mathbf{Z}(1)(x) = 0$ and the previous derivative becomes

$$j'(\mathbf{V}, \mathbf{W}) = \int_{C_{a,x}^V} \left[ -\Delta_C(\partial_H \bar{g}) + H\, g + \langle \nabla \bar{g}, \mathbf{n} \rangle \right] \langle \mathbf{Z}(1), \mathbf{n} \rangle \; \mathrm{d}C_{a,x}^V \qquad (3.85)$$

Eventually, we get the gradient of $j(\mathbf{V})$ using the expression furnished by equation (3.83) with

$$k = -\Delta_C(\partial_H \bar{g}) + H\, \bar{g} + \langle \nabla \bar{g}, \mathbf{n} \rangle \qquad (3.86)$$

# Chapter 4

## Shape differential equation and level set formulation

### 4.1 Introduction

In this chapter, we recall the concept of shape differential equation developed in [145],[147]. Here, we present a simplified version and some applications in dimension 2 which enable us to reach the time asymptotic result.

We consider a shape functional $J$ which is shape differentiable in $\mathcal{O}_k$ with respect to $\mathcal{V}_k$ to be specified later on. We denote $\nabla J(\Omega)$ its gradient, considered as a distribution in $\mathcal{A}_k^*$. For any $\Omega_0$ in $\mathcal{O}_k$ and $\mathbf{V}$ in $\mathcal{V}_k$, the absolute continuity of $J$ writes

$$J(\Omega_s(\mathbf{V})) - J(\Omega_0) = \int_0^s \langle \nabla J(\Omega_t(\mathbf{V})) , \mathbf{V}(t) \rangle_{\mathcal{A}_k^* \times \mathcal{A}_k} \, dt, \quad \forall s \geqslant 0 \quad (4.1)$$

Classical gradient based methods allow us to control the variations of $J$ with respect to the domain. Considering the problem,

$$\min_{\mathbf{V} \in \mathcal{V}_k} J(\Omega(\mathbf{V}))$$

we would like to elaborate a constructive method to decrease the functional *following the gradient*. This may be done by solving the non-linear equation for large evolution of the domain,

$$\nabla J(\Omega_t(\mathbf{V})) + \mathbb{A}(\mathbf{V}(t)) = 0, \quad \forall t \geqslant 0 \quad (4.2)$$

where $\mathbb{A}$ is an *ad-hoc* duality operator. This corresponds to the well-known steepest descent method.

From the structure theorem for shape gradient [135], we have[1],

$$\nabla J(\Omega) = \gamma_\Gamma^* \cdot (g \, \mathbf{n})$$

where the shape density gradient $g$ is a distribution on the boundary $\Gamma$. Usually it is a function on $\Gamma$ so that we consider *any extension* $\mathcal{G}$ of $g$ defined

---

[1]Under some regularity assumptions which are satisfied for a large class of problems.

in a neighbourhood of the boundary $\Gamma$. In this case the term $\gamma_\Gamma^* \cdot (g\,\mathbf{n})$ can be identified with $\mathcal{G}\,\nabla\chi$. Hence, the shape differential equation turns into an Hamilton-Jacobi equation for the characteristic function $\chi$,

$$\begin{cases} \partial_t\,\chi + \langle\nabla\chi, \mathbb{A}^{-1}\cdot(\mathcal{G}(\chi)\nabla\chi)\rangle = 0, \ (0,\tau) \\ \chi(0) = \chi_{\Omega_0}, \qquad\qquad\qquad\qquad \Omega_0 \subset D \end{cases} \tag{4.3}$$

We shall see in the sequel that the previous equation can be weakened using the *level set formulation* where we shall solve the following equation,

$$\begin{cases} \partial_t\,\Phi - \langle\nabla\Phi, \mathbb{A}^{-1}\cdot(\mathcal{G}(\chi_t(\Phi))\nabla\chi_t(\Phi))\rangle = 0, \ (0,\tau) \\ \Phi(0) = \Phi_0, \end{cases} \tag{4.4}$$

where $\chi_t(\Phi) = \{x \in D, \quad \Phi(t,x) > 0\}$.
We are going to recall the constructive proof of the existence of a $\mathbf{V}$ satisfying (4.2) and investigate the asymptotic behaviour of the method. The existence of a solution for this so-called *shape differential equation* has been proven in [147] inside a larger setting[2].

---

## 4.2   Classical shape differential equation setting

In this section, we recall the material introduced in [145, 146, 147] while solving the so-called shape differential equation. We denote $\mathcal{T}_k$ the subset of $C^k(\bar{D}, \mathbb{R}^d)$ whose elements are $C^k$-diffeomorphism of $\bar{D}$. It is endowed with the Courant metric[3] $\mathfrak{d}_k$ which is defined on the family of images of a given domain.
We fix a smooth bounded hold-all $D$ in $\mathbb{R}^d$ and a non-negative integer $k$. We denote by $\mathcal{O}_k$ the set of all open $C^k$-submanifold of $D$, and $\mathcal{O}_{\text{lip}}$ the set of Lipschitz open subset of $D$. We are going to use the following spaces

$$\mathcal{A}_k = \left\{\mathbf{V} \in C^k(D, \mathbb{R}^d)\ \middle|\ \langle\mathbf{V},\nu\rangle_{\mathbb{R}^d} = 0, \quad\text{on } \partial D\right\},$$
$$\mathcal{V}_k(I) = \left\{\mathbf{V} \in C^0(I, \mathcal{A}_k)\right\} \tag{4.5}$$

where $I$ is an interval of $\mathbb{R}^+$ which contains 0. For $I = \mathbb{R}^+$, we simply denote $\mathcal{V}_k = \mathcal{V}_k(\mathbb{R}^+)$.
The operator $\mathcal{A}_k^*$ stands for the dual space of distributions of order less than

---

[2]It holds for shape differentiable functional whose gradient is continuous and bounded on $\mathcal{O}_k$, endowed with the Courant's metric topology, ranging in a Sobolev space of Distributions.
[3]We refer the reader to the book [51].

$k$. In the case $k = 0$, it corresponds to the space of Radon measures. For any fixed domain $\Omega_0 \in \mathcal{O}_k$, we set

$$\mathcal{O}_k(\Omega_0) = \left\{ \Omega \in \mathcal{O}_k \ \middle| \ \exists T \in \mathcal{T}_k, \ \Omega = T(\Omega_0) \right\}$$

**REMARK 4.1** $(\mathcal{O}_k(\Omega_0), \partial_k)$ is a complete metric space. ▯

For a bounded universe $D$, the following compactness result holds,

**PROPOSITION 4.1 [51]**
*The inclusion* $(\mathcal{O}_{k+1}(\Omega_0), \partial_{k+1}) \hookrightarrow (\mathcal{O}_k(\Omega_0), \partial_k)$ *is compact.*

Also for a bounded universe $D$, the following continuity result holds,

**THEOREM 4.1 [147],[150]**
*The mapping*

$$\mathcal{V}_k \longrightarrow \mathcal{C}^0(I, \mathcal{O}_k(\Omega_0))$$
$$\mathbf{V} \longmapsto [t \mapsto \Omega_t(\mathbf{V}) = T_t(\mathbf{V})(\Omega_0)]$$

*is continuous and maps bounded subsets on equicontinuous parts.*

**LEMMA 4.1**
*The mappings*

$$\begin{array}{cc} \mathcal{V}_k(I) \to \mathcal{C}^1(I, \mathcal{C}^k(\bar{D}, \mathbb{R}^d)) & \mathcal{V}_k(I) \to \mathcal{C}^1(I, \mathcal{C}^k(\bar{D}, \mathbb{R}^d)) \\ \mathbf{V} \mapsto [t \mapsto T_t(\mathbf{V})] & \quad and \quad & \mathbf{V} \mapsto [t \mapsto T_t^{-1}(\mathbf{V})] \end{array}$$

*are continuous.*

Although $\mathcal{T}_k$ is not a vector-space, we will write, for shortness,

$$\mathcal{C}^1(I, \mathcal{T}_k) = \left\{ T \in \mathcal{C}^0(I, \mathcal{T}_k) \ \middle| \ T' \in \mathcal{C}^0(I, \mathcal{T}_k) \right\}$$

This space is endowed with the canonical norm

$$\|T\|_{\mathcal{C}^1(I, \mathcal{T}_k)} = \sup_{s \in I} \|T(s)\|_{\mathcal{T}_k} + \sup_{s \in I} \|T'(s)\|_{\mathcal{T}_k}$$

For $(\mathcal{T}_k, d_k)$, we have a result similar to theorem 4.1.

**THEOREM 4.2**
*The mapping*

$$\mathcal{V}_k(I) \to \mathcal{C}^1(I, \mathcal{T}_k)$$
$$\mathbf{V} \mapsto [t \mapsto T_t(\mathbf{V})]$$

*is surjective, continuous and maps bounded subsets on equicontinuous parts.*

We have the following characterization of the shape continuity,

### COROLLARY 4.1
*Let G be a shape function defined on $\mathcal{O}_k$ with values in a fixed Banach space $\mathcal{B}$. The followings properties are equivalent*

    *i) G is shape continuous with respect to $\mathcal{V}_k(I)$: for any initial domain $\Omega_0$, for all $\mathbf{V} \in \mathcal{V}_k(I)$, $s \mapsto G(\Omega_s(\mathbf{V}))$ belongs to $\mathcal{C}^0(I, \mathcal{B})$.*

    *ii) for any initial domain $\Omega_0$, for any $T$ in $\mathcal{C}^1(I, \mathcal{T}_k)$, $s \mapsto G\big(T(s)(\Omega_0)\big)$ belongs to $\mathcal{C}^0(I, \mathcal{B})$.*

It is important to notice how easy it is to characterize the shape continuity *via* the space $\mathcal{T}_k$. A characterization involving $\mathcal{O}_k$ would be more elegant, since the *real objects* are the domains, not the diffeomorphism. It is known (see [147, 51] for instance) that a shape functional $G$ defined on $\mathcal{O}_k$ with values in a fixed Banach space $\mathcal{B}$ is shape continuous (in the usual sense) as soon as $[s \mapsto G(\Omega(s))]$ is continuous (i.e., belongs to $\mathcal{C}^0(I, \mathcal{B})$) for any $[s \mapsto \Omega(s)] \in \mathcal{C}^0(I, \mathcal{O}_k)$.

---

## 4.3    The shape control problem

### 4.3.1    An existence result for the shape differential equation

This section aims at proving the following theorem, using a solution of equation (4.2).

### THEOREM 4.3
*Let J be a shape functional, differentiable in $\mathcal{O}_k$ with respect to $\mathcal{V}_k$. Let us assume the following conditions:*

    *(i) Both $J$ and $\nabla J$ are uniformly bounded on $\mathcal{O}_{k+1}$ (respectively in $\mathbb{R}$ and $\mathcal{A}^*_{k+1}$),*

    *(ii) $\nabla J$ is shape-continuous on $\mathcal{O}_{k+1}$, in $\mathcal{A}^*_k$, with respect to $\mathcal{V}_{k+1}$.*

*Then, there exists $\mathbf{V} \in \mathcal{V}_{k+1} \cap \mathrm{L}^2(\mathbb{R}^+; \mathcal{A}_{k+1})$ and $c > 0$ such that, for any $s \geqslant 0$,*

$$J(\Omega_s(\mathbf{V})) - J(\Omega_0(\mathbf{V})) = - \int_0^s \|\mathbf{V}(t)\|^2 \, dt = -c \int_0^s \|\nabla J(\Omega_t(\mathbf{V}))\|^2 \, dt$$

**PROOF**    Provided the duality operator $\mathbb{A}$ of equation (4.2) exists, a solution of this equation is convenient for the theorem. We are going to use a Sobolev space embedded in $\mathcal{A}_k$ to ensure the existence (and *good properties*) of the duality operator $\mathbb{A}$, and give a constructive proof of the existence of a solution of (4.2).

We fix $\kappa > 1$ such that the following space

$$\mathcal{H} = \left\{ \mathbf{V} \in \mathrm{H}^{\kappa}(D, \mathbb{R}^d) \ \middle| \ \langle \mathbf{V}, n \rangle_{\mathbb{R}^d} = 0 \text{ on } \partial D \right\}$$

satisfies the following embedding chain rule,

$$\mathcal{H} \hookrightarrow \mathcal{A}_{k+1} \hookrightarrow \mathcal{A}_k \tag{4.6}$$

We denote $\mathbb{A}$ the linear and continuous duality operator from $\mathcal{H}$ to its dual $\mathcal{H}^*$. We consider, the domain $\Omega_0$ being fixed in $\mathcal{O}_{k+1}$, an arbitrary interval $I$ of $\mathbb{R}^+$ which contains 0. Let $G_I$ be the mapping defined by

$$G_I(\mathbf{V}) : I \longrightarrow \mathcal{H}$$
$$s \longmapsto -\mathbb{A}^{-1}\big(\nabla J(\Omega_s(\mathbf{V}))\big) \tag{4.7}$$

for $\mathbf{V} \in \mathcal{C}^0(I, \mathcal{H}) \subset \mathcal{V}_{k+1}$.

Since we assumed the shape continuity of $\nabla J$, $G_I(\mathbf{V}) \in \mathcal{C}^0(I, \mathcal{H})$.

We are going to prove that $G_I$ has a fixed point, i.e., that equation (4.2) admits at least one solution. ⬚

### LEMMA 4.2
*There exists $m > 0$ such that*

$$B_{k,m} = \left\{ \mathbf{V} \in \mathcal{V}_{k+1} \ \middle| \ \sup_{s \in I} \|\mathbf{V}(s)\|_{\mathcal{A}_k} < m \right\} \supset G_I(B_{k,m})$$

**PROOF**    Due to the boundedness of $\nabla J$, there exists $m_1$ (which may depend on $\Omega_0$) such that for any $\Omega \in \mathcal{O}_k(\Omega_0)$,

$$\|\nabla J(\Omega)\|_{\mathcal{A}_k^*} \le m_1$$

It follows that

$$\|G_I(\mathbf{V})(s)\|_{\mathcal{H}} \le m_1 \|A^{-1}\|_{\mathcal{L}(\mathcal{H}^*, \mathcal{H})} = m_1$$

The choice $m = m_1$ is convenient. ⬚

### LEMMA 4.3
*The mapping $G_I$ is continuous. Provided $I$ is compact, $G$ is compact.*

**PROOF**    $G_I$ can be split using the following chain rule

$$G_I = G_3 \circ G_2 \circ G_1$$

where

$$G_1 : \mathcal{C}^0(I, \mathcal{H}) \longrightarrow \mathcal{C}^0(I, \mathcal{T}_k)$$
$$\mathbf{V} \; \mapsto \; [s \mapsto T_s(\mathbf{V})]$$

$$G_2 : \mathcal{C}^0(I, \mathcal{T}_k) \longrightarrow \mathcal{C}^0(I, \mathcal{A}_k^*)$$
$$T \; \mapsto \; \nabla J(T(.)(\Omega_0))$$

$$G_3 : \; \mathcal{C}^0(I, \mathcal{A}_k^*) \longrightarrow \mathcal{C}^0(I, \mathcal{H}^*) \longrightarrow \mathcal{C}^0(I, \mathcal{H})$$
$$g \qquad\qquad \mapsto \qquad g \qquad \mapsto \; -\mathbb{A}^{-1}g$$

Theorem 4.2 provides the continuity of $G_1$. Using Corollary 4.1, the continuity of $G_2$ is equivalent to the shape continuity of the $\nabla J$ and since the continuity of $G_3$ is clear, we deduce that $G_I$ is continuous.

We suppose $I$ is compact. By Theorem 4.2, a bounded subset $B \subset \mathcal{C}^0(I, \mathcal{A}_{k+1})$ is mapped by $G_1$ on a equicontinuous part of $\mathcal{C}^0(I, \mathcal{T}_{k+1}(\Omega_0))$. By Ascoli's theorem and the compactness of the inclusion of $\mathcal{T}_{k+1}(\Omega_0)$ in $\mathcal{T}_k(\Omega_0)$ (Theorem 4.1), the image of $B$ is para-compact in $\mathcal{C}^0(I, \mathcal{T}_k)$. Accordingly, $G_1$ is a compact mapping, and so is $G_I$.

Applying Leray-Schauder's fixed point theorem, we infer that for any initial domain $\Omega_0$ there exists $\mathbf{V}$ in $\mathcal{C}^0([0,1], \mathcal{H})$ with $G_{[0,1]}(\mathbf{V}) = \mathbf{V}$.    ☐

### 4.3.2   A constructive algorithm

Now, we shall build a sequence of converging domains. Let us define $\mathbf{V} \in \mathcal{C}^0(\mathbb{R}^+, \mathcal{H})$ and $(\Omega^n)_{n \in \mathbb{N}} \subset \mathcal{O}_{k+1}(\Omega_0)$ by

$$\forall n \in \mathbb{N}, \quad \begin{cases} G_{[n,n+1]}(\mathbf{V}\big|_{[n,n+1]}) = \mathbf{V}\big|_{[n,n+1]} \text{ on } D \\ \Omega_n(\mathbf{V}_n) = \Omega^n \\ \Omega_{n+1}(\mathbf{V}_n) = \Omega^{n+1} \\ \Omega_0(\mathbf{V}_0) = \Omega_0 \end{cases}$$

The continuity at integer points comes from equation (4.2). We have, for any $s \geqslant 0$, and any $n$,

$$J(\Omega^{n+1}) - J(\Omega^n) = -\int_n^{n+1} \|\mathbf{V}(t)\|_{\mathcal{H}}^2 \, dt = -\int_n^{n+1} \|\nabla J(\Omega_t(\mathbf{V}))\|_{\mathcal{H}^*}^2 \, dt \quad (4.8)$$

This so-built field $\mathbf{V}$ satisfies Theorem 4.3. Since we assumed $J$ is bounded, $s \mapsto J(\Omega^n)$ is bounded decreasing, hence has a limit and so does $\int_0^s \|\mathbf{V}(t)\|_{\mathcal{H}}^2 \, dt$, which proves $\mathbf{V} \in \mathrm{L}^2(\mathbb{R}^+, \mathcal{A}_{k+1})$.

## 4.4 The asymptotic behaviour

If $\mathbf{V}$ is given by Theorem 4.3, there exists a non-decreasing sequence $(s_n)_{n \geqslant 0}$ such that $\mathbf{V}(s_n) \to 0$ since $\mathbf{V} \in L^2(\mathbb{R}^+, \mathcal{A}_{k+1})$. We denote $\Omega^n = \Omega_{s_n}(\mathbf{V}) \subset \mathcal{O}_{k+1}(\Omega_0)$. The sequence $(\Omega^n)$ may not be bounded in $\mathcal{O}_k(\Omega_0)$ or $\mathcal{O}_{k+1}(\Omega_0)$, since the $L^2$ convergence of the speed given by this method is not sufficient in general. Nevertheless, we can use a weaker topology on the space of domains. We denote $\mathcal{O}_{\mathrm{op}}$ the family of all open subsets of $D$. In [51] it is proven to be a compact metric space for the Hausdorff-complementary metric

$$d(\Omega_1, \Omega_2) = \max \left\{ \sup_{x_1 \in D \setminus \Omega_1} \inf_{x_2 \in D \setminus \Omega_2} |x_1 - x_2| , \sup_{x_2 \in D \setminus \Omega_2} \inf_{x_1 \in D \setminus \Omega_1} |x_1 - x_2| \right\}$$

$$(4.9)$$

**LEMMA 4.4**

*Assume that*

*(i) the shape functional $J$ verifies the assumptions of Theorem 4.3,*

*(ii) the shape functional $J$ is defined and continuous in $\mathcal{O}_{\mathrm{op}}$,*

*(iii) the shape functional gradient $\nabla J$ is continuous for the Hausdorff complementary topology on $\mathcal{O}_k(\Omega_0)$.*

*Then $(\Omega^n)$ has cluster points in $\mathcal{O}_{\mathrm{op}}$ and if $\Omega^*$ is one of them, then*

$$\Omega^n \to \Omega^* \text{ in } \mathcal{O}_{\mathrm{op}} \text{ and } \nabla J(\Omega^n) \to 0 \text{ and } J(\Omega^n) \to J(\Omega^*)$$

**PROOF** The sequence $(\Omega^n)$ may be regarded as a sequence in the compact space $\mathcal{O}_{\mathrm{op}}$. Hence passing to a subsequence, it converges towards an open subset $\Omega^*$ of $D$. The gradient $\nabla J(\Omega^*)$ is not *a priori* defined, since the limit set has not enough regularity. Nevertheless since $\mathbf{V}$ satisfies (4.2), $\|\mathbf{V}(n)\|_{\mathcal{H}} = \|\nabla J(\Omega^n)\|$ hence $\nabla J(\Omega^n) \to 0$. ☐

In the next section, we shall furnish an example of a functional where the continuity hypothesis of Lemma (4.4) is satisfied.

## 4.5   Shape differential equation for the Laplace equation

### 4.5.1   The Laplace equation

In this section, we are given a family $g = (g_\Omega)_{\Omega \in \mathcal{O}_{\mathrm{lip}}}$ such that for any $\Omega \in \mathcal{O}_{\mathrm{lip}}$, $g_\Omega \in \mathrm{H}^{-1}(\Omega)$. We consider the Dirichlet problem

$$\mathcal{P}(\Omega, g) \quad \left\{ \begin{array}{rl} -\Delta y = g_\Omega & \text{in } \Omega \\ y = 0 & \text{on } \Gamma = \partial \Omega \end{array} \right.$$

which has a unique solution $y(\Omega, g)$ in the space $\mathrm{H}_0^1(\Omega)$ endowed with the norm $\|z\|_\Omega^2 = \int_\Omega |\nabla z|^2$.

**A priori estimates**

An *a priori* estimate for solution $y(\Omega, g)$ of $\mathcal{P}(\Omega, g)$ is derived from the variational formulation of the problem :
$y(\Omega, g)$ is the unique minimum of the functional $E_{\Omega,g}$ defined on $\mathrm{H}_0^1(\Omega)$ by

$$E_{\Omega,g}(z) = - \langle g_\Omega \, , \, z \rangle_{\mathrm{H}^{-1}(\Omega) \times \mathrm{H}_0^1(\Omega)} + \int_\Omega \frac{1}{2} |\nabla z|^2 \qquad (4.10)$$

where $|.|$ denotes the euclidean norm in $\mathbb{R}^d$. Accordingly,

$$E_{\Omega,g}(y(\Omega, g)) \leq 0$$

Thus it leads to $\dfrac{1}{2}\|y(\Omega, g)\|_\Omega^2 \leq \|g_\Omega\|_{\Omega,*}\|y(\Omega, g)\|_\Omega$ where $\|.\|_{\Omega,*}$ denotes the norm in the dual space $\mathrm{H}^{-1}(\Omega)$. This yields to

$$\|y(\Omega)\|_\Omega \leq 2\|g_\Omega\|_{\Omega,*} \qquad (4.11)$$

A mere consequence of this estimate is the following uniform boundedness result.

**LEMMA 4.5**
*Let O be a subset of $\mathcal{O}_{\mathrm{lip}}$ such that*

$$\{\|g_\Omega\|_{\Omega,*} | \Omega \in O\}$$

*is bounded. Then*

$$\{\|y(\Omega, g)\|_\Omega | \Omega \in O\}$$

*is bounded.*

In the sequel, we shall use the family $(f_{|\Omega})_{\Omega \in \mathcal{O}_{\text{lip}}}$ with $f \in L^2(D)$. Since

$$\|f_{|\Omega}\|_{\Omega,*} \leq \|f\|_{L^2(D)} \sup_{\substack{z \in H_0^1(\Omega) \\ \|z\|_{L^2(\Omega)} \leq 1}} \|z\|_{\Omega} \leq c_P(\Omega)\|f\|_{L^2(D)}$$

where $c_P(\Omega)$ is the Poincaré's constant[4] for the domain $\Omega$. Therefore, the uniform boundedness property of the solutions of $\mathcal{P}(\Omega, f)$[5] will arise from the following uniform boundedness of Poincaré's constant.

### LEMMA 4.6
*There exist a constant $c_P > 0$ such that*

$$\forall \Omega \in \mathcal{O}_{\text{lip}}, \quad c_p(\Omega) \leq c_P$$

**PROOF**　It is classical that for any $\Omega \in \mathcal{O}_{\text{lip}}$, there exists $z_\Omega$ in $H_0^1(\Omega)$ with $c_P(\Omega)^{-1} = \|\nabla z_\Omega\|_{L^2(\Omega)^d}$. Extending $z_\Omega$ by 0 provides a $\tilde{z}_\Omega$ in $H_0^1(D)$ such that $c_P(D)^{-1} \leq \|\nabla \tilde{z}_\Omega\|_{L^2(\Omega)^d}$. Thus $c_P(D) \geqslant c_P(\Omega)$ so $c_P = c_P(D)$ is convenient.　□

Eventually, we have proven the uniform boundedness of the solutions of $\mathcal{P}(\Omega, f)$ with respect to the domain.

### Strong shape continuity properties

Let $\Omega_0$ be a fixed initial domain in $\mathcal{O}_{\text{lip}}$ and let us assume the following,

i) The family $g_\Omega$ is *shape continuous*, i.e. for any $\mathbf{V}$ in $\mathcal{V}_k$, the mapping[6] $s \mapsto g_{\Omega_s(\mathbf{V})} \star T_s$ is continuous from $\mathbb{R}^+$ to $H^{-1}(\Omega_0)$

ii) For any $\mathbf{V}$ in $\mathcal{V}_k$,

$$\left[ s \mapsto \|g_{\Omega_s(\mathbf{V})}\|_{H^{-1}(\Omega_s)} \right] \in L_{\text{loc}}^\infty(\mathbb{R}^+) \tag{4.12}$$

---

[4]It may be defined *via* Rayleigh-quotient,

$$c_P(\Omega)^{-1} = \inf_{\substack{z \in H_0^1(\Omega) \\ z \neq 0}} \frac{\|\nabla z\|_{L^2(\Omega)^d}}{\|z\|_{L^2(\Omega)}}$$

[5]The accurate notation for this is $\mathcal{P}(\Omega, (f_{|\Omega})_{\Omega \in \mathcal{O}_{\text{lip}}})$.

[6]This mapping is defined as follows,
For any $z$ in $H_0^1(\Omega_0)$,

$$\left\langle g_{\Omega_s(\mathbf{V})} \star T_s, z \right\rangle_{H^{-1}(\Omega_0) \times H_0^1(\Omega_0)} = \left\langle g_{\Omega_s(\mathbf{V})}, \gamma_s^{-1} z \circ T_s^{-1} \right\rangle_{H^{-1}(\Omega_s(\mathbf{V})) \times H_0^1(\Omega_s(\mathbf{V}))}$$

where $\gamma_s = \det D T_s$.

**THEOREM 4.4**
*Under the last hypothesis, for $k \geqslant 2$, the following properties hold,*

*i) the transported solution of $\mathcal{P}(\Omega_0, g)$ is shape continuous:*

$$\forall \mathbf{V} \in V_k, \quad y(\Omega_s(\mathbf{V}), g) \circ T_s \xrightarrow{s \to 0} y(\Omega_0, g) \text{ in } H_0^1(\Omega_0)$$

*ii) the energy functional,*

$$E(\Omega_0, g) = E_{\Omega_0, g}(y(\Omega_0, g))$$

*is continuous,*

*iii) the extended solution of $\mathcal{P}(\Omega_0, g)$ is shape continuous:*

$$\forall \mathbf{V} \in V_k, \quad \tilde{y}(\Omega_s(\mathbf{V}), g) \xrightarrow{s \to 0} \tilde{y}(\Omega_0, g) \text{ in } H_0^1(D)$$

*Moreover, these properties are still valid for $k = 1$ provided $g_\Omega \in L^2(\Omega)$ for any $\Omega \in \mathcal{O}_{\mathrm{lip}}$.*

**PROOF**     For the sake of simplicity, we denote $y_s = y(\Omega_s(\mathbf{V}), g)$ and $y^s = y_s \circ T_s$. We have $y_0 = y^0$. Due to the local boundedness property (4.12), there exists $\varepsilon$ such that $(\|y_s\|_{\Omega_s})_{0 \leq s \leq \varepsilon}$ is uniformly bounded. Since

$$\|y^s\|_{\Omega_0}^2 \leq \|\gamma(s)^{-1} | D(T_s^{-1})^{-1} |\|_{L^\infty(D)} \|y_s\|_{\Omega_s}$$

the family $(y^s)_{0 \leq s \leq \varepsilon}$ is uniformly bounded in $H_0^1(\Omega_0)$. Hence we can extract a subsequence $(y^{s_n})_{n \geq 0}$ where $0 \leq s_n \leq \varepsilon$ and $s_n \to 0$ that converges towards an element $y^*$, weakly in $H_0^1(\Omega_0)$.
We denote $E^{\Omega_s, g}$ the functional defined on $H_0^1(\Omega_0)$ by the following identity,

$$E^{\Omega_s, g}(z) \stackrel{\mathrm{def}}{=} E_{\Omega_s, g}(z \circ T_s^{-1})$$
$$= - \langle g_{\Omega_s} \star T_s, \gamma_s z \rangle_{H^{-1}(\Omega_0) \times H_0^1(\Omega_0)} + \int_{\Omega_0} \frac{1}{2} |^* D T_s^{-1} \cdot \nabla z|^2 \gamma_s$$

and we have

$$\min E^{\Omega_s, g} = E^{\Omega_s, g}(y^s) = E_{\Omega_s, g}(y_s) = \min E_{\Omega_s, g} \qquad (4.13)$$

Since $k \geqslant 2$ the Jacobian $\gamma_{s_n}$ converges towards $\gamma(0) \equiv 1$ in $\mathcal{C}^1(D, \mathbb{R}^d)$. This is sufficient for

$$\langle g_{\Omega_{s_n}} \star T_{s_n}, \gamma(s_n) y^{s_n} \rangle_{H^{-1}(\Omega_0) \times H_0^1(\Omega_0)} \xrightarrow{s \to 0} \langle g_{\Omega_0}, y^0 \rangle_{H^{-1}(\Omega_0) \times H_0^1(\Omega_0)} \qquad (4.14)$$

Furthermore, the convergence of $\gamma_{s_n} (D T_{s_n})^{-1}$ towards $I$ in $\mathcal{C}^0(D, \mathbb{R}^d)$ induces the weak convergence of $(\gamma_{s_n} {}^* D T_{s_n}^{-1}) \cdot \nabla y^{s_n}$ towards $\nabla y^*$ in $L^2(\Omega_0)^d$. Using the weak-lower semi-continuity of the $L^2$-norm, we have

$$\int_{\Omega_0} \frac{1}{2} |\nabla y^*|^2 \leq \liminf_{n \to \infty} \int_{\Omega_0} \frac{1}{2} |^* D T_{s_n}^{-1} \cdot \nabla y^{s_n}|^2 \gamma_{s_n}$$

Thus, we have proven the weak-lower semi-continuity of the mapping $(s, z) \mapsto E^{\Omega_s, g}(z)$ on $\mathbb{R}^+ \times H_0^1(\Omega_0)$ at $(0, z)$ for any $z$. This proves that for any $z$

$$E^{\Omega_0, g}(y^*) \leq \liminf_{n \to \infty} E^{\Omega_{s_n}, g}(y^{s_n}) \leq \liminf_{n \to \infty} E^{\Omega_{s_n}, g}(z)$$

But $s \mapsto E^{\Omega_s, g}(z)$ is continuous for any $z$. Hence $y^* = y^0 = y_0$, which proves a weak shape-continuity for the transported solutions of $\mathcal{P}(\Omega, g)$.

The strong continuity will arise from a continuity of the norms, *via* the so-called compliance equality, and a compactness argument for the $y_s$. For any $\Omega$ in $\mathcal{O}_{\text{lip}}$, the necessary (and sufficient) condition of optimality for $E_{\Omega, g}$ is written

$$\forall z \in H_0^1(\Omega), \quad \int_\Omega \langle \nabla y(\Omega, g), \nabla z \rangle_{\mathbb{R}^d} = \langle g_\Omega, z \rangle_{H^{-1}(\Omega) \times H_0^1(\Omega)}$$

Choosing $z = y(\Omega, g)$ in the above equation, we come to

$$E_{\Omega, g}(y(\Omega, g)) = -\frac{1}{2} \|y(\Omega, g)\|_\Omega^2 = -\frac{1}{2} \langle g_\Omega, y(\Omega, g) \rangle_{H^{-1}(\Omega) \times H_0^1(\Omega)} \quad (4.15)$$

which leads together with equation (4.14) and the weak continuity of $s \mapsto y^s$, that $s \mapsto E_{\Omega, g}(y(\Omega_s, g))$ is continuous.

Since the sequence $(\|y_{s_n}\|_\Omega) = (\|\tilde{y}_{s_n}\|_D)$, where $\tilde{\cdot}$ denote the extension to $D$ with 0, there exists a $y_*$ in $H_0^1(D)$ such that $\tilde{y}_{s_n} \rightharpoonup y_*$ in $H_0^1(D)$. Since the sequence $(\Omega_{s_n})$ converges towards $\Omega_0$ for the Hausdorff complementary topology, $y_*$ has support in $\overline{\Omega_0}$ and may be written $\tilde{y}_\sharp$ with $y_\sharp \in H_0^1(\Omega_0)$, since the boundary $\partial \Omega_0$ has non-zero capacity. Using arguments similar to the ones which established (4.14), we have

$$\tilde{y}_{s_n} \circ T_{s_n}^{-1} = \widetilde{y_{s_n} \circ T_{s_n}^{-1}} \rightharpoonup \tilde{y}_\sharp \text{ in } H_0^1(D)$$

Accordingly, $y^{s_n} \rightharpoonup y_\sharp$ in $H_0^1(\Omega_0)$ and $y_\sharp = y_0$. This eventually proves that

$$\tilde{y}_{s_n} \rightharpoonup \tilde{y}_0 \text{ in } H_0^1(D)$$

But due to the continuity of $s \mapsto E_{\Omega_s, g}(y(\Omega_s, g))$ and equation (4.15),

$$\|\tilde{y}_{s_n}\|_D = \|y_{s_n}\|_{\Omega_{s_n}} \to \|\tilde{y}_0\|_D = \|y_0\|_{\Omega_0}$$

and this is sufficient for

$$\tilde{y}_{s_n} \to \tilde{y}_0 \text{ (strongly) in } H_0^1(D)$$

and

$$y^{s_n} \to y^0 \text{ (strongly) in } H_0^1(\Omega_0)$$

□

**REMARK 4.2**    In this proof, the assumption $k \geq 2$ is needed to prove the convergence result (4.14). In the case where $g_\Omega \in L^2(\Omega)$ for any $\Omega \in \mathcal{O}_{\mathrm{lip}}$, this assumption is not needed anymore.                                      ☐

The following differentiability result is well known when the right-hand side is fixed in $H^1(D)$. It may easily be extended to a domain-dependent right-hand side [135].

**THEOREM 4.5**
*Assume that for any $\Omega_0$ in $\mathcal{O}_k$, $g_{\Omega_0} \in L^2(\Omega_0)$ and for any $\mathbf{V}$ in $\mathcal{V}_k$ the mapping $s \mapsto g_{\Omega_s(\mathbf{V})} \circ T_s$ is strongly differentiable at $s = 0$ in $H^{-1}(\Omega_0)$ with derivative $\dot{g}_{(\Omega_0;\mathbf{V})}$.*
*Then the solution $y(\Omega)$ of problem $\mathcal{P}(\Omega, f)$ has a material derivative $\dot{y}(\Omega; \mathbf{V})$ in $H_0^1(\Omega)$ for any velocity field $\mathbf{V} \in \mathcal{V}_k$. Moreover, for any $\phi \in H_0^1(\Omega)$,*

$$\int_\Omega \langle \nabla \dot{y}(\Omega; \mathbf{V}) , \nabla \phi \rangle =$$
$$\int_\Omega -\left\langle [\tfrac{1}{2} \operatorname{div} \mathbf{V}(0) \, \mathrm{I} - \varepsilon(\mathbf{V}(0))] \cdot \nabla y(\Omega) , \nabla \phi \right\rangle_{\mathbb{R}^d} + [g_\Omega \operatorname{div}(\mathbf{V}(0)) + \dot{g}_{\Omega_0;\mathbf{v}}] \, \phi$$
(4.16)

### 4.5.2    The shape control problem

We would like to minimize the following shape functional,

$$J(\Omega) = \int_\Omega (y(\Omega) - y_{\mathrm{d}})^2 \qquad (4.17)$$

where $y_{\mathrm{d}} \in H_0^1(D)$ is a given target.

**The shape gradient**

**PROPOSITION 4.2**
*For any domain $\Omega$ in $\mathcal{O}_{\mathrm{lip}}$, for any field $\mathbf{V} \in \mathcal{V}_k$ ($k \geq 1$), the functional $J$ admits an Eulerian derivative,*

$$dJ(\Omega; \mathbf{V}) = \int_\Omega (y - y_{\mathrm{d}}) \left[ \operatorname{div} \mathbf{V}(0)(y - y_{\mathrm{d}}) - \mathrm{D}\, y_{\mathrm{d}} \cdot \mathbf{V}(0) \right]$$
$$+ 2 \langle A' \nabla y , \nabla p \rangle - 2 \langle \operatorname{div}(f \, \mathbf{V}) , p \rangle \quad (4.18)$$

*where $p$ is the solution of the adjoint problem $\mathcal{P}(\Omega, y(\tilde{\Omega}) - y_{\mathrm{d}})$.*

**PROOF** If $T_s$ is the flow-mapping associated to the velocity field $\mathbf{V}$, a change of variable gives

$$J(\Omega_s) = \int_{\Omega_s} (y_s - y_{\mathrm{d}})^2$$

$$= \int_{\Omega} \gamma_s \, (y^s - y_{\mathrm{d}} \circ T_s)^2$$

Since $s \mapsto T_s$ is of class $\mathcal{C}^1$, we get

$$\partial_s J(\Omega_s) = \int_{\Omega} \gamma'(s)(y^s - y_{\mathrm{d}} \circ T_s)^2 - 2\gamma(s)(y^s - y_{\mathrm{d}} \circ T_s)(\partial_s y^s - \mathrm{D}\, y \circ T_s \partial_s T_s)$$

Therefore,

$$dJ(\Omega; \mathbf{V}) = \int_{\Omega} [\operatorname{div} \mathbf{V}(0)] \, (y - y_{\mathrm{d}})^2 - 2(y - y_{\mathrm{d}}) \, (\dot{y} - \mathrm{D}\, y_{\mathrm{d}} \cdot \mathbf{V}(0))$$

We consider the adjoint problem,

$$\mathcal{P}(\Omega, y(\tilde{\Omega}) - y_{\mathrm{d}}) \quad \begin{cases} -\Delta p = y - y_{\mathrm{d}} & \text{on } \Omega \\ \quad p = 0 & \text{on } \Gamma = \partial \Omega \end{cases} \tag{4.19}$$

Thanks to Theorem 4.5, the strong material derivative of $y(\Omega)$ satisfies

$$\forall \phi \in \mathrm{H}_0^1(\Omega) \,, \quad \int_{\Omega} \langle \nabla \dot{y} \,, \, \nabla \phi \rangle = - \int_{\Omega} \langle A'(\mathbf{V}) . \nabla y \,, \, \nabla \phi \rangle \, + \langle \operatorname{div}(f \, \mathbf{V}(0)) \,, \, \phi \rangle$$

with

$$A'(\mathbf{V}(0)) \overset{\text{def}}{=} \operatorname{div} \mathbf{V}(0) \, \mathrm{I} - 2\, \varepsilon(\mathbf{V}(0))$$

Then, we deduce

$$dJ(\Omega; \mathbf{V}) = \int_{\Omega} \operatorname{div} \mathbf{V}(0)(y - y_{\mathrm{d}})^2 + 2(\Delta p) \, (\dot{y} - \mathrm{D}\, y_{\mathrm{d}} \cdot \mathbf{V}(0))$$

$$= \int_{\Omega} \operatorname{div} \mathbf{V}(0)(y - y_{\mathrm{d}})^2 - \mathrm{D}\, y_{\mathrm{d}} \cdot \mathbf{V}(0) - 2\nabla p \cdot \nabla \dot{y}$$

$$= \int_{\Omega} (y - y_{\mathrm{d}})[\operatorname{div} \mathbf{V}(0)(y - y_{\mathrm{d}}) - \mathrm{D}\, y_{\mathrm{d}} \cdot \mathbf{V}(0)]$$
$$+ 2 \, \langle A'(\mathbf{V}(0))\nabla y \,, \, \nabla p \rangle - 2 \, \langle \operatorname{div}(f \, \mathbf{V}) \,, \, p \rangle$$

When the domain $\Omega$ is fixed, the mapping $\mathbf{V} \mapsto dJ(\Omega; \mathbf{V})$ is linear and continuous. We consider the element $\nabla J(\Omega)$ of $\mathcal{A}_k^*$ given by

$$\langle \nabla J(\Omega) \,, \, \mathbf{V} \rangle_{\mathcal{A}_k^* \times \mathcal{A}_k} = \int_{\Omega} (y(\Omega) - y_{\mathrm{d}})[\operatorname{div} \mathbf{V}(y(\Omega) - y_{\mathrm{d}}) - \mathrm{D}\, y_{\mathrm{d}} \cdot \mathbf{V}]$$

$$+ 2 \, \langle A'(\mathbf{V})\nabla y(\Omega) \,, \, \nabla p \rangle - 2 \int_{\Omega} \langle \operatorname{div}(f \, \mathbf{V}) \,, \, p(\Omega) \rangle \quad (4.20)$$

$\square$

**Uniform boundedness**

### PROPOSITION 4.3
*There exists a constant $M > 0$ such that*

$$\|\nabla J(\Omega)\|_{\mathcal{A}_k^*} \leq M, \quad \forall \Omega \in \mathcal{O}_{\text{lip}}$$

**PROOF**

$$\|\nabla J(\Omega)\|_{\mathcal{A}_k^*} = \sup_{\mathbf{V} \in \mathcal{A}_k} \langle \nabla J(\Omega), \mathbf{V} \rangle_{\mathcal{A}_k^* \times \mathcal{A}_k}$$

$$\leq \|\operatorname{div} \mathbf{V}\|_{\text{L}^\infty} \|y - y_{\text{d}}\|_{\text{L}^2}^2 + \|D\, y_{\text{d}} \cdot (y - y_{\text{d}})\|_{\text{L}^1} \|\mathbf{V}\|_{\text{L}^\infty}$$

$$+ 2\|A'(\mathbf{V})\|_{\text{L}^\infty} \|\langle \nabla y, \nabla p \rangle\|_{\text{L}^1} + 2 \|f|\nabla p|\|_{\text{L}^1} \|\mathbf{V}\|_{\text{L}^\infty}$$

Then, there exists a constant $m > 0$ such that

$$\|\nabla J(\Omega)\|_{\mathcal{A}_k^*} \leq m\|y\|_\Omega \|p\|_\Omega \|\mathbf{V}\|_{\mathcal{A}_k}$$

and the uniform boundedness of $\|y\|_\Omega$ and $\|p\|_\Omega$ gives the correct result. ⬛

**Shape continuity properties of the gradient**

### THEOREM 4.6
*The gradient distribution $\nabla J$ is continuous in $\mathcal{A}_k^*$.*

**PROOF**   The gradient $\nabla J(\Omega)$ is a continuous function $G(\chi_\Omega, \tilde{y}(\Omega), \tilde{p}(\Omega))$ which is continuous from $\text{L}^2(D) \times \text{H}_0^1(D) \times \text{H}_0^1(D)$ to $\mathcal{A}_k^*$ with

$$\langle G(\chi_\Omega, y, p), \mathbf{V} \rangle_{\mathcal{A}_k^* \times \mathcal{A}_k} =$$

$$\int_D \chi_\Omega (y - y_{\text{d}})[\operatorname{div} \mathbf{V} \cdot (y - y_{\text{d}}) - D\, y_{\text{d}}.\mathbf{V}] + 2\chi_\Omega \langle A'(\mathbf{V})\nabla y, \nabla p \rangle$$

$$- 2\chi_\Omega \langle \operatorname{div}(f\, \mathbf{V}), p \rangle$$

Theorem 4.4 provides that the mapping $s \mapsto \tilde{y}(\Omega_s(\mathbf{V}))$, $s \mapsto \tilde{p}(\Omega_s(\mathbf{V}))$ are continuous for any $\mathbf{V}$ in $\mathcal{V}_k$. Consequently, $s \mapsto \nabla J(\Omega_s(\mathbf{V}))$ is continuous. ⬛

### 4.5.3   An asymptotic result in the 2D case

We shall apply theorem 4.3 to the Laplace equation $\mathcal{P}(\Omega, g)$.

### PROPOSITION 4.4
*For any $\Omega_0$ in $\mathcal{O}_{k+1}$, there exists a $\mathbf{V} \in \mathcal{V}_{k+1}$ and an open subset $\Omega_*$ of $D \subset \mathbb{R}^2$ such that,*

i) $J(\Omega_s(\mathbf{V})) - J(\Omega_0) = \int_0^s \|\mathbf{V}(t)\|^2 \, \mathrm{d}t$

ii) *for any sequence $(s_n)_{n \geqslant 0}$ with $s_n \to \infty$, $\Omega_{s_n}(\mathbf{V}) \to \Omega_*$ for Hausdorff complementary topology, $J(\Omega_{s_n}(\mathbf{V})) \to J(\Omega_*)$ and $\nabla J(\Omega_{s_n}(\mathbf{V})) \to 0$.*

The general asymptotic behaviour of section 4.4 may be developed in the case $d = 2$. Indeed, the continuity of $\Omega \mapsto \tilde{y}(\Omega)$ for Hausdorff-complementary topology does not hold in general. Nevertheless, this continuity holds under capacity constraints( [23], [19]). In the 2-dimensional case, Sverak has proven in [137],[136] the convergence of $\tilde{y}(\Omega_n)$ towards $\tilde{y}(\Omega)$, provided $(\complement\Omega_n)_n$ converges to $\complement\Omega$ for Hausdorff topology, with $\sharp\complement\Omega_n$ uniformly bounded.

---

## 4.6 Shape differential equation in $\mathbb{R}^{d+1}$

We come back to the optimal dynamical evolution of a geometrical domain, a moving domain, say $\Omega_t$, where $t$ is the time parameter. That optimality is built in the context of the minimization of some cost functional. For each tube

$$Q = \bigcup_{0 \leq t \leq \tau} \{t\} \times \Omega_t \subset \mathbb{R}^{d+1}$$

with lateral boundary

$$\Sigma = \bigcup_{0 \leq t \leq \tau} \{t\} \times \partial\Omega_t \subset \mathbb{R}^{d+1}$$

We define as previously the unitary normal field to $\Sigma$ pointing outside $Q$,

$$\boldsymbol{\nu}(t, x) = (1 + v(t, x)^2)^{-1/2} \left( -v(t, x), \mathbf{n}_t(x) \right)$$

where $\mathbf{n}_t(x) \subset \mathbb{R}^d$ is the unitary normal field to $\Gamma_t \stackrel{\text{def}}{=} \partial\Omega_t$, pointing outside the moving domain $\Omega_t$.

Let us consider a cost functional of the tracking type,

$$J(Q) = \frac{1}{2} \int_Q |y_Q - Y_g|^2 \, \mathrm{d}x \, \mathrm{d}t$$

where $Y_g$ is a given element, e.g., $Y_g \in H^1((0, \tau) \times D)$ where $D$ is a given bounded open domain in $\mathbb{R}^{d+1}$ such that the cylindrical domain $]0, \tau[\times D$ will contain the potential evolution of the moving tubes $Q_s$. The element $y_Q \in W(Q)$ is a state variable which is defined as an element of some functional space built on the tube $Q$.

As an example of such a state variable, we can consider the non-cylindrical wave equation, where for a given rhs $F \in L^2(0, \tau; L^2(D))$, the state variable $y_Q \in L^2(0, \tau, H_0^1(\Omega_t))$ satisfies the following equation,

$$\partial_{tt} y - \Delta y = F, \text{ in } Q \tag{4.21}$$

The basic "Optimal Moving Domain" problem consists in finding the "best" dynamical time evolution for the moving domain $\Omega_t$ in order to minimize the functional $J$. In other words, we look for the best tube $Q \subset \mathbb{R}^{d+1}$ which minimizes the functional $J$. For that purpose, we just consider the tube $Q$ as a moving domain in $\mathbb{R}^{d+1}$ and we consider an "evolution" parameter $s$ to "follow" the virtual evolution of the tube $Q_s$.
Given a smooth enough vector field

$$\tilde{\mathbf{Z}}(s; t, x) \overset{\text{def}}{=} \left( 0, \mathbf{Z}(s; t, x) \right) \in \mathbb{R}_t \times \mathbb{R}_x^d$$

we consider the associated flow mapping

$$\mathbf{T}_s(\tilde{\mathbf{Z}})(t, x) \overset{\text{def}}{=} \left( t, \mathbf{T}_s(\mathbf{Z})(t, x) \right)$$

The specific structure of that flow is obviously deriving from the "horizontal" character of the vector field $\tilde{\mathbf{Z}}$ itself. That choice does not produce any generality loss[7] in the potential deformations of the cylinder $Q$. A special role will be played by the $d$ dimensional vector field

$$\mathbf{Z}(t)(s, x) := \mathbf{Z}(s; t, x)$$

From the classical shape analysis, we derive the $\mathbb{R}^{d+1}$-shape gradient associated to the shape functional $J(Q)$ as follows:

$$J(Q_s) = \min_{\phi \in L^2(0, \tau, H_0^1(\Omega_t))} \max_{\psi \in L^2(0, \tau, H_0^1(\Omega_t))} \mathcal{L}(s; \phi, \psi)$$

where the Lagrangian functional is given by the following expression,

$$\mathcal{L}(s; \phi, \psi) = \frac{1}{2} \int_{Q_s} |\phi(t) \circ \mathbf{T}_s(\mathbf{Z})^{-1} - Y_g|^2 \, dx \, dt$$

$$+ \int_{Q_s} \left[ \partial_t (\phi(t) \circ \mathbf{T}_s(\mathbf{Z})^{-1}) \partial_t (\psi(t) \circ T_s(Z)^{-1}) \right.$$

$$+ \nabla(\phi(t) \circ \mathbf{T}_s(\mathbf{Z})^{-1}) \cdot \nabla(\psi(t) \circ \mathbf{T}_s(\mathbf{Z})^{-1}) \, dx \, dt$$

$$- \int_{Q_s} F \, \psi(t) \circ \mathbf{T}_s(\mathbf{Z})^{-1}, dx \, dt - \int_{\Omega_0} y_1 \, \psi(0) \, dx$$

---

[7] Indeed a general form for the vector field $\tilde{\mathbf{Z}}$ could induce a change in the time variable $t$ which would create a "time extension" of the tube.

Using the structure theorem, we can state that the eulerian directional derivative of $J$ is only supported by the lateral boundary $\Sigma$ and is given by the following expression,

$$dJ(Q,\tilde{\mathbf{Z}}) \overset{\text{def}}{=} \left(\frac{\partial}{\partial s}J(Q_s)\right)_{s=0} = \int_\Sigma g\,\tilde{\mathbf{Z}}\cdot\boldsymbol{\nu}\,d\Sigma$$

Notice that

$$\int_\Sigma f\,d\Sigma = \int_0^\tau \int_{\Gamma_t} f(t,x)\,\sqrt{1+v(t,x)^2}\;dt\,d\Gamma(x)$$

Moreover

$$\langle\tilde{\mathbf{Z}}(t,x),\boldsymbol{\nu}(t,x)\rangle = \frac{1}{\sqrt{1+v(t,x)^2}}\,\langle\mathbf{Z}(t,x),\mathbf{n}_t(x)\rangle$$

So that

$$dJ(Q,\tilde{\mathbf{Z}}) = \int_0^\tau \int_{\Gamma_t} g(t,x)\,\mathbf{Z}(t,x)\cdot\mathbf{n}_t(x)\,d\Gamma_t(x)\,dt$$

That is

$$dJ(Q,\tilde{\mathbf{Z}}) = \langle G(Q),\tilde{\mathbf{Z}}(0)\rangle_{\mathcal{D}'(]0,\tau[\times D,\mathbb{R}^{d+1})\times\mathcal{D}(]0,\tau[\times D,\mathbb{R}^{d+1})}$$

Thus the "time-space" vectorial distribution is given by the following expression,

$$G(Q) \overset{\text{def}}{=} {}^*\gamma_\Sigma(g\,\boldsymbol{\nu})$$

where the density gradient $g$ is an element of $L^1(\Sigma)$ (in general a $(d+1)$- vector measure on $\Sigma$ with zero transverse order) and its expression involves, in the non-cylindrical wave equation case, the normal derivatives at the boundary of the state and co-state solution.

Now we apply, in the $(d+1)$-dimensional case, the *Shape Differential Equation* framework developed in the previous section of this chapter. As a consequence, we derive the existence of a solution $\tilde{\mathbf{Z}} = (0,\mathbf{Z})$ to the shape differential equation,

$$\tilde{\mathbf{Z}} = \left(0,\mathcal{A}^{-1}.G(\mathbf{T}_s(\tilde{\mathbf{Z}}))\right) \tag{4.22}$$

where $\mathcal{A}$ stands for an *ad hoc* duality positive operator as described previously. So we obtain

$$J(Q_{s_0}) \le J(Q) - \alpha\int_0^{s_0}\|\tilde{\mathbf{Z}}(s)\|^2_{H(D)}\,ds$$

for some "pivot" Hilbertian Sobolev vector space $H(D)$ over the hold-all domain $D$.

The construction of the tube $Q_{s_0} \overset{\text{def}}{=} \mathbf{T}_s(\tilde{\mathbf{Z}})(Q)$ from the field $\tilde{\mathbf{Z}}$ solution to the previous $\mathbb{R}^{d+1}$-shape differential equation consists in finding the $\mathbb{R}^d$-dimensional vector field $\mathbf{V}(t,x)$ whose flow mapping $\mathbf{T}_t(\mathbf{V})$ builds the tube

$Q_{s_0}$ from the initial domain $\Omega_0$.
Let us assume that the initial tube $Q$ is the cylindrical domain

$$Q = ]0, \tau[ \times \Omega_0$$

That tube is obviously built by the vector speed $\mathbf{V} = 0$. Now we look for a vector field $\mathbf{V}^{s_0}(t, x)$ such that its flow mapping $\mathbf{T}_t(\mathbf{V}^{s_0})$ builds the "optimal" tubes $Q_{s_0}$ from the initial domain $\Omega_0$. This means that the following identity holds true,

$$\mathbf{T}_t(\mathbf{V}^{s_0}) = \mathbf{T}_{s_0}(\mathbf{Z}(t))(\Omega_0), \forall 0 < t < \tau \qquad (4.23)$$

where $\mathbf{Z}(t)$ stands for the $\mathbb{R}^{d+1}$-dimensional vector field $\mathbf{Z}(t)(s, x) = \mathbf{Z}(s, t, x)$. Since, by definition, we have

$$\mathbf{T}_{s_0}(\tilde{\mathbf{Z}})(t, x) = \left( t, x + \int_0^{s_0} \mathbf{Z}(\sigma; t, \mathbf{T}_\sigma(\tilde{\mathbf{Z}})(t, x))) d\sigma \right)$$

we get

$$\frac{\partial}{\partial t} \left( \mathbf{T}_{s_0}(\tilde{\mathbf{Z}}) \right)(t, x) = \big( 1,$$
$$\int_0^{s_0} \frac{\partial}{\partial t} \left( \mathbf{Z}(t, \mathbf{T}_\sigma(\tilde{\mathbf{Z}})(t, x)) + D_x \, \mathbf{Z}(\sigma; t, \mathbf{T}_\sigma(\tilde{\mathbf{Z}})(t, x)) \cdot \frac{\partial}{\partial t}(\mathbf{T}_\sigma(\tilde{\mathbf{Z}})(t, x)) \right) d\sigma \big)$$

By definition, the $\mathbb{R}^d$-dimensional speed vector $\mathbf{V}^{s_0}$ is furnished by the following expression,

$$\left( 1, \mathbf{V}^{s_0}(t, x) \right) \stackrel{\text{def}}{=} \frac{\partial}{\partial t} \mathbf{T}_{s_0}(\tilde{\mathbf{Z}})(t, \mathbf{T}_{s_0}(\tilde{\mathbf{Z}})^{-1}(t, x)) \qquad (4.24)$$

Now, since the $\mathbb{R}^{d+1}$-dimensional flow is horizontal, i.e.,

$$\mathbf{T}_{s_0}(\tilde{\mathbf{Z}})(t, x) = (t, \mathbf{T}_{s_0}(\mathbf{Z}(t))(x)) \qquad (4.25)$$

then we obtain

$$\mathbf{V}^{s_0}(t, x) = \frac{\partial}{\partial t} \left( \mathbf{T}_{s_0}(\mathbf{Z}(t)) \right) \left( \mathbf{T}_{s_0}(\mathbf{Z}(t))^{-1}(x) \right) \qquad (4.26)$$

Also the previous differential equation can be rewritten in term of $\mathbf{Z}(t)$ as follows,

$$\mathbf{T}_{s_0}(\mathbf{Z}(t))(x) = x + \int_0^{s_0} \mathbf{Z}(t)(\sigma, \mathbf{T}_\sigma(\mathbf{Z}(t))(x)) \, d\sigma \qquad (4.27)$$

so that

$$\frac{\partial}{\partial t}[\mathbf{T}_{s_0}(\mathbf{Z}(t))](x) = \int_0^{s_0} \{ \frac{\partial}{\partial t} \mathbf{Z}(t)(\sigma, \mathbf{T}_\sigma(\mathbf{Z}(t)))(x)$$
$$+ D \, \mathbf{Z}(t)(\sigma, \mathbf{T}_\sigma(\mathbf{Z}(t)))(x) \cdot \frac{\partial}{\partial t} \mathbf{T}_\sigma(\mathbf{Z}(t))(x) \} \, d\sigma$$

Then,

$$\mathbf{V}^{s_0}(t,x) = \int_0^{s_0} \{ \frac{\partial}{\partial t}\mathbf{Z}(t)(\sigma, \mathbf{T}_\sigma(\mathbf{Z}(t)) \circ \mathbf{T}_{s_0}(\mathbf{Z}(t))^{-1})(x)$$
$$+ D\,\mathbf{Z}(t)(\sigma, \mathbf{T}_\sigma(\mathbf{Z}(t)) \circ \mathbf{T}_{s_0}(\mathbf{Z}(t))^{-1})(x)) \cdot \mathbf{V}^\sigma(t, \mathbf{T}_\sigma(\mathbf{Z}(t)) \circ \mathbf{T}_{s_0}(\mathbf{Z}(t))^{-1}(x))\,d\sigma$$

## 4.7   The level set formulation

### 4.7.1   Introduction

In [149],[135],[51],[93] we have considered domains family parametrized as follows:

$$\Omega_t = \Omega_t(\Phi) = \{x \in D \mid \Phi(t,x) > 0\},\ \Gamma_t = \{x \in D \mid \Phi(t,x) = 0\}\ \ (4.28)$$

where $\Phi(t,.)$ is a function defined on $D$ verifying a negative condition at the boundary of $D$, say $\Phi = -1$ on the boundary $\partial D$. The singular points are those in $D$ at which the gradient of $\Phi(t,.)$ vanishes. We assume that no such points lie on $\Gamma_t$ so that in a neighbourhood of the lateral boundary $\Sigma$ of the tube, the following vector field

$$\mathbf{V}(t,x) \stackrel{\text{def}}{=} -\partial_t\,\Phi(t,x)\,\frac{\nabla\Phi(t,x)}{\|\nabla\Phi(t,x)\|^2} \tag{4.29}$$

is always defined. If $\mathbf{V}$ is smooth enough we get

$$\Omega_t = T_t(\mathbf{V})(\Omega_0)$$

In this case, the shape differential equation writes

$$\mathbb{A}^{-1}.\nabla J(\Omega_t(\mathbf{V})) - \partial_t\,\Phi(t,x)\,\frac{\nabla\Phi(t,x)}{\|\nabla\Phi(t,x)\|^2} = 0 \tag{4.30}$$

which would imply, with the notation $\Omega_t(\Phi) = \Omega_t(\mathbf{V})$ ($\mathbf{V}$ previously defined), the following Hamilton-Jacobi equation for the function $\Phi$:

$$\partial_t\,\Phi(t,x) - \langle \nabla\Phi(t,x),\ \mathbb{A}^{-1} \cdot \nabla J(\Omega_t(\Phi)) \rangle = 0 \tag{4.31}$$

Notice that the Hamilton-Jacobi versions (4.30) and (4.31) are not equivalent. They would be merely equivalent if $\mathbb{A}^{-1} \cdot \nabla J(\Omega_t(\Phi))$ was proportional to $\nabla\Phi(t,x)$. In order to bypass that point, we consider the *scalar shape differential equation*. From the general structure theorem for shape gradient [135], we have

$$G = \nabla J(\Omega) = \gamma_\Gamma^*.(g\,\mathbf{n}) = \mathcal{G}(\Omega)\,\nabla\chi_\Omega$$

where $\mathbf{n}$ is the normal field to the boundary $\Gamma$, while $g$ is a scalar distribution on the boundary. In almost classical regular problems $g$ turns to be a function defined on the boundary with $g \in L^2(\Gamma)$. We choose to work with the *normal scalar shape differential equation* obtained by setting

$$v \overset{\text{def}}{=} \langle \mathbf{V}(t,.), \mathbf{n}_t \rangle = -\mathbb{A}_{\Gamma_t}^{-1} \cdot g(\Gamma_t) \quad \text{on} \quad \Gamma_t \tag{4.32}$$

In terms of level set modeling, with $\mathbf{n}_t = \dfrac{\nabla \Phi(t.)}{\|\nabla \Phi(t,.)\|}$ on $\Gamma_t$, it leads to the following *normal-level* shape differential equation :

$$\frac{\partial_t \Phi}{\|\nabla \Phi\|} = -\mathbb{A}_{\Gamma_t}^{-1} \cdot g(\Gamma_t) \tag{4.33}$$

It leads to the following *normal Hamilton-Jacobi* level set equation in the whole domain $D$ :

$$\partial_t \Phi + \mathbb{A}_{\Gamma_t}^{-1} \cdot \mathcal{G} \, \|\nabla \Phi\| = 0 \tag{4.34}$$

Assuming $\Phi$ is a solution to the previous *normal Hamilton-Jacobi*, in order to derive a solution to the previous normal scalar shape differential equation, we need to divide by $\|\nabla \Phi\|$. For that reason we now focus on a class of functions $\Phi$ without step so that $\|\nabla \Phi\|$ is different from zero almost everywhere in $D$.

## 4.7.2   Solutions without step

We are interested in functions $\Phi(t,.)$ without steps.

**DEFINITION 4.1**   *We say that a function $f$ defined on a set $D$ has a step $t$ if*

$$\text{meas}(\{x \in D \ | \ f(x) = t \ \}) > 0$$

Now we shall describe a construction of function without step, derived from a method, we have introduced for the modeling of free boundary value problems arising in plasma physics [8].
Consider, for any $\varepsilon > 0$ and for $g \in H^{-1}(D)$, the variational problem

$$z \in \underset{u \in H_0^1(D)}{\arg\min} \int_D \left[ \frac{1}{2} \|\nabla u\|^2 + (\varepsilon|D| - g) \, u \right] dx - \frac{\varepsilon}{2} \int_D \int_D (u(x) - u(y))^+ \, dx \, dy$$

The associated Euler-Lagrange equation writes

$$-\Delta z = \varepsilon \, \beta(z) + g, \quad \text{in } D \tag{4.35}$$

---

[8]The so-called Harold Grad Adiabatic equation of plasmas at equilibrium in the Tokomak [148], [149].

with $\beta(z)(x) = \text{meas}\,(\{y \in D \mid z(y) < z(x)\})$, together with the extra *no step* condition :

$$\text{meas}\,(\{x \in D \mid z(x) = t\}) = 0, \quad \forall\, t \in \mathbb{R}^+ \tag{4.36}$$

**REMARK 4.3**  This method can be limited to *zero step* functions :

$$z \in \underset{u \in H_0^1(D)}{\arg\min} \int_D \left[\frac{1}{2}\|\nabla u\|^2 - g\,u\right]\,dx - \frac{\varepsilon}{2}\int_D (u(x))^+\,dx$$

The associated Euler-Lagrange equation writes

$$-\Delta z = \varepsilon\,\beta_0(z) + g, \quad \text{in } D \tag{4.37}$$

with $\beta_0(z)(x) = \chi_{\{x \in D\,|\,z(x)>0\}}$, together with the extra *no step* condition :

$$\text{meas}\,(\{x \in D \mid z(x) = 0\}) = 0 \tag{4.38}$$

☐

## 4.7.3  Iterative Scheme

Obviously the function $\chi$ must satisfy the constraint $\chi^2 = \chi$ while the *level set* function $\Phi$ does not. Then, immediately from the previous study, we understand that in the second Hamilton-Jacobi equation (4.31), the velocity vector field $\mathbf{V} = \mathbb{A}^{-1}\cdot\nabla J(\Omega_t(\Phi))$ only needs to be in $L^2(0,\tau,L^2(D)^3)$ together with its divergence too. Indeed, in order to perform a fixed point argument inside the following iterative approximation scheme,

$$\partial_t \Phi^n(t,x) - \langle \nabla\Phi^n(t,x), \mathbf{V}^{n-1}\rangle = 0$$

with

$$\mathbf{V}^{n-1} \overset{\text{def}}{=} \mathbb{A}^{-1}\cdot\nabla J(\Omega_t(\Phi^{n-1}))$$

we only need $\mathbf{V}^n \in L^2(0,\tau;L^2(D)^3)$ and div $\mathbf{V}^n \in L^2(0,\tau;L^\infty(D))$. The idea is to choose a possibly non-linear operator $\mathbb{A}$ *powerfull enough* so that its inverse $\mathbb{A}^{-1}$ would map compactly and continuously the functional space where lies the cost function gradient into the previous velocity space.
In almost all *smooth problems*, i.e., following the general assumptions of the structure derivative theorem, the cost gradient takes the following form,

$$G = \nabla J(\Omega_t(\Phi)) = \mathcal{G}(\chi_t(\Phi))\nabla\chi_t(\Phi)$$

So that $G$ is a distribution, with support on the boundary $\Gamma$, zero transverse order and lying in some negative Sobolev space over $D$. Furthermore, $\mathcal{G} \in W^{1,1}(D)$ and is non-uniquely determined. Actually, only its trace[9] $g$ on the

---

[9] $g$ is called the density gradient. It is a scalar distribution on the *manifold* $\Gamma$, when $\Gamma$ is smooth enough in order to make sense.

boundary $\Gamma$, if smooth enough, is intrinsically determined.
We consider the following family of without step functions,

$$(A_\varepsilon)^{-1}(g) = \{\ z \text{ solutions to } (4.35), (4.36)\ \}$$

$$(A_\varepsilon^0)^{-1}(g) = \{\ z \text{ solutions to } (4.37), (4.38)\ \}$$

**LEMMA 4.7**
*Let $g_n \longrightarrow g$ in $H^{-1}(D)$ and $z_n \in (A_\varepsilon)^{-1}(g_n)$ (resp. $z_n \in (A_\varepsilon^0)^{-1}(g_n)$). Then there exists a subsequence and a limit element $z$ such that:*

$$z_{n_k} \longrightarrow z \text{ weakly in } H_0^1(D)$$

*Moreover any such element $z$ verifies $z \in (A_\varepsilon)^{-1}(g)$ (resp. $z \in (A_\varepsilon^0)^{-1}(g)$.*

The most important result concerning such elements without zero step is the following,

**LEMMA 4.8**
*Let $\Phi_n$, $\Phi \in L^2(D)$. Assume that $\Phi_n \longrightarrow \Phi$ strongly in $L^2(D)$ and $\Phi$ is without zero step :*

$$\text{meas}(\{x \in D \ | \ \Phi(x) = 0\ \}) = 0$$

*Then let $\Omega_n = \{x \in D \ | \ \Phi_n(x) > 0\ \}$, $\Omega = \{x \in D \ | \ \Phi(x) > 0\ \}$. We denote by $\chi_n$ and $\chi$ the respective characteristic functions of those subsets. Then $\chi_n \longrightarrow \chi$ strongly in $L^2(D)$.*

**PROOF**    Obviously, we have

$$\chi_n \Phi_n = (\Phi_n)^+$$

There exists a subsequence and an element $\zeta$, $0 \leq \zeta \leq 1$ such that $\chi_n$ weakly converges to $\zeta$ in $L^2(D)$. In the limit case, as $|(\Phi_n)^+(x) - (\Phi)^+(x)| \leq |\Phi_n(x) - \Phi(x)|$, we get :

$$\zeta \Phi = (\Phi)^+$$

So that $\zeta = 1$ a.e. in $\Omega$, and $\zeta = 0$ a.e. in $D - \Omega$. Since $\Phi$ has no zero step, we conclude that $\zeta = \chi$, so that the sequence converges strongly in $L^2(D)$. □

At that point, it is obvious that with some continuity assumption on the gradient $G(\chi)$ (that hypothesis will be satisfied in the following example), with respect to the characteristic function, the previous iterative construction $\mathbf{V}^n$ will converge and we will derive the existence of solutions to the Hamilton-Jacobi equation for the Level set function associated to the multi-valued operators $A_\varepsilon$ and $A_\varepsilon^0$.

### 4.7.4 An example: the transverse magnetic like inverse problem

As an illustration for the construction of solution to the Hamilton-Jacobi equation, we consider a simplified version of the famous transverse magnetic inverse problem. The analysis is greatly simplified since we consider a bounded universe $D$. Let $\Omega \subset D$ and $y \overset{\text{def}}{=} y(\chi) \in H_0^1(D)$ be the solution to the following problem,

$$-\Delta y + k\,\chi\,y = f$$

where $k$ is the contrast parameter while $f$ is given in $L^2(D)$ and $\chi \overset{\text{def}}{=} \chi_\Omega$. We introduce the observability functional,

$$J(\chi) = 1/2 \int_E (y - y_{\text{d}})^2 \, dx$$

where $E \subset D$.

**The classical shape derivative**

Here, we would like to compute the derivative of $J(\chi)$ with respect to $\chi$. It is given by

$$G(\Omega) = \gamma_\Gamma^*(y\,p\,\mathbf{n}) \tag{4.39}$$

where $p \in H_0^1(D)$ is the solution of the adjoint equation :

$$-\Delta p + k\,\chi\,p = \chi_E\,(y - y_{\text{d}}) \tag{4.40}$$

The associated density gradient is given by

$$g(\Gamma) = (y\,p)|_\Gamma$$

Let us consider the extension $\mathcal{G}$ of $g$ inside $D$,

$$\mathcal{G}(\chi) = y\,p \in W^{2,1}(D)$$

We easily check the continuity of the following mapping,

$$\chi \in L^2(D) \longrightarrow \mathcal{G}(\chi) \in W^{2,1}(D)$$

**Topological derivative versus set derivative**

At that point, it is very interesting, on that example, to understand that the shape derivative of that functional $J(\Omega)$ can be relaxed to a *set derivative setting* which coincide in a smooth situation with the classical shape derivative and shape gradient analysis.

We introduce the usual Lagrangian functional,

$$\mathcal{L}(\Omega, \phi, \psi) = \int_D (\nabla\phi.\nabla\psi + k\,\chi_\Omega\,\phi\,\psi - f\,\psi + \chi_E\,1/2\,(\phi - y_{\text{d}})^2\,)\,dx \tag{4.41}$$

Then

$$J(\Omega) = \min_{\phi \in H^2(D)} \max_{\psi \in H^2(D)} \mathcal{L}(\Omega, \phi, \psi)$$

with unique saddle point $(y, p)$. The Lagrangian $\mathcal{L}$ being partially concave-convex, weakly lower semi-continuous and upper weakly semi-continuous, considering any evolutive characteristic function $\chi(t)$, we get (see([42]) :

$$\frac{\partial}{\partial t} J(\chi)|_{\{t=0\}} = \frac{\partial}{\partial t} \mathcal{L}(\chi(t), y_0, p_0))|_{\{t=0\}}$$

where $y_0, p_0$ are the solution at $t = 0$. That derivative does make sense as soon as the right hand side derivative does. This technique enables us to manage holes creation inside $D$.

Assume for example that $\Omega_t$ is a vanishing sequence, as $t \to 0$ of measurable subsets in $D \subset \mathbb{R}^d$ verifying:

$$\Omega_n \to \{x_0\} \text{ in Hausdorff topology}$$

$$(\text{meas}(\Omega_t) / t^d) \to 0 \text{ as } t \to 0$$

Then, assuming $d = 2$ so that $y, p \in H^2(D) \subset C^0(\bar{D})$, we get :

$$\frac{\partial}{\partial t} J(\chi(t))_{\{t=0\}} = k (yp)(x_0)$$

This set derivative setting works thanks to the simple fact that $\frac{\chi(t)}{t} \to \delta_{x_0}$ (Dirac measure at point $x_0 \in D$), as $t \to 0$. Hence, we understand that the important concept in that situation is the derivative, in a measure space of the mapping $t \to \chi(t)$ at $t = 0$. Following this line, we shall consider, when it exists, the following element:

$$\dot{\chi}(t) = \lim_{\varepsilon \to 0} \frac{(\chi(t + \varepsilon) - \chi(t))}{\varepsilon}, \quad \text{in } \mathcal{M}(D)$$

if the saddle point $(y_t, p_t)$ satisfies the following continuity property:

$$t \longrightarrow (y_t, p_t) \text{ continuous in } H(D) \subset C^0(\bar{D})$$

We get the notion of *set derivative* for the set functional $J(\Omega)$, defined as follows,

$$\frac{\partial}{\partial t} J(\chi(t)) = < \dot{\chi}(t), \ \mathcal{G}(\chi(t)) > \tag{4.42}$$

In the previous example, we had $\mathcal{G}(\chi) = y\, p$. Such an example is considered in ([129]) for a transverse magnetic *inverse shape problem* in buried obstacle reconstruction.

## Hole-differentiability

Here we shall illustrate the various dependencies of the topological derivative in the case of the Laplace equation with Dirichlet boundary conditions.
Let us consider the domain

$$\Omega = \{x \in \mathbb{R}^2 \mid \|x\| < 1\}$$

together with the *perforated* subdomain

$$\Omega_\varepsilon = \{x \in \Omega \mid \|x\| > \varepsilon\}$$

Let $y_\varepsilon$ be the solution to the following Dirichlet problem,

$$\begin{cases} \Delta y_\varepsilon = 0, \ \Omega_\varepsilon, \\ y_\varepsilon(x) = 0, \ \|x\| = \varepsilon, \\ y_\varepsilon(x) = 1, \ \|x\| = 1 \end{cases} \tag{4.43}$$

Obviously, we get

$$y_\varepsilon(x) = -\frac{\ln r}{\ln \varepsilon} + 1, \quad \|x\| = r$$

and $y_0(x) = 1$.
Let us first consider the functional

$$j^1(\varepsilon) = \int_{\|x\|=1} (\frac{\partial}{\partial n} y_\varepsilon(s))^2 \, ds$$

We get

$$j^1(\varepsilon) - j^1(0) = \frac{2\pi}{(\ln \varepsilon)^2} \to 0, \ \varepsilon \to 0$$

Setting $a_\varepsilon^1 = (\frac{1}{\ln \varepsilon})^2$, we get

$$\lim_{\varepsilon \to 0} \frac{j^1(\varepsilon) - j^1(0)}{a_\varepsilon^1} = 2\pi$$

This limit is usually referred to as the topological derivative of the functional
$j^1(\varepsilon)$.
Now, let us choose a different functional,

$$j^2(\varepsilon) = \int_{\Omega_\varepsilon} |\nabla y_\varepsilon(x)|^2 \, dx$$

We find

$$j^2(\varepsilon) - j^2(0) = -\frac{2\pi}{\ln \varepsilon}$$

Setting $a_\varepsilon^2 = -(\frac{1}{\ln \varepsilon})$, we get

$$\lim_{\varepsilon \to 0} \frac{j^2(\varepsilon) - j^2(0)}{a_\varepsilon^2} = 2\pi$$

Hence, the choice of the scaling parameter $a_\varepsilon$ depends not only on the type
of state equation, but also on the type of functional.

# Chapter 5

## Dynamical shape control of the Navier-Stokes equations

## 5.1 Introduction

This chapter deals with the analysis of an inverse dynamical shape problem involving a fluid inside a moving domain. This type of inverse problem happens frequently in the design and the control of many industrial devices such as aircraft wings, cable-stayed bridges, automobile shapes, satellite reservoir tanks and more generally of systems involving fluid-solid interactions.
The control variable is the shape of the moving domain, and the objective is to minimize a given cost functional that may be chosen by the designer.
On the theoretical level, early works concerning optimal control problems for general parabolic equations written in non-cylindrical domains have been considered in [43], [29], [30], [142], [2]. In [140], [151], [152], the stabilization of structures using the variation of the domain has been addressed. The basic principle is to define a map sending the non-cylindrical domain into a cylindrical one. This process leads to the mathematical analysis of non-autonomous PDE's systems.
Recently, a new methodology to obtain Eulerian derivatives for non-cylindrical functionals has been introduced in [157], [156], [58]. This methodology was applied in [59] to perform dynamical shape control of the non-cylindrical Navier-Stokes equations where the evolution of the domain is the control variable. Hence the classical optimal shape optimization theory has been extended to deal with non-cylindrical domains.
The aim of this chapter is to review several results on the dynamical shape control of the Navier-Stokes system and suggest an alternative treatment using the Min-Max principle [45, 46]. Despite its lack of rigorous mathematical justification in the case where the Lagrangian functional is not convex, we shall show how this principle allows, at least formally, to bypass the tedious computation of the state differentiability with respect to the shape of the moving domain.

## 5.2   Problem statement

Let us consider a moving domain $\Omega_t \in \mathbb{R}^d$. We introduce a diffeomorphic map sending a fixed reference domain $\Omega_0$ into the physical configuration $\Omega_t$ at time $t \geq 0$.

Without loss of generality, we choose the reference configuration to be the physical configuration at initial time $\Omega_{t=0}$.

Hence we define a map $T_t \in \mathcal{C}^1(\overline{\Omega_0})$ such that

$$\overline{\Omega_t} = T_t(\overline{\Omega_0}),$$
$$\overline{\Gamma_t} = T_t(\overline{\Gamma_0})$$

We set $\Sigma \equiv \bigcup_{0<t<T} (\{t\} \times \Gamma_t)$, $Q \equiv \bigcup_{0<t<T} (\{t\} \times \Omega_t)$. The map $T_t$ can be actually defined as the flow of a particular vector field, as described in the following lemma :

### THEOREM 5.1 [147]

$$\overline{\Omega_t} = T_t(V)(\overline{\Omega_0}),$$
$$\overline{\Gamma_t} = T_t(V)(\overline{\Gamma_0})$$

*where $T_t(V)$ is solution the of the following dynamical system :*

$$T_t(V) : \Omega_0 \longrightarrow \Omega$$
$$x_0 \longmapsto x(t, x_0) \equiv T_t(V)(x_0)$$

*with*

$$\frac{\mathrm{d}\,x}{\mathrm{d}\,\tau} = V(\tau, x(\tau)), \ \tau \in [0, T] \tag{5.1}$$
$$x(\tau = 0) = x_0, \quad in \ \Omega_0$$

The fluid filling $\Omega_t$ is assumed to be a viscous incompressible newtonian fluid. Its evolution is described by its velocity $u$ and its pressure $p$. The couple $(u, p)$ satisfies the classical Navier-Stokes equations written in non-conservative form,

$$\begin{cases} \partial_t u + \mathrm{D}\,u \cdot u - \nu \Delta u + \nabla p = 0, & Q(V) \\ \mathrm{div}(u) = 0, & Q(V) \\ u = V, & \Sigma(V) \\ u(t = 0) = u_0, & \Omega_0 \end{cases} \tag{5.2}$$

where $\nu$ stands for the kinematic viscosity.

The quantity $\sigma(u, p) = -p\,\mathrm{I} + \nu(\mathrm{D}\,u + {}^*\mathrm{D}\,u)$ stands for the fluid stress tensor inside $\Omega_t$, with $(\mathrm{D}\,u)_{i,j} = \partial_j u_i$.

We are interested in solving the following minimization problem :

$$\min_{V \in \mathcal{U}} j(V) \tag{5.3}$$

where $j(V) = J_V(u(V), p(V))$ with $(u(V), p(V))$ is a weak solution of problem (5.2) and $J_V(u, p)$ is a real functional of the following form :

$$J_V(u, p) = \frac{\alpha}{2} \|\mathcal{B}\,u\|_{Q(V)}^2 + \frac{\gamma}{2} \|\mathcal{K}\,V\|_{\Sigma(V)}^2 \tag{5.4}$$

where $\mathcal{B} \in \mathcal{L}(\mathcal{H}, \mathcal{H}^*)$ is a general linear differential operator satisfying the following identity,

$$\langle \mathcal{B}\,u, v \rangle + \langle u, \mathcal{B}^* v \rangle = \langle \mathcal{B}_\Sigma\,u, v \rangle_{L^2(\Sigma)} \tag{5.5}$$

where $\mathcal{H} = \{v \in L^2(0, T; (H_0^1(\mathrm{div}, \Omega_t(V)))^d)\}$ and $\mathcal{K} \in \mathcal{L}(\mathcal{U}, L^2(\Sigma(V)))$ is a general linear differential operator satisfying the following identity,

$$\langle \mathcal{K}\,u, v \rangle_{L^2(\Sigma)} + \langle u, \mathcal{K}^* v \rangle_{L^2(\Sigma)} = \langle \mathcal{K}_\Sigma\,u, v \rangle_{L^2(\Sigma)} \tag{5.6}$$

The main difficulty in dealing with such a minimization problem is related to the fact that integrals over the domain $\Omega_t(V)$ depend on the control variable $V$. This point will be solved by using the Arbitrary Lagrange-Euler (ALE) map $T_t(V)$ introduced previously. The purpose of this chapter is to prove using several methods the following result,

**MAIN RESULT:** For $V \in \mathcal{U}$ and $\Omega_0$ of class $\mathcal{C}^2$, the functional $j(V)$ possesses a gradient $\nabla j(V)$ which is supported on the moving boundary $\Gamma_t(V)$ and can be represented by the following expression,

$$\nabla j(V) = -\lambda\,n - \sigma(\varphi, \pi) \cdot n + \alpha\,\mathcal{B}_\Sigma \mathcal{B}\,u + \gamma\,[-\mathcal{K}^*\mathcal{K}\,V + \mathcal{K}_\Sigma \mathcal{K}\,V] \tag{5.7}$$

where $(\varphi, \pi)$ stands for the adjoint fluid state solution of the following system,

$$\begin{cases} -\partial_t\varphi - \mathrm{D}\,\varphi \cdot u + {}^*\mathrm{D}\,u \cdot \varphi - \nu\Delta\varphi + \nabla\pi = -\alpha\,\mathcal{B}^*\mathcal{B}\,u, & Q(V) \\ \mathrm{div}(\varphi) = 0, & Q(V) \\ \varphi = 0, & \Sigma(V) \\ \varphi(T) = 0, & \Omega_T \end{cases} \tag{5.8}$$

and $\lambda$ is the adjoint transverse boundary field, solution of the tangential dynamical system,

$$\begin{cases} -\partial_t\lambda - \nabla_\Gamma \lambda \cdot V - (\mathrm{div}\,V)\,\lambda = f, & (0, T) \\ \lambda(T) = 0, & \Gamma_T(V) \end{cases} \tag{5.9}$$

with

$$f = [-(\sigma(\varphi, \pi) \cdot n) + \alpha\,\mathcal{B}_\Sigma \mathcal{B}\,u] \cdot (\mathrm{D}\,V \cdot n - \mathrm{D}\,u \cdot n) + \frac{1}{2}\,[\alpha|\mathcal{B}\,u|^2 + \gamma\,H|\mathcal{K}\,V|^2] \tag{5.10}$$

### Example 5.1
We set

$$(\mathcal{B}, \mathcal{B}^*, \mathcal{B}_\Sigma) = (\mathrm{I}, -\mathrm{I}, 0)$$
$$(\mathcal{K}, \mathcal{K}^*, \mathcal{K}_\Sigma) = (\mathrm{I}, -\mathrm{I}, 0)$$

This means that we consider the cost functional,

$$J_V(u, p) = \frac{\alpha}{2}\|u\|^2_{L^2(Q(V))} + \frac{\gamma}{2}\|V\|^2_{L^2(\Sigma(V))} \qquad (5.11)$$

Then its gradient is given by

$$\nabla j(V) = -\lambda\, n - \sigma(\varphi, \pi) \cdot n + \gamma\, V \qquad (5.12)$$

where $(\varphi, \pi)$ stands for the adjoint fluid state solution of the following system,

$$\begin{cases} -\partial_t\varphi - \mathrm{D}\,\varphi \cdot u + {}^*\mathrm{D}\,u \cdot \varphi - \nu\Delta\varphi + \nabla\pi = \alpha\, u, & Q(V) \\ \mathrm{div}(\varphi) = 0, & Q(V) \\ \varphi = 0, & \Sigma(V) \\ \varphi(T) = 0, & \Omega_T \end{cases} \qquad (5.13)$$

and $\lambda$ is the adjoint transverse boundary field, solution of the tangential dynamical system,

$$\begin{cases} -\partial_t\lambda - \nabla_\Gamma\,\lambda \cdot V - (\mathrm{div}\,V)\,\lambda = f, & (0, T) \\ \lambda(T) = 0, & \Gamma_T(V) \end{cases} \qquad (5.14)$$

with

$$f = -\nu(\mathrm{D}\,\varphi \cdot n) \cdot (\mathrm{D}\,V \cdot n - \mathrm{D}\,u \cdot n) + \frac{1}{2}(\alpha + \gamma\, H)|V|^2 \qquad (5.15)$$

⬜

### Example 5.2
We set

$$(\mathcal{B}, \mathcal{B}^*, \mathcal{B}_\Sigma) = (\mathrm{curl}, \mathrm{curl}, \wedge n)$$
$$(\mathcal{K}, \mathcal{K}^*, \mathcal{K}_\Sigma) = (\mathrm{I}, -\mathrm{I}, 0)$$

$$J_V(u, p) = \frac{\alpha}{2}\|\,\mathrm{curl}\,u\|^2_{L^2(Q(V))} + \frac{\gamma}{2}\|V\|^2_{L^2(\Sigma(V))} \qquad (5.16)$$

Then its gradient is given by

$$\nabla j(V) = -\lambda\, n - \sigma(\varphi, \pi) \cdot n + \alpha\,(\operatorname{curl} u) \wedge n + \gamma\, V \qquad (5.17)$$

where $(\varphi, \pi)$ stands for the adjoint fluid state solution of the following system,

$$\begin{cases} -\partial_t\varphi - D\varphi \cdot u + {}^*\!D\,u \cdot \varphi - \nu\Delta\varphi + \nabla\pi = -\alpha\,\Delta u, & Q(V) \\ \operatorname{div}(\varphi) = 0, & Q(V) \\ \varphi = 0, & \Sigma(V) \\ \varphi(T) = 0, & \Omega_T \end{cases} \qquad (5.18)$$

and $\lambda$ is the adjoint transverse boundary field, solution of the tangential dynamical system,

$$\begin{cases} -\partial_t\lambda - \nabla_\Gamma\,\lambda \cdot V - (\operatorname{div} V)\,\lambda = f, & (0, T) \\ \lambda(T) = 0, & \Gamma_T(V) \end{cases} \qquad (5.19)$$

with

$$f = [-\nu\, D\varphi \cdot n + \alpha\,(\operatorname{curl} u) \wedge n] \cdot (D\,V \cdot n - D\,u \cdot n)$$
$$+ \frac{1}{2}\left[\alpha|\operatorname{curl} u|^2 + \gamma\, H|V|^2\right]$$

□

In the next section, we introduce several concepts closely related to shape optimization tools for moving domain problems. We also recall elements of tangential calculus that will be used through this chapter. Then we treat successively the following points,

1. In section (5.5), we choose to prove the differentiability of the fluid state $(u, p)$ with respect to the design variable $V$. The directional shape derivative $(u', p')(V) \cdot W$ is then used to compute the directional derivative $j'(V) \cdot W$ of the cost functional $j(V)$. Using the adjoint state $(\varphi, \pi)(V)$ associated to $(u', p')(V)$ and the adjoint field $\Lambda$ associated to the transverse field $Z_t$ introduced in section (5.3), we are able to furnish an expression of the gradient $\nabla j(V)$ which is a distribution supported by the moving boundary $\Gamma_t(V)$.

2. In section (5.6), we choose to bypass the computation of the state shape derivative $(u', p')(V) \cdot W$, by using a Min-Max formulation of problem (5.3) and a transport technique. The state and multiplier spaces are chosen in order to be independent on the scalar perturbation parameter used in the computation of the derivative of the Lagrangian functional with respect to $V$. This method directly furnishes the fluid state and transverse field adjoint systems and the resulting gradient $\nabla j(V)$.

3. In section (5.7), we again use a Min-Max strategy coupled with a state and multiplier functional space embedding. This means that the state and multiplier variables live in the hold-all domain $D$. Hence the derivative of the Lagrangian functional with respect to $V$ only involves terms coming from the flux variation through the moving boundary $\Gamma_t(V)$. This again leads to the direct computation of the fluid state and transverse field adjoints and consequently to the gradient $\nabla j(V)$.

## 5.3   Elements of non-cylindrical shape calculus

This section introduces several concepts that will be intensively used through this chapter. It concerns the differential calculus of integrals defined on moving domains or boundaries with respect to their support.

### 5.3.1   Non-cylindrical speed method

In this paragraph, we are interested in differentiability properties of integrals defined over moving domains,

$$J_1(\Omega_t) = \int_{\Omega_t} f(\Omega_t) \, d\Omega$$

$$J_2(\Gamma_t) = \int_{\Gamma_t} g(\Gamma_t) \, d\Gamma$$

The behaviour of $J_1$ and $J_2$ while perturbing their moving support highly depends on the regularity in space and time of the domains. In this work, we choose to work with domains $\Omega_t$ that are images of a fixed domain $\Omega_0$ through an ALE map $T_t(V)$ as introduced in the first section. Hence, the design parameter is no more the support $\Omega_t$ but rather the velocity field $V \in \mathcal{U} \overset{\text{def}}{=} C^0([0,T]; (W^{k,\infty}(D))^d)$ that builds the support. This technique has the advantage to transform shape calculus into classical differential calculus on vector spaces [157],[59]. For an other choice based on the non-cylindrical identity perturbation, the reader is referred to the next chapter. Before stating the main result of this section, we recall the notion of transverse field.

**Transverse applications**

**DEFINITION 5.1**   *The transverse map $T_\rho^t$ associated to two vector fields $(V, W) \in \mathcal{U}$ is defined as follows,*

$$T_\rho^t : \overline{\Omega_t} \longrightarrow \overline{\Omega_t^\rho} \overset{\text{def}}{=} \overline{\Omega_t(V + \rho W)}$$
$$x \mapsto T_t(V + \rho W) \circ T_t(V)^{-1}$$

**REMARK 5.1**    The transverse map allows us to perform sensitivity analysis on functions defined on the unperturbed domain $\Omega_t(V)$. ⬚

The following result states that the transverse map $T^t_\rho$ can be considered as a dynamical flow with respect to the perturbation variable $\rho$,

**THEOREM 5.2 [156]**
*The Transverse map $T^t_\rho$ is the flow of a transverse field $\mathcal{Z}^t_\rho$ defined as follows*

$$\mathcal{Z}^t_\rho \stackrel{\text{def}}{=} \mathcal{Z}^t(\rho,.) = \left(\frac{\partial T^t_\rho}{\partial \rho}\right) \circ (T^t_\rho)^{-1} \tag{5.20}$$

*i.e., is the solution of the following dynamical system :*

$$T^\rho_t(\mathcal{Z}^t_\rho) : \overline{\Omega_t} \longrightarrow \overline{\Omega^\rho_t}$$

$$x \longmapsto x(\rho, x) \equiv T^\rho_t(\mathcal{Z}^t_\rho)(x)$$

*with*

$$\frac{\mathrm{d}\, x(\rho)}{\mathrm{d}\, \rho} = \mathcal{Z}^t(\rho, x(\rho)), \ \rho \geq 0$$
$$x(\rho = 0) = x, \qquad in \ \Omega_t(V) \tag{5.21}$$

Since, we will mainly consider derivatives of perturbed functions at point $\rho = 0$, we set $Z_t \stackrel{\text{def}}{=} \mathcal{Z}^t_{\rho=0}$. A fundamental result lies in the fact that $Z_t$ can be obtained as the solution of a linear time dynamical system depending on the vector fields $(V, W) \in \mathcal{U}$,

**THEOREM 5.3 [59]**
*The vector field $Z_t$ is the unique solution of the following Cauchy problem,*

$$\begin{cases} \partial_t Z_t + [Z_t, V] = W, \ D \times (0, T) \\ Z_{t=0} = 0, \qquad D \end{cases} \tag{5.22}$$

*where $[Z_t, V] \stackrel{\text{def}}{=} DZ_t \cdot V - DV \cdot Z_t$ stands for the Lie bracket of the pair $(Z_t, V)$.*

**Shape derivative of non-cylindrical functionals**

The main theorem of this section uses the notion of a non-cylindrical material derivative that we recall here,

**DEFINITION 5.2**    *The derivative with respect to $\rho$ at point $\rho = 0$ of the following composed function,*

$$f^\rho : [0, \rho_0] \rightarrow H(\Omega_t(V))$$
$$\rho \mapsto f(V + \rho W) \circ T^t_\rho$$

$\dot{f}(V;W)$ *is called the non-cylindrical material derivative of* $f(V)$ *at point* $V \in$ $\mathcal{U}$ *in the direction* $W \in \mathcal{U}$. *We shall use the notation,*

$$\dot{f}(V) \cdot W = \dot{f}(V;W) \overset{\text{def}}{=} \frac{d}{d\rho} f^{\rho} \bigg|_{\rho=0}$$

With the above definition, we can state the differentiability properties of non-cylindrical integrals with respect to their moving support,

**THEOREM 5.4 [59]**
*For a bounded measurable domain* $\Omega_0$ *with boundary* $\Gamma_0$, *let us assume that for any direction* $W \in \mathcal{U}$ *the following hypothesis holds,*

 i) $f(V)$ *admits a non-cylindrical material derivative* $\dot{f}(V) \cdot W$

*then* $J_1(.)$ *is Gâteaux differentiable at point* $V \in \mathcal{U}$ *and its derivative is given by the following expression,*

$$J_1'(V) \cdot W = \int_{\Omega_t(V)} \left[ \dot{f}(V) \cdot W + f(V) \operatorname{div} Z_t \right] d\Omega \tag{5.23}$$

*Futhermore, if*

 ii) $f(V)$ *admits a non-cylindrical shape derivative given by the following expression,*

$$f'(V) \cdot W = \dot{f}(V) \cdot W - \nabla f(V) \cdot Z_t \tag{5.24}$$

*then*

$$J_1'(V) \cdot W = \int_{\Omega_t(V)} [f'(V) \cdot W + \operatorname{div}(f(V) Z_t)] d\Omega \tag{5.25}$$

*Furthermore, if* $\Omega_0$ *is an open domain with a Lipschitzian boundary* $\Gamma_0$, *then*

$$J_1'(V) \cdot W = \int_{\Omega_t(V)} f'(V) \cdot W d\Omega + \int_{\Gamma_t(V)} f(V) \langle Z_t, n \rangle d\Gamma \tag{5.26}$$

**REMARK 5.2**  The last identity will be of great interest while trying to prove a gradient structure result for general non-cylindrical functionals. ⬚

It is also possible to establish a similar result for integrals over moving boundaries. For that purpose, we need to define the non-cylindrical tangential material derivative,

**DEFINITION 5.3**  *The derivative with respect to* $\rho$ *at point* $\rho = 0$ *of the following composed function,*

$$g^{\rho} : [0, \rho_0] \to H(\Gamma_t(V))$$
$$\rho \mapsto g(V + \rho W) \circ T_{\rho}^t$$

is called the non-cylindrical material derivative of the function $g(V) \in H(\Gamma_t(V))$ in the direction $W \in \mathcal{U}$. We shall use the notation,

$$\dot{g}(V) \cdot W = \dot{g}(V; W) \overset{\text{def}}{=} \frac{d}{d\rho} g^\rho \Big|_{\rho=0}$$

This concept is involved in the differentiability property of boundary integrals,

### THEOREM 5.5
*For a bounded measurable domain $\Omega_0$ with boundary $\Gamma_0$, let us assume that for any direction $W \in U$ the following hypothesis holds,*

*i) $g(V)$ admits a non-cylindrical material derivative $\dot{g}(V) \cdot W$*

*then $J_2(.)$ is Gâteaux differentiable at point $V \in \mathcal{U}$ and its derivative is given by the following expression,*

$$J_2'(V) \cdot W = \int_{\Gamma_t(V)} [\dot{g}(V) \cdot W + g(V) \operatorname{div}_\Gamma Z_t] \, d\Gamma \tag{5.27}$$

*Futhermore, if*

*ii) $g(V)$ admits a non-cylindrical shape derivative given by the following expression,*

$$g'(V) \cdot W = \dot{g}(V) \cdot W - \nabla_\Gamma g(V) \cdot Z_t \tag{5.28}$$

*then*

$$J_2'(V) \cdot W = \int_{\Gamma_t(V)} [\tilde{g}'(V) \cdot W + H \, g(V) \langle Z_t, n \rangle] \, d\Gamma \tag{5.29}$$

*where $H$ stands for the additive curvature (Def. (5.4)). Furthermore, if $g(V) = \tilde{g}(V)|_{\Gamma_t(V)}$ with $\tilde{g} \in H(\Omega_t(V))$, then*

$$J_2'(V) \cdot W = \int_{\Gamma_t(V)} [g'(V) \cdot W + (\nabla \tilde{g}(V) \cdot n + H \, g(V)) \langle Z_t, n \rangle] \, d\Gamma \tag{5.30}$$

### Adjoint transverse field

It is possible to define the solution of the adjoint transverse system,

### THEOREM 5.6 [58]
*For $F \in L^2(0, T; (H^1(D))^d)$, there exists a unique field*

$$\Lambda \in \mathcal{C}^0([0, T]; (L^2(D))^d)$$

*solution of the backward dynamical system,*

$$\begin{cases} -\partial_t \Lambda - D\, \Lambda \cdot V - {}^* D V \cdot \Lambda - (\operatorname{div} V)\Lambda = F, \, (0, T) \\ \Lambda(T) = 0 \end{cases} \tag{5.31}$$

**REMARK 5.3**　　The field $\Lambda$ is the dual variable associated to the transverse field $Z_t$ and is the solution of the adjoint problem associated to the transverse dynamical system.　　　　　　　　　　　　　　　　　　　　　　　　　　　　　□

In this chapter, we shall deal with a specific right-hand side $F$ of the form $F(t) = {}^{*}\gamma_{\Gamma_t(V)}(f(t)n)$. In this case, the adjoint field $\Lambda$ is supported on the moving boundary $\Gamma_t(V)$ and has the following structure,

### THEOREM 5.7 [59]

*For $F(t) = {}^{*}\gamma_{\Gamma_t(V)}(f(t)n)$, with $f \in L^2(0,T; L^2(\Gamma_t(V)))$, the unique solution $\Lambda$ of the problem is given by the following identity,*

$$\Lambda = (\lambda \circ p)\,\nabla\,\chi_{\Omega_t(V)} \in \mathcal{C}^0([0,T]; (H^1(\Gamma_t))^d) \tag{5.32}$$

*where $\lambda \in \mathcal{C}^0([0,T]; H^1(\Gamma_t))$ is the unique solution of the following boundary dynamical system,*

$$\begin{cases} -\partial_t \lambda - \nabla_\Gamma \lambda \cdot V - (\operatorname{div} V)\lambda = f, & (0,T) \\ \lambda(T) = 0, & \Gamma_t(V) \end{cases} \tag{5.33}$$

*$p$ is the canonical projection on $\Gamma_t(V)$ and $\chi_{\Omega_t(V)}$ is the characteristic function of $\Omega_t(V)$ inside $D$.*

### Gradient of non-cylindrical functionals

In the next sections, we will often deal with boundary integrals of the following forms,

$$K = \int_0^T \int_{\Gamma_t(V)} E\,\langle Z_t, n\rangle$$

with $E \in L^2(0,T;\Gamma_t(V))$ and $Z_t$ is the solution of the transverse equation (5.22). The following result allows us to eliminate the auxiliary variable $Z_t$ inside the functional $K$,

### THEOREM 5.8 [59]

*For any $E \in L^2(0,T;\Gamma_t(V))$ and $(V,W) \in \mathcal{U}$, the following identity holds,*

$$\int_0^T \int_{\Gamma_t(V)} E\,\langle Z_t, n\rangle = -\int_0^T \int_{\Gamma_t(V)} \lambda\,\langle W, n\rangle \tag{5.34}$$

*where $\lambda \in \mathcal{C}^0([0,T]; H^1(\Gamma_t))$ is the unique solution of problem (5.33) with $f = E$.*

## 5.4 Elements of tangential calculus

In this section, we review basic elements of differential calculus on a $\mathcal{C}^k$-submanifold with $k \geq 2$ of codimension one in $\mathbb{R}^d$. The following approach avoids the use of local bases and coordinates by using the intrinsic tangential derivative.

### 5.4.1 Oriented distance function

Let $\Omega$ be an open domain of class $\mathcal{C}^k$ in $\mathbb{R}^d$ with compact boundary $\Gamma$. We define the oriented distance function to be as follows,

$$b_\Omega(x) = \begin{cases} d_\Gamma(x), & x \in \mathbb{R}^d \setminus \overline{\Omega} \\ -d_\Gamma(x), & x \in \Omega \end{cases}$$

where $d_\Gamma(x) = \min_{y \in \Omega} |y - x|$.

### PROPOSITION 5.1 [51]

*Let $\Omega$ be an open domain of class $\mathcal{C}^k$ for $k \geq 2$ in $\mathbb{R}^d$ with compact boundary $\Gamma$. There exists a neighbourhood $U(\Gamma)$ of $\Gamma$, such that $b \in \mathcal{C}^k(U(\Gamma))$. Furthermore, we have the following properties,*

*i) $\nabla b|_\Gamma = n$, where $n$ stands for the unit exterior normal on $\Gamma$,*

*ii) $\mathrm{D}^2 b : T_{p(x)}\Gamma \to T_{p(x)}\Gamma$ coincides with the second fundamental form on $\Gamma$, where*

$$p : U(\Gamma) \to \Gamma$$
$$x \mapsto x - b(x) \cdot \nabla b(x)$$

*stands for the projection mapping and $T_{p(x)}\Gamma$ stands for the tangent plane.*

*iii) $(0, \beta_1, \ldots, \beta_{d-1})$ are the eigenvalues of $\mathrm{D}^2 b$ associated to the eigenfunctions*

$$(n, \mu_1, \ldots, \mu_{d-1})$$

*where $(\beta_i, \mu_i)_{1 \leq i \leq d-1}$ are the mean curvatures and principal direction of curvatures of $\Gamma$.*

### PROPOSITION 5.2 [51]

*For $\Gamma$ of class $\mathcal{C}^2$, the projection mapping $p$ is differentiable and its derivative* has the following properties,

$$^*\mathrm{D}\,p = \mathrm{D}\,p = \mathrm{I} - \nabla b \cdot {}^*\nabla b - b\,\mathrm{D}^2\,b$$
$$\mathrm{D}\,p \cdot \tau = \tau, \qquad on\ \Gamma,$$
$$\mathrm{D}\,p \cdot n = 0, \qquad on\ \Gamma \tag{5.35}$$

**DEFINITION 5.4 [51]**    *For $\Gamma$ of class $C^2$, the additive curvature $H$ of $\Gamma$ is defined as the trace of the second order fundamental form :*

$$H = \mathrm{Tr}\,\mathrm{D}^2\,b = \Delta b = (d-1)\bar{H}, \qquad on\ \Gamma \tag{5.36}$$

*and $\bar{H}$ stands for the mean curvature of $\Gamma$.*

## 5.4.2   Intrinsic tangential calculus

Using arbitrary smooth extensions of functions defined on $\Gamma$ to $\Omega \in \mathbb{R}^d$ is the most classical way of defining tangential operators. Hence the differential calculus on manifolds can be reduced to classical differential calculus in $\mathbb{R}^d$. In this section we recall standard formulas for differential tangential operators using arbitrary extensions. We also emphasize the particular case where the extension is of the canonical type $(f \circ p)$. This is the basis of a simple differential calculus in the neighbourhood of $\Gamma$.

**DEFINITION 5.5**    *For $\Gamma$ of class $C^2$, given any extension $F \in C^1(U(\Gamma))$ of $f \in C^1(\Gamma)$, the tangential gradient of $f$ is defined as*

$$\nabla_\Gamma f \stackrel{\mathrm{def}}{=} \nabla F|_\Gamma - (\partial_n F)\,n \tag{5.37}$$

*where $\partial_n F = \nabla F \cdot n$.*

**PROPOSITION 5.3 [51]**
*Assume that $\Gamma$ of class $C^2$ is compact and $f \in C^1(\Gamma)$, then*

  *i)*

$$\nabla_\Gamma f = (P\nabla F)|_\Gamma$$
$$n \cdot \nabla_\Gamma f = \nabla b \cdot \nabla_\Gamma f = 0 \tag{5.38}$$

  *where $P \stackrel{\mathrm{def}}{=} \mathrm{I} - \nabla b {}^*\nabla b$ is the orthogonal projection operator onto the tangent plane $T_{p(x)}\Gamma$.*

  *ii)*

$$\nabla(f \circ p) = \left[\mathrm{I} - b\,\mathrm{D}^2\,b\right] \nabla_\Gamma f \circ p$$
$$\nabla(f \circ p)|_\Gamma = \nabla_\Gamma f \tag{5.39}$$

Hence $(f \circ p)$ plays the role of a canonical extension in the neighbourhood $U(\Gamma)$ and its gradient is tangent to the level sets of $b$. Consequently, we can define in an intrinsic way the tangential gradient,

**DEFINITION 5.6** *For $\Gamma$ of class $\mathcal{C}^2$ and $f \in \mathcal{C}^1(\Gamma)$, the tangential gradient of $f$ is defined as*

$$\nabla_\Gamma f = \nabla(f \circ p)|_\Gamma \tag{5.40}$$

In the sequel, we shall use the above definition for the tangential gradient whenever the function under derivation is intrinsically defined on $\Gamma$. We now define the other classical tangential operators,

**DEFINITION 5.7** *For $\Gamma$ of class $\mathcal{C}^2$,*

*i) for $v \in (\mathcal{C}^1(\Gamma))^d$, and $\tilde{v} \in (\mathcal{C}^1(U(\Gamma)))^d$ an arbitrary extension, the tangential jacobian is defined as follows,*

$$\begin{aligned} D_\Gamma v &\overset{def}{=} D\tilde{v}|_\Gamma - (D\tilde{v} \cdot n)^* n \\ &= D\tilde{v}|_\Gamma - D\tilde{v} \cdot (n \otimes n) \end{aligned} \tag{5.41}$$

*Furthermore,*

$$\begin{aligned} D(v \circ p) &= D_\Gamma v \circ p \left[ I - b\, D^2\, b \right] \\ D_\Gamma v &= D(v \circ p)|_\Gamma \end{aligned} \tag{5.42}$$

*ii) for $v \in (\mathcal{C}^1(\Gamma))^d$, and $\tilde{v} \in (\mathcal{C}^1(U(\Gamma)))^d$ an arbitrary extension, the tangential divergence is defined as follows,*

$$\text{div}_\Gamma\, v \overset{def}{=} \text{div}\, \tilde{v}|_\Gamma - (D\tilde{v} \cdot n) \cdot n \tag{5.43}$$

*Furthermore,*

$$\text{div}_\Gamma\, v = \text{div}(v \circ p)|_\Gamma = \text{Tr}(D_\Gamma\, v) \tag{5.44}$$

*iii) for $f \in \mathcal{C}^2(\Gamma)$, and $F \in \mathcal{C}^2(U(\Gamma))$ an arbitrary extension, the tangential Laplace-Beltrami operator is defined as follows,*

$$\Delta_\Gamma f = \Delta F|_\Gamma - H \partial_n F - \partial_n^2 F \tag{5.45}$$

*with $\partial_n^2 F = (D^2\, F \cdot n) \cdot n$. Furthermore,*

$$\Delta_\Gamma f = \text{div}_\Gamma(\nabla_\Gamma\, f) = \Delta(f \circ p)|_\Gamma \tag{5.46}$$

In some cases, it may be interesting to use a splitting of the function $v$ onto a normal and a tangential component,

**DEFINITION 5.8**  *For $v \in (\mathcal{C}^1(\Gamma))^d$, we define the tangential component $v_\Gamma \in (\mathcal{C}^1(\Gamma))^d$ and the normal component $v_n \in \mathcal{C}^1(\Gamma)$ such that*

$$v = v_\Gamma + v_n\, n \tag{5.47}$$

Using the above definition, we obtain the following identities,

**PROPOSITION 5.4**
*For $v \in (\mathcal{C}^1(\Gamma))^d$, we have*

$$\mathrm{D}_\Gamma\, v = \mathrm{D}_\Gamma\, v_\Gamma + v_n\, \mathrm{D}^2\, b + n \cdot {}^*\nabla_\Gamma\, v_n \tag{5.48}$$

$$\nabla_\Gamma\, v_n = {}^*\mathrm{D}_\Gamma\, v \cdot n + \mathrm{D}^2\, b \cdot v_\Gamma \tag{5.49}$$

$$\mathrm{div}_\Gamma\, v = \mathrm{div}_\Gamma\, v_\Gamma + H\, v_n \tag{5.50}$$

### 5.4.3  Tangential Stokes formula

In order to perform integration by parts on $\Gamma$, we will use the following tangential Stokes identity,

**PROPOSITION 5.5**
*Let $\Gamma$ be a $\mathcal{C}^2$-submanifold in $\mathbb{R}^d$; for $E \in H^1(\Gamma; \mathbb{R}^d)$ and $\psi \in H^1(\Gamma; \mathbb{R})$ the following identity holds*

$$\int_\Gamma \langle E, \nabla_\Gamma\, \psi \rangle_{\mathbb{R}^d} + \int_\Gamma (\mathrm{div}_\Gamma\, E)\psi = \int_\Gamma H\, \psi\, \langle E, n \rangle_{\mathbb{R}^d} \tag{5.51}$$

---

## 5.5  State derivative strategy

In this section, we shall prove the main theorem of this chapter using an approach based on the differentiability of the solution of the Navier-Stokes system (Eq. (5.2)) with respect to the velocity field $V$. First, we introduce a weak formulation for Eq. (5.2) and recall the associated classical solvability result. Then, using the weak implicit function theorem, we will prove the existence of a weak material derivative. Finally, introducing the adjoint equations associated to the linearized fluid and transverse systems, we will be able to express the gradient of the functional $j(V)$. For the sake of simplicity, we shall only prove the main theorem in the case of example (5.1)-1 with free divergence control velocity fields.

### 5.5.1 Weak formulation and solvability

In order to take into account the non-homogeneous Dirichlet boundary condition on $\Gamma_t(V)$, we use the following change of variable $\tilde{u} = u - V$, where $\tilde{u}$ satisfies the following homogeneous Dirichlet Navier-Stokes system,

$$\begin{cases} \partial_t \tilde{u} + D\,\tilde{u} \cdot \tilde{u} + D\,\tilde{u} \cdot V + D V \cdot \tilde{u} - \nu \Delta \tilde{u} + \nabla p = F(V), & Q(V) \\ \operatorname{div}(\tilde{u}) = 0, & Q(V) \\ \tilde{u} = 0, & \Sigma(V) \\ \tilde{u}(0) = u_0 - V(0), & \Omega_0 \end{cases} \qquad (5.52)$$

with $F(V) = -\partial_t V - D V \cdot V + \nu \Delta V$.

We consider the following classical functional spaces [97], [139],

$$H(D) = \left\{ v \in (L^2(D))^d, \quad \operatorname{div} v = 0, \text{ in } D, \quad v \cdot n = 0 \text{ on } \partial D \right\}$$

$$H_0^1(\operatorname{div}, D) = \left\{ v \in (H_0^1(D))^d, \quad \operatorname{div} v = 0, \text{ in } D \right\}$$

$$\mathcal{H} = \left\{ v \in L^2(0, T; (H_0^1(\operatorname{div}, \Omega_t(V)))^d) \right\}$$

$$V = \left\{ v \in \mathcal{H}, \quad \partial_t v \in L^2(0, T; (H_0^1(\Omega_t(V)))^d) \right\}$$

In the sequel, we shall use the notation $u$ instead of $\tilde{u}$, keeping in mind that the original variable is obtained by translation.

**DEFINITION 5.9**   *The function $u \in V$ is called a weak solution of problem (5.52), if it satisfies the following identity,*

$$\langle e_V(u), v \rangle = \langle [e_V^1(u), e_V^2(u)], v \rangle = [0, 0], \quad \forall v \in \mathcal{H} \qquad (5.53)$$

*with*

$$\langle e_V^1(u), v \rangle =$$

$$\int_0^T \int_{\Omega_t(V)} [(\partial_t u + D u \cdot u + D u \cdot V + D V \cdot u) \cdot v + \nu D u \cdot\cdot D v] \qquad (5.54)$$

$$- \int_0^T \int_{\Omega_t(V)} F(V) \cdot v$$

$$\langle e_V^2(u), v \rangle = \int_{\Omega_0} (u(0) - \tilde{u}_0) \cdot v(0) \qquad (5.55)$$

We set

$$\mathcal{U} = \left\{ V \in H^1(0, T; (H^m(D))^d), \quad \operatorname{div} V = 0 \text{ in } D, \quad V \cdot n = 0 \text{ on } \partial D \right\}$$

with $m > 5/2$.

**THEOREM 5.9 [58]**

*We assume the domain $\Omega_0$ to be of class $\mathcal{C}^1$. For $V \in \mathcal{U}$ and $u_0 \in H(D)$ such that $u_0|_{\Omega_0} \in H(\Omega_0)$,*

*1. it exists at least a weak solution of problem (5.52) with*

$$u \in \mathcal{H} \cup L^\infty(0, T; H),$$

*2. if $u_0 \in (H^2(D))^d \cup H_0^1(\text{div}, D)$ and $\nu$ is large or $u_0$ is a small data, then the uniqueness of a weak solution is guaranteed, and we have $\partial_t u \in \mathcal{H} \cup L^\infty(0, T; H(\Omega_t))$,*

*3. if $\Omega$ is of class $\mathcal{C}^2$, $u \in L^\infty(0, T; (H^2(\Omega_t))^d \cup H_0^1(\text{div}, \Omega_t))$.*

### 5.5.2   The weak Piola material derivative

We are interested in solving the following minimization problem :

$$\min_{V \in \mathcal{U}} j(V) \qquad (5.56)$$

with

$$j(V) = \frac{\alpha}{2} \int_0^T \int_{\Omega_t(V)} |u(V)|^2 + \frac{\gamma}{2} \int_0^T \int_{\Omega_t(V)} |V|^2 \qquad (5.57)$$

In order to derive first-order optimality conditions for problem (5.56), we need to analyse the derivability of the state $u(V)$ with respect to $V \in \mathcal{U}$. There exist at least two methods in order to establish such a differentiability result:

- Limit analysis of the differential quotient,

$$\dot{u}(V; W) = \lim_{\rho \to 0} \frac{1}{\rho}(u(V + \rho W) \circ T_t(V + \rho W) - u(V) \circ T_t(V))$$

- Application of the weak implicit function theorem and deduction of the local differentiability of the solution $u(V)$ associated to the implicit equation $\langle e_V(u), v \rangle = 0, \quad \forall v \in \mathcal{H}$.

We recall here how the weak implicit function theorem can be applied following the result obtained in [59].

In order to work with divergence free functions, we need to introduce the Piola transform that preserves the free divergence condition.

***LEMMA 5.1 [13]***
*The Piola transform,*

$$P_I : H_0^1(\text{div}, \Omega_t(V)) \longrightarrow H_0^1(\text{div}, \Omega_t^\rho)$$
$$v \mapsto (D\, T_\rho^t \cdot v) \circ (T_\rho^t)^{-1}$$

*is an isomorphism.*

We consider the solution $u_\rho = u(V + \rho W)$ defined on $\Omega_t^\rho$ of the following implicit equation,

$$\langle e_{(V+\rho W)}(u), v \rangle = 0, \quad \forall v \in \mathcal{H}^\rho$$

with

$$\mathcal{H}^\rho = \{v \in L^2(0, T; (H_0^1(\mathrm{div}, \Omega_t(V + \rho W)))^d)\}$$

We introduce the element $\hat{u}_\rho = (\mathrm{D}\, T_\rho^t)^{-1} \cdot (u_\rho \circ T_\rho^t)$ defined on $\Omega_t(V)$ (i.e., $u_\rho = P_I(\hat{u}_\rho)$).

## LEMMA 5.2

*The element $u_\rho$ satisfies the following identity,*

$$\langle e_{(V+\rho W)}(u), v \rangle = 0, \quad \forall v \in \mathcal{H}^\rho$$

*if and only if $\hat{u}_\rho$ satisfies the following identity,*

$$\langle e^\rho(\hat{u}_\rho), \hat{v} \rangle = 0, \quad \forall \hat{v} \in \mathcal{H}$$

*with*

$$\langle e_1^\rho(v), w \rangle =$$

$$\int_0^T \int_{\Omega_t(V)} [(\partial_t(\mathrm{D}\, T_\rho^t \cdot v)) \cdot (\mathrm{D}\, T_\rho^t \cdot w)$$

$$-(\mathrm{D}(\mathrm{D}\, T_\rho^t \cdot v) \cdot (\mathrm{D}\, T_\rho^t)^{-1} \cdot (\partial_t T_\rho^t)) \cdot (\mathrm{D}\, T_\rho^t \cdot w) + (\mathrm{D}(\mathrm{D}\, T_\rho^t \cdot v) \cdot v) \cdot (\mathrm{D}\, T_\rho^t \cdot w)$$

$$+(\mathrm{D}(\mathrm{D}\, T_\rho^t \cdot v) \cdot (\mathrm{D}\, T_\rho^t)^{-1} \cdot ((V + \rho W) \circ T_\rho^t)) \cdot (\mathrm{D}\, T_\rho^t \cdot w)$$

$$+(\mathrm{D}((V + \rho W) \circ T_\rho^t) \cdot v) \cdot (\mathrm{D}\, T_\rho^t \cdot w)$$

$$+\nu (\mathrm{D}(\mathrm{D}\, T_\rho^t \cdot v) \cdot (\mathrm{D}\, T_\rho^t)^{-1}) \cdot \cdot (\mathrm{D}(\mathrm{D}\, T_\rho^t \cdot w) \cdot (\mathrm{D}\, T_\rho^t)^{-1})$$

$$-(F(V + \rho W) \circ T_\rho^t) \cdot (\mathrm{D}\, T_\rho^t \cdot w)]$$

$$\langle e_2^\rho(v), w \rangle = \int_\Omega (v(0) - \hat{u}_0) \cdot w$$

*with*

$$F(V) = -\partial_t V - \mathrm{D}\, V \cdot V + \nu \Delta V$$

*and*

$$\partial_t T_\rho^t = (V + \rho W) \circ T_\rho^t - D T_\rho^t \cdot V$$

**PROOF**    We consider the solution $u_\rho$ of the perturbed state equation $e_{(V+\rho W)} = 0$, with

$$\langle e_{(V+\rho W)}(u), v\rangle = \int_0^T \int_{\Omega_t^\rho} \left[ (\partial_t u + D u \cdot u + D u \cdot (V + \rho W) \right.$$
$$\left. + D(V + \rho W) \cdot u) \cdot v + \nu D u \cdot\cdot D v - F(V + \rho W) \cdot v \right]$$

with $v \in \mathcal{H}^\rho$.
We introduce the variables $(\hat{u}, \hat{v})$ defined in $\Omega_t(V)$ such that

$$[u, v] = \left[ (D T_\rho^t \cdot \hat{u}) \circ (T_\rho^t)^{-1}, (D T_\rho^t \cdot \hat{v}) \circ (T_\rho^t)^{-1} \right]$$

We replace this new representation inside the state equation and we use a back transport in $\Omega_t(V)$. This leads to the following identity,

$$\langle e_1^\rho(\hat{u}), \hat{v}\rangle = \int_{Q(V)} \left[ (\partial_t((D T_\rho^t \cdot \hat{u}) \circ (T_\rho^t)^{-1}) \right.$$
$$+ D((D T_\rho^t \cdot \hat{u}) \circ (T_\rho^t)^{-1}) \cdot (D T_\rho^t \cdot \hat{u}) \circ (T_\rho^t)^{-1}$$
$$+ D((D T_\rho^t \cdot \hat{u}) \circ (T_\rho^t)^{-1}) \cdot (V + \rho W)$$
$$+ D(V + \rho W) \cdot (D T_\rho^t \cdot \hat{u}) \circ (T_\rho^t)^{-1}) \circ T_\rho^t \cdot (D T_\rho^t \cdot \hat{v})$$
$$+ \nu D((D T_\rho^t \cdot \hat{u}) \circ (T_\rho^t)^{-1}) \circ T_\rho^t \cdot\cdot (D((D T_\rho^t \cdot \hat{v}) \circ (T_\rho^t)^{-1})) \circ T_\rho^t$$
$$\left. - F(V + \rho W) \circ T_\rho^t \cdot (D T_\rho^t \cdot \hat{v}) \right]$$

$$\square$$

**LEMMA 5.3**
$$D((T_\rho^t)^{-1}) \circ T_\rho^t = (D T_\rho^t)^{-1} \tag{5.58}$$
$$\partial_t((T_\rho^t)^{-1}) \circ T_\rho^t = -(D T_\rho^t)^{-1} \cdot \partial_t T_\rho^t \tag{5.59}$$
$$\partial_t((D T_\rho^t \cdot \hat{u}) \circ (T_\rho^t)^{-1}) \circ T_\rho^t = \partial_t(D T_\rho^t \circ \hat{u}) - D(D T_\rho^t \circ \hat{u}) \cdot (D T_\rho^t)^{-1} \cdot \partial_t T_\rho^t \tag{5.60}$$

**PROOF**    Using the identity,

$$(T_\rho^t)^{-1} \circ T_\rho^t = I$$

we get

$$D((T_\rho^t)^{-1} \circ T_\rho^t) = I$$
$$D((T_\rho^t)^{-1}) \circ T_\rho^t \cdot D T_\rho^t = I$$

By differentiation with respect to time t, we also get

$$\partial_t((T^t_\rho)^{-1} \circ T^t_\rho) = 0$$
$$\partial_t((T^t_\rho)^{-1}) \circ T^t_\rho + D((T^t_\rho)^{-1}) \circ T^t_\rho \cdot \partial_t T^t_\rho = 0$$
$$\partial_t((T^t_\rho)^{-1}) \circ T^t_\rho + (D\,T^t_\rho)^{-1} \cdot \partial_t T^t_\rho = 0$$

Using the chain rule, we deduce

$$[\partial_t((D\,T^t_\rho \cdot \hat{u}) \circ (T^t_\rho)^{-1})] \circ T^t_\rho = [\partial_t(D\,T^t_\rho \cdot \hat{u}) \circ (T^t_\rho)^{-1}$$
$$+ D(D\,T^t_\rho \cdot \hat{u}) \circ (T^t_\rho)^{-1} \cdot \partial_t((T^t_\rho)^{-1})] \circ T^t_\rho$$
$$= \partial_t(D\,T^t_\rho \cdot \hat{u}) + D(D\,T^t_\rho \cdot \hat{u}) \cdot \partial_t((T^t_\rho)^{-1}) \circ T^t_\rho$$
$$= \partial_t(D\,T^t_\rho \cdot \hat{u}) - D(D\,T^t_\rho \cdot \hat{u}) \cdot (D\,T^t_\rho)^{-1} \cdot \partial_t T^t_\rho$$

▯

In order to get the correct state operator, we need also the following identities,

**LEMMA 5.4**

$$D(\phi \circ (T^t_\rho)^{-1}) \circ T^t_\rho = D(\phi) \cdot (D\,T^t_\rho)^{-1} \tag{5.61}$$

$$D(V + \rho W) \circ T^t_\rho \cdot (D\,T^t_\rho) = D((V + \rho W) \circ T^t_\rho) \tag{5.62}$$

We shall apply the first identity with $\phi = (D\,T^t_\rho \cdot \hat{u})$. Finally, using all the identities proven above, we deduce the expression of $e^1_\rho(\hat{u}, \hat{v})$. Now, we simply need to prove the following lemma in order to conclude the proof,

**LEMMA 5.5**

$$\partial_t T^t_\rho = (V + \rho W) \circ T^t_\rho - D\,T^t_\rho \cdot V \tag{5.63}$$

**PROOF**    We use the definition of the Transverse map,

$$\partial_t(T^t_\rho) = \partial_t(T_t(V + \rho W) \circ T_t(V)^{-1})$$
$$= \partial_t(T_t(V + \rho W)) \circ T_t(V)^{-1}$$
$$+ D(T_t(V + \rho W)) \circ T_t(V)^{-1} \cdot \partial_t(T_t(V)^{-1})$$
$$= ((V + \rho W) \circ T_t(V + \rho W)) \circ T_t(V)^{-1}$$
$$- D(T_t(V + \rho W)) \circ T_t(V)^{-1} \cdot (D\,T_t^{-1}(V)) \cdot \partial_t(T_t(V)) \circ T_t^{-1}(V)$$
$$= (V + \rho W) \circ T^t_\rho - D(T_t(V + \rho W) \circ T_t(V)^{-1}) \cdot \partial_t(T_t(V)) \circ T_t^{-1}(V)$$
$$= (V + \rho W) \circ T^t_\rho - D(T^t_\rho) \cdot V$$

▯

We now consider the application,

$$[0, \rho_0] \times V \to \mathcal{H}^* \times H_0^1(\text{div}, \Omega_0)$$
$$(\rho, v) \mapsto e^\rho(v) \tag{5.64}$$

and

$$[0, \rho_0] \to \mathcal{H}$$
$$\rho \mapsto \hat{u}_\rho = (\mathrm{D}\, T_\rho^t)^{-1} \cdot (u_\rho \circ T_\rho^t) \tag{5.65}$$

where $\hat{u}_\rho \in V$ is the solution of the state equation,

$$\langle e^\rho(v), w \rangle = 0, \quad \forall w \in \mathcal{H} \tag{5.66}$$

**LEMMA 5.6 [135]**
*For any $F \in H^s(D)$, with $s \geq 1$,*

$$\frac{1}{\rho}(F \circ T_\rho^t - F) \overset{\rho \to 0}{\longrightarrow} \nabla F \cdot Z_t \tag{5.67}$$

*strongly in $H^{s-1}(D)$. In the case $s < 1$, the convergence only holds weakly in $H^{s-1}(D)$.*

In order to prove the differentiability of $\hat{u}_\rho$ with respect to $\rho$ in a neighbourhood of $\rho = 0$, we cannot use the classical implicit function theorem, since it requires strong differentiability results in $H^{-1}$ for our application. Then we shall use the weak implicit function theorem, recalled below,

**THEOREM 5.10 [147]**
*Let $X, Y^*$ be two Banach spaces, $I$ an open bounded set in $\mathbb{R}$, and consider the following mapping,*

$$e : I \times X \to Y^*$$
$$(\rho, x) \mapsto e(\rho, x)$$

Let us assume the following hypothesis,

a)  (i) the application $\rho \mapsto \langle e(\rho, x), y \rangle$ is continuously differentiable for any $y \in Y$,

   (ii) the application $(\rho, x) \mapsto \langle \partial_\rho e(\rho, x), y \rangle$ is continuous.

b) It exists $u \in X$ such that

$$u \in \mathcal{C}^{0,1}(I; X)$$
$$e(\rho, u(\rho)) = 0, \ \forall \rho \in I$$

c) $x \mapsto e(\rho, x)$ is differentiable and $(\rho, x) \mapsto \partial_x e(\rho, x)$ is continuous.

d) It exists $\rho_0 \in I$ such that $\partial_x e(\rho, x)|_{(\rho_0, x(\rho_0))} \in \mathrm{ISOM}(X, Y^*)$.

Then the mapping

$$u(.) : I \to X$$
$$\rho \mapsto u(\rho)$$

is differentiable at point $\rho = \rho_0$ for the weak topology in $X$ and its weak derivative $\dot{u}(\rho)$ is the solution of the following linearized equation,

$$\langle \partial_x e(\rho_0, u(\rho_0)) \cdot \dot{u}(\rho_0), y \rangle + \langle \partial_\rho e(\rho_0, u(\rho_0)), y \rangle = 0, \quad \forall y \in Y \qquad (5.68)$$

In order to apply the above theorem to Eq. (5.66), we need to state the following properties,

**LEMMA 5.7**
*The mapping,*

$$[0, \rho_0] \to \mathbb{R} \qquad \cdot$$
$$\rho \mapsto \langle e^\rho(v), w \rangle \qquad (5.69)$$

*is $C^1$ for any $(v, w) \in V \times H$ and its derivative is given by the following expression,*

$$\langle \partial_\rho e_1^\rho(v), w \rangle =$$

$$\int_{Q(V)} \left[ (\partial_t (\mathrm{D}(\mathcal{Z}_\rho^t \cdot \mathcal{T}_\rho^t) \cdot v)) + (\mathrm{D}(\mathrm{D}(\mathcal{Z}_\rho^t \cdot \mathcal{T}_\rho^t) \cdot v) \cdot V + (\mathrm{D}(\mathrm{D}(\mathcal{Z}_\rho^t \cdot \mathcal{T}_\rho^t) \cdot v) \cdot v) \right.$$

$$+ \mathrm{D} \left[ (\mathrm{D}(V + \rho W) \cdot \mathcal{Z}_\rho^t) \circ \mathcal{T}_\rho^t + W \circ \mathcal{T}_\rho^t \right] \cdot v - \partial_\rho (F(V + \rho W) \circ \mathcal{T}_\rho^t) \right] \cdot (\mathrm{D} \mathcal{T}_\rho^t \cdot w)$$

$$+ \left[ (\partial_t (\mathrm{D} \mathcal{T}_\rho^t \cdot v)) + (\mathrm{D}(\mathrm{D} \mathcal{T}_\rho^t \cdot v) \cdot V + (\mathrm{D}(\mathrm{D} \mathcal{T}_\rho^t \cdot v) \cdot v) \right.$$

$$+ (\mathrm{D}((V + \rho W) \circ \mathcal{T}_\rho^t) \cdot v) - (F(V + \rho W) \circ \mathcal{T}_\rho^t] \cdot (\mathrm{D}(\mathcal{Z}_\rho^t \cdot \mathcal{T}_\rho^t) \cdot w)$$

$$+ \nu (\mathrm{D}(\mathrm{D}(\mathcal{Z}_\rho^t \circ \mathcal{T}_\rho^t) \cdot v) \cdot (\mathrm{D} \mathcal{T}_\rho^t)^{-1}) \cdot \cdot (\mathrm{D}(\mathrm{D} \mathcal{T}_\rho^t \cdot w) \cdot (\mathrm{D} \mathcal{T}_\rho^t)^{-1})$$

$$- \nu (\mathrm{D}(\mathrm{D} \mathcal{T}_\rho^t \cdot v) \cdot (\mathrm{D} \mathcal{T}_\rho^t)^{-1}) \cdot \mathrm{D}(\mathcal{Z}_\rho^t \circ \mathcal{T}_\rho^t) \cdot (\mathrm{D} \mathcal{T}_\rho^t)^{-1}) \cdot \cdot (\mathrm{D}(\mathrm{D} \mathcal{T}_\rho^t \cdot w) \cdot (\mathrm{D} \mathcal{T}_\rho^t)^{-1})$$

$$+ \nu (\mathrm{D}(\mathrm{D} \mathcal{T}_\rho^t \cdot v) \cdot (\mathrm{D} \mathcal{T}_\rho^t)^{-1}) \cdot \cdot (\mathrm{D}(\mathrm{D}(\mathcal{Z}_\rho^t \circ \mathcal{T}_\rho^t) \cdot w) \cdot (\mathrm{D} \mathcal{T}_\rho^t)^{-1})$$

$$- \nu (\mathrm{D}(\mathrm{D} \mathcal{T}_\rho^t \cdot v) \cdot (\mathrm{D} \mathcal{T}_\rho^t)^{-1}) \cdot \cdot (\mathrm{D}(\mathrm{D} \mathcal{T}_\rho^t \cdot w) \cdot (\mathrm{D} \mathcal{T}_\rho^t)^{-1}) \cdot \mathrm{D}(\mathcal{Z}_\rho^t \circ \mathcal{T}_\rho^t) \cdot (\mathrm{D} \mathcal{T}_\rho^t)^{-1})$$

**PROOF**    We first simplify the expression of the weak state operator, using that

$$\partial_t T_\rho^t = (V + \rho W) \circ T_\rho^t - D\,T_\rho^t \cdot V$$

and we get

$$\langle e_1^\rho(v), w \rangle =$$
$$\int_0^T \int_{\Omega_t(V)} \left[ (\partial_t (D\,T_\rho^t \cdot v)) + (D(D\,T_\rho^t \cdot v) \cdot V + (D(D\,T_\rho^t \cdot v) \cdot v) \right.$$
$$+ (D((V + \rho W) \circ T_\rho^t) \cdot v) - (F(V + \rho W) \circ T_\rho^t] \cdot (D\,T_\rho^t \cdot w)$$
$$\left. + \nu (D(D\,T_\rho^t \cdot v) \cdot (D\,T_\rho^t)^{-1}) \cdot \cdot (D(D\,T_\rho^t \cdot w) \cdot (D\,T_\rho^t)^{-1}) \right.$$

We use the expression of the weak state operator and the following identities,

$$\partial_\rho T_\rho^t = \mathcal{Z}_\rho^t \circ T_\rho^t$$

$$\partial_\rho (D\,T_\rho^t)^{-1} = -(D\,T_\rho^t)^{-1} \cdot D(\mathcal{Z}_\rho^t \circ T_\rho^t) \cdot (D\,T_\rho^t)^{-1}$$

□

### LEMMA 5.8

*The mapping,*

$$[0, \rho_0] \times V \to \mathcal{H}^*$$
$$(\rho, v) \mapsto \partial_\rho e^\rho(v) \tag{5.70}$$

*is weakly continuous.*

**PROOF**    We can prove that for $(V, W) \in V$, the associated flow $T_\rho^t \in \mathcal{C}^1([0, \rho_0[; \mathcal{C}^2(D, \mathbb{R}^3))$, and the weak continuity follows easily.    □

In order to apply the implicit function derivative identity, we need to express the derivative $\partial_\rho e^\rho(v)$ at point $\rho = 0$,

**LEMMA 5.9**

$\langle \partial_\rho e_1^\rho|_{\rho=0}(v), w \rangle =$

$$\int_Q [\partial_t(\mathrm{D}\, Z_t \cdot v) + \mathrm{D}(\mathrm{D}\, Z_t \cdot v) \cdot V + \mathrm{D}(\mathrm{D}\, Z_t \cdot v) \cdot v + \mathrm{D}\, [\mathrm{D}\, V \cdot Z_t + W] \cdot v] \cdot w$$

$$+ [\partial_t v + \mathrm{D}\, v \cdot V + \mathrm{D}\, v \cdot v + \mathrm{D}\, V \cdot v] \cdot (\mathrm{D}\, Z_t \cdot w) + \nu\, \mathrm{D}(\mathrm{D}\, Z_t \cdot v) \cdot\cdot \mathrm{D}\, w$$

$$- \nu (\mathrm{D}\, v \cdot \mathrm{D}\, Z_t) \cdot\cdot \mathrm{D}\, w + \nu\, \mathrm{D}\, v \cdot\cdot \mathrm{D}(\mathrm{D}\, Z_t \cdot w) - \nu\, \mathrm{D}\, v \cdot\cdot (\mathrm{D}\, w \cdot \mathrm{D}\, Z_t)$$

$$+ [\partial_t W + \mathrm{D}\, W \cdot V + \mathrm{D}\, V \cdot W - \nu \Delta W] \cdot w$$

$$+ (\mathrm{D}\, [\partial_t V + \mathrm{D}\, V \cdot V - \nu \Delta V] \cdot Z_t) \cdot w + [\partial_t V + \mathrm{D}\, V \cdot V - \nu \Delta V] \cdot (\mathrm{D}\, Z_t \cdot w)$$

**PROOF**    We set $\rho = 0$ in the expression of $\langle \partial_\rho e^\rho(v), w \rangle$ and we use the following identities,

$$\mathcal{T}_{\rho=0}^t = \mathrm{I}$$

$$\mathcal{Z}_\rho^t|_{\rho=0} \overset{\text{def}}{=} Z_t$$

▯

**LEMMA 5.10**
*The mapping,*

$$V \to \mathcal{H}^*$$
$$v \mapsto e^\rho(v) \tag{5.71}$$

*is differentiable for any $\rho \in [0, \rho_0]$ and its derivative is given by the following expression,*

$\langle \partial_v e_1^\rho(v) \cdot \delta v, w \rangle =$

$$\int_{Q(V)} [(\partial_t(\mathrm{D}\, \mathcal{T}_\rho^t \cdot \delta v)) + \mathrm{D}(\mathrm{D}\, \mathcal{T}_\rho^t \cdot \delta v) \cdot V + \mathrm{D}(\mathrm{D}\, \mathcal{T}_\rho^t \cdot \delta v) \cdot v + \mathrm{D}(\mathrm{D}\, \mathcal{T}_\rho^t \cdot v) \cdot \delta v$$

$$+ (\mathrm{D}((V + \rho W) \circ \mathcal{T}_\rho^t) \cdot \delta v)] + \nu (\mathrm{D}(\mathrm{D}\, \mathcal{T}_\rho^t \cdot \delta v) \cdot (\mathrm{D}\, \mathcal{T}_\rho^t)^{-1}) \cdot\cdot (\mathrm{D}(\mathrm{D}\, \mathcal{T}_\rho^t \cdot w) \cdot (\mathrm{D}\, \mathcal{T}_\rho^t)^{-1})$$

*and the mapping,*

$$[0, \rho_0] \times V \to \mathcal{L}(V; \mathcal{H}^*)$$
$$(\rho, v) \mapsto \partial_v e^\rho(v) \tag{5.72}$$

*is continuous.*

**LEMMA 5.11**
*The mapping,*

$$V \to \mathcal{F}$$
$$\delta v \mapsto \partial_v e^{\rho=0}(v) \cdot \delta v \tag{5.73}$$

*is an isomorphism and its expression is furnished by the following identity,*

$$\langle \partial_v e_1^{\rho=0}(v) \cdot \delta v, w \rangle =$$
$$\int_{Q(V)} [(\partial_t \delta v) \cdot w + (D \, \delta v \cdot v) \cdot w + (D \, v \cdot \delta v) \cdot w + (D \, \delta v \cdot V) \cdot w$$
$$+ (D \, V \cdot \delta v) \cdot w + \nu \, D \, \delta v \cdot \cdot \, D \, w]$$

**PROOF**    This result follows from the uniqueness result for the Navier-Stokes system under regularity and smallness assumptions (see Th. (5.9) and [139]). Indeed, for $u_1$ and $u_2$ solutions of the Navier-Stokes equations, it is proven that the element $y = u_1 - u_2$ satisfying the following identity,

$$\int_{Q(V)} [(\partial_t y) \cdot w + (D \, y \cdot u_1) \cdot w + (D \, u_2 \cdot y) \cdot w + (D \, y \cdot V) \cdot w + (D \, V \cdot y) \cdot w$$
$$+ \nu \, D \, y \cdot \cdot \, D \, w] = 0, \quad \forall w \in \mathcal{H}$$

*exists and is identically equal to the null function. Similar a priori estimates hold for $\delta v$ and the unique solvability of the linearized system is established.*
◻

**LEMMA 5.12**
*The solution $\hat{u}_\rho \in V$ of the implicit equation,*

$$\langle e^\rho(v), w \rangle = 0, \quad \forall w \in \mathcal{H} \tag{5.74}$$

*is Lipschitz with respect to $\rho$.*

**PROOF**    We need the identity satisfied by $\hat{u}_{\rho_1} - \hat{u}_{\rho_2}$ and we shall follow the same steps described in [56] (pp. 31).    ◻

Hence the hypothesis of Th. (5.10) is satisfied by the Eq. (5.66) and we can state the following differentiability result,

**THEOREM 5.11**
*The Piola material derivative $\dot{u}^P = \partial_\rho(\hat{u}_\rho)|_{\rho=0}$ exists and is characterized by the linear tangent equation,*

$$\langle \partial_v e^{\rho=0}(v)|_{v=\hat{u}} \cdot \dot{u}^P, w \rangle + \langle \partial_\rho e^\rho(\hat{u})|_{\rho=0}, w \rangle = 0, \quad \forall\, w \in \mathcal{H} \qquad (5.75)$$

*which possesses the following structure,*

$$\int_{Q(V)} \left[ (\partial_t \dot{u}^P) \cdot w + (\mathrm{D}\,\dot{u}^P \cdot u) \cdot w + (\mathrm{D}\,u \cdot \dot{u}^P) \cdot w + (\mathrm{D}\,\dot{u}^P \cdot V) \cdot w \right.$$
$$\left. + (\mathrm{D}\,V \cdot \dot{u}^P) \cdot w + \nu\,\mathrm{D}\,\dot{u}^P \cdot\cdot\,\mathrm{D}\,w \right] = \langle L(u, Z_t, V, W), w \rangle$$

*with*

$$\langle L(u, Z_t, V, W), w \rangle =$$

$$-\int_Q \left[ \partial_t(\mathrm{D}\,Z_t \cdot u) + \mathrm{D}(\mathrm{D}\,Z_t \cdot u) \cdot V + \mathrm{D}(\mathrm{D}\,Z_t \cdot u) \cdot u + \mathrm{D}\left[\mathrm{D}\,V \cdot Z_t + W\right] \cdot u \right] \cdot w$$

$$- \left[ \partial_t u + \mathrm{D}\,u \cdot V + \mathrm{D}\,u \cdot u + \mathrm{D}\,V \cdot u \right] \cdot (\mathrm{D}\,Z_t \cdot w) - \nu\,\mathrm{D}(\mathrm{D}\,Z_t \cdot u) \cdot\cdot\,\mathrm{D}\,w$$

$$+ \nu(\mathrm{D}\,u \cdot \mathrm{D}\,Z_t) \cdot\cdot\,\mathrm{D}\,w - \nu\,\mathrm{D}\,u \cdot\cdot\,\mathrm{D}(\mathrm{D}\,Z_t \cdot w) + \nu\,\mathrm{D}\,u \cdot\cdot(\mathrm{D}\,w \cdot \mathrm{D}\,Z_t)$$

$$+ \left[ -\partial_t W - \mathrm{D}\,W \cdot V - \mathrm{D}\,V \cdot W + \nu\Delta W \right] \cdot w$$

$$- (\mathrm{D}\left[\partial_t V + \mathrm{D}\,V \cdot V - \nu\Delta V\right] \cdot Z_t) \cdot w - \left[\partial_t V + \mathrm{D}\,V \cdot V - \nu\Delta V\right] \cdot (\mathrm{D}\,Z_t \cdot w)$$

### 5.5.3   Shape derivative

In the last section, we have proven that the solution $u(V)$ of the moving Navier-Stokes system is differentiable with respect to the velocity $V$. We have also characterized the linearized system satisfied by the Piola material derivative $\dot{u}^P(V) \cdot W$. In this paragraph, we will identify the shape derivative $u'(V) \cdot W$ under some regularity assumptions.
Let us consider the weak solution $\tilde{u}$ of Eq. (5.52), i.e.,

$$\langle e_V(\tilde{u}), v \rangle = \langle [e_V^1(u), e_V^2(u)], v \rangle = [0, 0], \quad \forall\, v \in \mathcal{H} \qquad (5.76)$$

with

$$\langle e_V^1(\tilde{u}), v \rangle = \int_{Q(V)} \left[ (\partial_t \tilde{u} + \mathrm{D}\,\tilde{u} \cdot \tilde{u} + \mathrm{D}\,\tilde{u} \cdot V + \mathrm{D}\,V \cdot \tilde{u}) \cdot v \right.$$
$$\left. + 2\nu\varepsilon(\tilde{u}) \cdot\cdot\,\varepsilon(v) - F(V) \cdot v \right]$$

$$\langle e_V^2(\tilde{u}), v \rangle = \int_{\Omega_0} (\tilde{u}(0) - u_0) \cdot v(0) \tag{5.77}$$

where $\varepsilon(v) = \frac{1}{2}(\mathrm{D}\,v + {}^*\mathrm{D}\,v)$ stands for the symmetrical deviation tensor. This definition is motivated by the following lemma,

**LEMMA 5.13**

$$-\nu \int_{\Omega_t} \Delta u \cdot v = 2\nu \int_{\Omega_t} \varepsilon(u) \cdot \cdot \varepsilon(v) - 2\nu \int_{\Gamma_t} \langle \varepsilon(u) \cdot n, v \rangle, \quad \forall v \in H^1(\mathrm{div}, \Omega_t) \tag{5.78}$$

**THEOREM 5.12**

*For $\Omega_0$ of class $\mathcal{C}^2$, the shape derivative $\tilde{u}' = \dot{\tilde{u}} - \mathrm{D}\,\tilde{u} \cdot Z_t$ exists and is characterized as the solution of the following linearized system,*

$$\begin{cases} \partial_t \tilde{u}' + \mathrm{D}\,\tilde{u}' \cdot \tilde{u} + \mathrm{D}\,\tilde{u} \cdot \tilde{u}' + \mathrm{D}\,\tilde{u}' \cdot V + \mathrm{D}\,V \cdot \tilde{u}' - \nu\Delta\tilde{u}' + \nabla p' = L(V, W), & Q \\ \mathrm{div}(\tilde{u}') = 0, & Q \\ \tilde{u}' = -(\mathrm{D}\,\tilde{u} \cdot n)\langle Z_t, n \rangle, & \Sigma \\ \tilde{u}'(0) = 0, & \Omega_0 \end{cases} \tag{5.79}$$

*with*

$$L(V, W) = -\partial_t W - \mathrm{D}\,W \cdot V - \mathrm{D}\,V \cdot W + \nu\Delta W - \mathrm{D}\,\tilde{u} \cdot W - \mathrm{D}\,W \cdot \tilde{u} \tag{5.80}$$

**PROOF**  In order to state such a result, we use Th. (5.4) and we get

$$\frac{d}{dV}\left(\int_{\Omega_t(V)} G(V)dx\right) \cdot W = \int_{\Omega_t(V)} G'(V) \cdot W dx + \int_{\Gamma_t(V)} G\langle Z_t, n \rangle \tag{5.81}$$

where $G'(V) \cdot W$ stands for non-cylindrical shape derivative of $G$ and $Z_t$ is the transverse vector field solution of the Transverse Equation (Eq. (5.22)) with

$$G = [(\partial_t \tilde{u} + \mathrm{D}\,\tilde{u} \cdot \tilde{u} + \mathrm{D}\,\tilde{u} \cdot V + \mathrm{D}\,V \cdot \tilde{u}) \cdot v + \nu\varepsilon(\tilde{u}) \cdot \cdot \varepsilon(v) - F(V) \cdot v]$$

We assume that $v$ has a compact support, then $G|_{\Gamma_t(V)} = 0$.

**LEMMA 5.14**

$$G'(V) \cdot W = [(\partial_t \tilde{u}' + \mathrm{D}\,\tilde{u}' \cdot \tilde{u} + \mathrm{D}\,\tilde{u} \cdot \tilde{u}' + \mathrm{D}\,\tilde{u}' \cdot V + \mathrm{D}\,\tilde{u} \cdot W$$
$$+ \mathrm{D}\,W \cdot \tilde{u} + \mathrm{D}\,V \cdot \tilde{u}') \cdot v + \nu\varepsilon(\tilde{u}') \cdot \cdot \varepsilon(v) - F'(V) \cdot W \cdot v]$$

*with*

$$F'(V) \cdot W = -\partial_t W - \mathrm{D}\,W \cdot V - \mathrm{D}\,V \cdot W + \nu\Delta W$$

Finally we obtain

$$\frac{d}{dV}\langle e_1(\tilde{u}) \cdot W, \cdot v\rangle = \int_{Q(V)} [(\partial_t\tilde{u}' + D\,\tilde{u}' \cdot \tilde{u} + D\,\tilde{u} \cdot \tilde{u}' + D\,\tilde{u}' \cdot V + D\,\tilde{u} \cdot W$$
$$+ D\,W \cdot \tilde{u} + D\,V \cdot \tilde{u}') \cdot v + \nu\varepsilon(\tilde{u}') \cdot \cdot\varepsilon(v) - F'(V) \cdot W \cdot v]$$

for any $v \in \mathcal{H}$ with compact support.

Using integration by parts for the term $\int_{Q(V)} \nu\varepsilon(\tilde{u}') \cdot \cdot\varepsilon(v)$, we recover the correct strong formulation of the linearized equation (Eq. (5.79)) satisfied by the shape derivative $u'(V) \cdot W$.

The boundary condition comes from the fact that the shape derivative of the condition $\tilde{u} = 0$ on $\Gamma_t(V)$ is given by

$$\tilde{u}' = -D\,u \cdot Z_t, \quad \text{on } \Gamma_t(V)$$

Since $u = 0$ on $\Gamma_t(V)$, we have $D\,u|_{\Gamma_t} = D\,u \cdot (n \otimes n)$ which gives

$$\tilde{u}' = -(D\,u \cdot n)\langle Z_t, n\rangle, \quad \text{on } \Gamma_t(V)$$

☐

The shape derivative $u'(V){\cdot}W$ of the solution $u$ of the original non-homogeneous Dirichlet boundary problem (Eq. (5.2)) is given by the expression

$$u'(V) \cdot W = \tilde{u}'(V) \cdot W + W \tag{5.82}$$

**COROLLARY 5.1**
*The shape derivative $u'(V){\cdot}W$ of the solution $u$ of Eq. (5.2) exists and satisfies the following linearized problem,*

$$\begin{cases} \partial_t u' + D\,u' \cdot u + D\,u \cdot u' - \nu\Delta u' + \nabla p' = 0, & Q \\ \text{div}(u') = 0, & Q \\ u' = W + (D\,V \cdot n - D\,u \cdot n)\langle Z_t, n\rangle, & \Sigma \\ u'(0) = 0, & \Omega_0 \end{cases} \tag{5.83}$$

**PROOF**    We simply set in Eq. (5.79), $\tilde{u}' = u' - W$ and $\tilde{u} = u - V$.    ☐

**REMARK 5.4**    If we choose $V = (V \circ p)$ the canonical extension of $V$ in Eq. (5.83), then we get the simpler boundary condition,

$$u' = W - (D\,u \cdot n)\langle Z_t, n\rangle, \quad \text{on } \Gamma_t(V) \tag{5.84}$$

☐

### 5.5.4    Extractor Identity

In the last section, we have established the structure of the system satisfied by the non-cylindrical shape derivative $u'(V) \cdot W$ of the solution $u(V)$ of the Navier-Stokes problem in the moving domain $\Omega_t(V)$. This linearized system has been obtained independently of the system satisfied by the non-cylindrical material derivative $\dot{\tilde{u}}^P(V) \cdot W$. However, there exists an explicit relation between the original shape $u'$ and the Piola material derivative $\dot{\tilde{u}}^P(V) \cdot W$ of the shift state $\tilde{u} = u - V$.

### LEMMA 5.15
*Let $u(V)$ stand for the weak solution of the non-homogeneous Navier-Stokes equations (Eq. (5.2)) in moving domain, $u'(V) \cdot W$ stands for its shape derivative and $\dot{\tilde{u}}^P(V) \cdot W$ stands for the Piola material derivative of the shift flow $\tilde{u} = u(V) - V$ in the direction $W$. Then the following identity holds,*

$$\dot{\tilde{u}}^P(V) \cdot W = \tilde{u}'(V) \cdot W + [\tilde{u}(V), Z_t] \tag{5.85}$$
$$= u'(V) \cdot W + [u(V), Z_t] - [V, Z_t] - W \tag{5.86}$$

*where $[X, Y] = D\, X \cdot Y - D\, Y \cdot X$.*

This relation can be fruitful in order to obtain an identity concerning the solution $\tilde{u}(V)$ inside $\Omega_t(V)$.

### PROPOSITION 5.6
*We consider $\Omega_0$ of class $\mathcal{C}^2$, $\tilde{u}$ solution of the homogeneous Navier-Stokes equations Eq. (5.52) and $Z_t$ solution of Eq. (5.22), the following identity holds for all $(V, W) \in \mathcal{U}$,*

$$\int_{Q(V)} \{[\partial_t(D\, \tilde{u} \cdot Z_t) + D(D\, \tilde{u} \cdot Z_t) \cdot \tilde{u} + D\, \tilde{u} \cdot (D\, \tilde{u} \cdot Z_t) + D(D\, \tilde{u} \cdot Z_t) \cdot V$$

$$+ D\, V \cdot (D\, \tilde{u} \cdot Z_t)] \cdot w - [D\, \tilde{u} \cdot (D\, Z_t \cdot \tilde{u}) + D\, V \cdot (D\, Z_t \cdot \tilde{u}) - D(D\, V \cdot Z_t) \cdot \tilde{u}] \cdot w$$

$$+ [\partial_t \tilde{u} + D\, \tilde{u} \cdot V + D\, \tilde{u} \cdot \tilde{u} + D\, V \cdot \tilde{u}] \cdot (D\, Z_t \cdot w) - D\, \tilde{u} \cdot W \cdot w$$

$$+ \nu\, D(D\, \tilde{u} \cdot Z_t) \cdot\cdot D\, w - \nu(D\, \tilde{u} \cdot D\, Z_t) \cdot\cdot D\, w + \nu\, D\, \tilde{u} \cdot\cdot D(D\, Z_t \cdot w) - \nu\, D\, \tilde{u} \cdot\cdot (D\, w \cdot D\, Z_t)$$

$$+ (D\, [\partial_t V + D\, V \cdot V - \nu \Delta V] \cdot Z_t) \cdot w + [\partial_t V + D\, V \cdot V - \nu \Delta V] \cdot (D\, Z_t \cdot w)\} = 0,$$
$$\forall\, w \in \mathcal{H}$$

**PROOF**  We recall that the shape derivative $\tilde{u}'$ satisfies the following identity,

$$\int_{Q(V)} [\partial_t \tilde{u}' + D\,\tilde{u}' \cdot \tilde{u} + D\,\tilde{u} \cdot \tilde{u}' + D\,\tilde{u}' \cdot V + D\,V \cdot \tilde{u}'] \cdot w + \nu\, D\,\tilde{u}' \cdot\cdot\, D\,w = \langle \ell_1, w \rangle$$

with

$$\langle \ell_1, w \rangle = \int_{Q(V)} [-\partial_t W - D\,W \cdot V - D\,V \cdot W + \nu \Delta W - D\,\tilde{u} \cdot W - D\,W \cdot \tilde{u}] \cdot w$$

Then we set $\tilde{u}'(V) = \dot{\tilde{u}}^P - [\tilde{u}, Z_t] = \dot{\tilde{u}}^P - D\,\tilde{u} \cdot Z_t + D\,Z_t \cdot \tilde{u}$. This leads to the following identity,

$$\int_{Q(V)} \left[ (\partial_t \dot{\tilde{u}}^P) \cdot w + (D\,\dot{\tilde{u}}^P \cdot u) \cdot w + (D\,u \cdot \dot{\tilde{u}}^P) \cdot w + (D\,\dot{\tilde{u}}^P \cdot V) \cdot w \right.$$
$$\left. + (D\,V \cdot \dot{\tilde{u}}^P) \cdot w + \nu\, D\,\dot{\tilde{u}}^P \cdot\cdot\, D\,w \right] = \langle \ell_2, w \rangle$$

with

$$\langle \ell_2, w \rangle =$$
$$\int_{Q(V)} \Big\{ \big[ \partial_t(D\,\tilde{u} \cdot Z_t) + D(D\,\tilde{u} \cdot Z_t) \cdot \tilde{u} + D\,\tilde{u} \cdot (D\,\tilde{u} \cdot Z_t) + D(D\,\tilde{u} \cdot Z_t) \cdot V$$
$$+ D\,V \cdot (D\,\tilde{u} \cdot Z_t) \big] \cdot w - \big[ \partial_t(D\,Z_t \cdot \tilde{u}) + D(D\,Z_t \cdot \tilde{u}) \cdot \tilde{u} + D\,\tilde{u} \cdot (D\,Z_t \cdot \tilde{u})$$
$$+ D(D\,Z_t \cdot \tilde{u}) \cdot V + D\,V \cdot (D\,Z_t \cdot \tilde{u}) \big] \cdot w + \big[ \nu\, D(D\,\tilde{u} \cdot Z_t) \cdot\cdot\, D\,w - \nu\, D(D\,Z_t \cdot \tilde{u}) \cdot\cdot\, D\,w \big]$$
$$+ \big[ -\partial_t W - D\,W \cdot V - D\,V \cdot W + \nu \Delta W - D\,\tilde{u} \cdot W - D\,W \cdot \tilde{u} \big] \cdot w \Big\}$$

Using Theorem (5.11), we deduce that

$$\langle \ell_2, w \rangle = \langle L, w \rangle, \quad \forall w \in \mathcal{H} \tag{5.87}$$

with

$$\langle L, w \rangle =$$
$$-\int_{Q(V)} \Big\{ \big[ \partial_t(D\,Z_t \cdot \tilde{u}) + D(D\,Z_t \cdot \tilde{u}) \cdot \tilde{u} + D(D\,Z_t \cdot \tilde{u}) \cdot V + D(D\,V \cdot Z_t) \cdot \tilde{u}$$

$$+ D\,W \cdot \tilde{u} \big] \cdot w - [\partial_t \tilde{u} + D\,\tilde{u} \cdot V + D\,\tilde{u} \cdot \tilde{u} + D\,V \cdot \tilde{u}] \cdot (D\,Z_t \cdot w)$$

$$- \nu\, D(D\,Z_t \cdot \tilde{u}) \cdot\cdot\, D\,w + \nu(D\,\tilde{u} \cdot D\,Z_t) \cdot\cdot\, D\,w - \nu\, D\,\tilde{u} \cdot\cdot\, D(D\,Z_t \cdot w)$$

$$+ \nu\, D\,\tilde{u} \cdot\cdot (D\,w \cdot D\,Z_t) + [-\partial_t W - D\,W \cdot V - D\,V \cdot W + \nu \Delta W] \cdot w$$

$$- (D\,[\partial_t V + D\,V \cdot V - \nu \Delta V] \cdot Z_t) \cdot w - [\partial_t V + D\,V \cdot V - \nu \Delta V] \cdot (D\,Z_t \cdot w) \Big\}$$

The sequence

$$[-\partial_t(\mathrm{D}\,Z_t \cdot \tilde{u}) - \mathrm{D}(\mathrm{D}\,Z_t \cdot \tilde{u}) \cdot \tilde{u} - \mathrm{D}(\mathrm{D}\,Z_t \cdot \tilde{u}) \cdot V - \mathrm{D}\,W \cdot \tilde{u}] \cdot w$$
$$- \nu\,\mathrm{D}(\mathrm{D}\,Z_t \cdot \tilde{u}) \cdot\cdot \mathrm{D}\,w + [-\partial_t W - \mathrm{D}\,W \cdot V - \mathrm{D}\,V \cdot W + \nu\Delta W] \cdot w$$

cancels and it remains the following terms,

$$\int_{Q(V)} \big\{ [\partial_t(\mathrm{D}\,\tilde{u} \cdot Z_t) + \mathrm{D}(\mathrm{D}\,\tilde{u} \cdot Z_t) \cdot \tilde{u} + \mathrm{D}\,\tilde{u} \cdot (\mathrm{D}\,\tilde{u} \cdot Z_t) + \mathrm{D}(\mathrm{D}\,\tilde{u} \cdot Z_t) \cdot V$$

$$+ \mathrm{D}\,V \cdot (\mathrm{D}\,\tilde{u} \cdot Z_t)] \cdot w - [\mathrm{D}\,\tilde{u} \cdot (\mathrm{D}\,Z_t \cdot \tilde{u}) + \mathrm{D}\,V \cdot (\mathrm{D}\,Z_t \cdot \tilde{u}) - \mathrm{D}(\mathrm{D}\,V \cdot Z_t) \cdot \tilde{u}] \cdot w$$

$$+ [\partial_t \tilde{u} + \mathrm{D}\,\tilde{u} \cdot V + \mathrm{D}\,\tilde{u} \cdot \tilde{u} + \mathrm{D}\,V \cdot \tilde{u}] \cdot (\mathrm{D}\,Z_t \cdot w) - \mathrm{D}\,\tilde{u} \cdot W \cdot w$$

$$+ \nu\,\mathrm{D}(\mathrm{D}\,\tilde{u} \cdot Z_t) \cdot\cdot \mathrm{D}\,w - \nu(\mathrm{D}\,\tilde{u} \cdot \mathrm{D}\,Z_t) \cdot\cdot \mathrm{D}\,w + \nu\,\mathrm{D}\,\tilde{u} \cdot\cdot \mathrm{D}(\mathrm{D}\,Z_t \cdot w) - \nu\,\mathrm{D}\,\tilde{u} \cdot\cdot (\mathrm{D}\,w \cdot \mathrm{D}\,Z_t)$$

$$+ (\mathrm{D}\,[\partial_t V + \mathrm{D}\,V \cdot V - \nu\Delta V] \cdot Z_t) \cdot w + [\partial_t V + \mathrm{D}\,V \cdot V - \nu\Delta V] \cdot (\mathrm{D}\,Z_t \cdot w) \big\} = 0$$

$$\square$$

**REMARK 5.5**    If we set $\tilde{u} = u - V$, we can obtain an identity only involving the 4-uplet $(u, Z_t, V, W)$. $\quad\square$

### 5.5.5    Adjoint system and cost function shape derivative

We are now coming back to the original problem of computing the gradient of the cost function $j(V)$. Let us first state a differentiability property,

**PROPOSITION 5.7**
*For $\Omega_0$ of class $\mathcal{C}^2$, the functional $j(V)$ is Gâteaux differentiable at point $V \in \mathcal{U}$ and its directional derivative has the following expression,*

$$\langle j'(V), W \rangle = \int_{Q(V)} \alpha\,u(V) \cdot u'(V) \cdot W$$

$$+ \int_{\Sigma(V)} \left[ \gamma\,V \cdot W + \frac{1}{2}(\alpha + \gamma\,H)|V|^2 \langle Z_t, n \rangle \right], \quad \forall W \in \mathcal{U} \qquad (5.88)$$

*where $u'(V) \cdot W$ is the solution of the shape derivative system (Eq. (5.83)).*

**PROOF**    We recall that

$$j(V) = \frac{\alpha}{2} \int_0^T \int_{\Omega_t(V)} |u(V)|^2 + \frac{\gamma}{2} \int_0^T \int_{\Gamma_t(V)} |V|^2 \qquad (5.89)$$

The differentiability property is an easy consequence of the differentiability of $J_V(u)$ with respect to $(u, V)$ and the shape differentiability of $u(V)$ with respect to $V$. The expression of the directional derivative is a direct consequence of Th. (5.4) and Th. (5.5). □

Using the fluid adjoint state and the adjoint transverse field, it is possible to identify the gradient distribution associated to the functional $j(V)$,

**THEOREM 5.13**
*For $V \in \mathcal{U}$ and $\Omega_0$ of class $\mathcal{C}^2$, the functional $j(V)$ possesses a gradient $\nabla j(V)$ which is supported on the moving boundary $\Gamma_t(V)$ and can be represented by the following expression,*

$$\nabla j(V) = -\lambda \, n - \sigma(\varphi, \pi) \cdot n + \gamma \, V \tag{5.90}$$

*where $(\varphi, \pi)$ stands for the adjoint fluid state solution of the following system,*

$$\begin{cases} -\partial_t \varphi - D \varphi \cdot u + {}^* D u \cdot \varphi - \nu \Delta \varphi + \nabla \pi = \alpha \, u, & Q(V) \\ \operatorname{div}(\varphi) = 0, & Q(V) \\ \varphi = 0, & \Sigma(V) \\ \varphi(T) = 0, & \Omega_T \end{cases} \tag{5.91}$$

*and $\lambda$ is the adjoint transverse boundary field and is the solution of the tangential dynamical system,*

$$\begin{cases} -\partial_t \lambda - \nabla_\Gamma \, \lambda \cdot V = f, & (0, T) \\ \lambda(T) = 0, & \Gamma_T(V) \end{cases} \tag{5.92}$$

*with $f = -(\sigma(\varphi, \pi) \cdot n) \cdot (D V \cdot n - D u \cdot n) + \frac{1}{2}(\alpha + \gamma \, H)|V|^2$.*

**PROOF**  We need the following identity,

$$\int_0^T \int_{\Omega_t(V)} [\partial_t u' + D u' \cdot u + D u \cdot u' - \nu \Delta u' + \nabla p'] v - \int_0^T \int_{\Omega_t(V)} q \operatorname{div} u'$$

$$= \int_0^T \int_{\Omega_t(V)} [-\partial_t v - D v \cdot u + {}^* D u \cdot v - \nu \Delta v + \nabla q] u' - \int_0^T \int_{\Omega_t(V)} p' \operatorname{div} v$$

$$+ \int_0^T \int_{\Gamma_t(V)} [p' \, n \cdot v - \nu v \cdot \partial_n u' + \nu u' \cdot \partial_n v - u' \cdot q \, n] \tag{5.93}$$

We define $(\varphi, \pi)$ to be the solution of the adjoint system (Eq. (5.8)), and we set $(v, q) = (\varphi, \pi)$ in Eq. (5.93). We get

$$\int_0^T \int_{\Omega_t(V)} \alpha \, u \cdot u' = -\int_0^T \int_{\Gamma_t(V)} \langle \sigma(\varphi, \pi) \cdot n, u' \rangle \tag{5.94}$$

We use the boundary condition on $\Gamma_t(V)$ for the linearized state $u'$, i.e.,

$$u' = W + (\mathrm{D}\,V \cdot n - \mathrm{D}\,u \cdot n)\langle Z_t, n \rangle, \quad \text{on } \Gamma_t(V) \tag{5.95}$$

Thus,

$$
\begin{aligned}
\langle j'(V), W \rangle = \int_0^T \int_{\Gamma_t(V)} & \big[ -(\sigma(\varphi, \pi) \cdot n) \cdot (\mathrm{D}\,V \cdot n - \mathrm{D}\,u \cdot n) \\
& + \frac{1}{2}(\alpha + \gamma\,H)|V|^2 \big]\langle Z_t, n \rangle + \int_0^T \int_{\Gamma_t(V)} [-\sigma(\varphi, \pi) \cdot n + \gamma V] \cdot W
\end{aligned}
$$

Then we use Th. (5.8) with

$$E = -(\sigma(\varphi, \pi) \cdot n) \cdot (\mathrm{D}\,V \cdot n - \mathrm{D}\,u \cdot n) + \frac{1}{2}(\alpha + \gamma\,H)|V|^2$$

and we get the correct result. $\qquad\qquad \Box$

**REMARK 5.6**    Actually, we have $\pi(\mathrm{D}(V - u) \cdot n) \cdot n = \pi \operatorname{div} V|_{\Gamma_t}$ using the formula,

$$(\mathrm{D}(V - u) \cdot n) \cdot n = \operatorname{div}(V - u)|_{\Gamma_t} - \operatorname{div}_\Gamma(V - u)$$

and the fact that $V - u = 0$ on $\Gamma_t$. Furthermore, we have considered free divergence field $V$; then this term is null and we get that

$$f = -\nu(\mathrm{D}\,\varphi \cdot n) \cdot (\mathrm{D}\,V \cdot n - \mathrm{D}\,u \cdot n) + \frac{1}{2}(\alpha + \gamma\,H)|V|^2$$

$$\Box$$

## 5.6    Min-Max and function space parametrization

In the previous section, we have been using the differentiability of the fluid state with respect to the Eulerian velocity $V$ as a sufficient condition in order to derive first-order optimality conditions, involving the adjoint of the linearized state. Actually, the tedious computation of the state differentiability is not necessary in many cases, and even if the state is not differentiable, it can happen that first-order optimality conditions still hold. This is a consequence of a fundamental result in optimal control theory, the so-called Maximum Principle.

Avoiding the differentiation of the state equations with respect to the design variable $V$ is of great interest for shape optimization problems, especially if we deal with a moving domain system.

In this section, we are concerned with the function space parametrization, which consists in transporting the different quantities defined in the perturbed moving domain back into the reference moving domain that does not depend on the perturbation parameter. Thus, differential calculus can be performed since the functions involved are defined in a fixed domain with respect to the perturbations.

In the first part, we define the saddle point formulation of the fluid state equations and the Lagrangian functional associated to the cost functional. Then, we perform a sensitivity analysis of the Lagrangian thanks to the transverse field and the fundamental Min-Max principle. This allows us to derive the expression of the cost function gradient involving the fluid and transverse field adjoints.

### 5.6.1   Saddle point formulation of the fluid state system

In the next paragraphs, we shall describe how to build an appropriate Lagrangian functional that can take into account all the constraints imposed by the mechanical problem, such as the divergence free condition or the non-homogeneous Dirichlet boundary conditions.

**Null divergence condition**

The divergence free condition coming from the fact that the fluid has an homogeneous density and evolves as an incompressible flow is difficult to impose on the mathematical and numerical point of view. We suggest at least 3 possible choices to handle this condition in our Min-Max formulation,

1. It can be taken into account in the state and multipliers spaces. In this case, the divergence free condition must be invariant with respect to the use of transport map during the derivation of optimality condition for the Lagrangian functional. This reduces the choice of appropriate maps and indeed the ALE map $T_t$ does not satisfy this invariance condition. It is well known that the Piola transform does preserve the divergence quantity. Indeed we have the following property:

**LEMMA 5.16 [13]**

*The Piola transform*

$$P_t : H_0^1(\mathrm{div}, \Omega_0) \longrightarrow H_0^1(\mathrm{div}, \Omega_t)$$
$$\varphi \longmapsto ((J_t)^{-1} \, \mathrm{D} \, T_t \cdot \varphi) \circ T_t^{-1} \tag{5.96}$$

*is an isomorphism.*

This new transform introduces additional mathematical and computational efforts, but it seems to be the best approach in order to get rigorous mathematical justifications of the Lagrangian framework in the context of non-cylindrical and free boundary problems.

2. One way to avoid the use of this transform is the penalization of the divergence free condition inside the Navier-Stokes system. Let $\varepsilon > 0$ be a small parameter. We may consider the new penalized system :

$$\begin{cases} \partial_t u + \mathrm{D}\, u \cdot u - \nu \Delta u - \dfrac{1}{\varepsilon} \nabla(\mathrm{div}\, u) = 0, & Q \\ u = V, & \Sigma \\ u(t = 0) = u_0, & \Omega_0 \times \mathbb{R}^2 \end{cases} \qquad (5.97)$$

with $\sigma^\varepsilon(u) = \frac{1}{\varepsilon} \mathrm{div}(u)\, \mathrm{I} + \nu(\mathrm{D}\, u + {}^*\mathrm{D}\, u)$.

We may work with such a modified system, derive the optimality conditions of the penalized Lagrangian functional and finally perform an asymptotic analysis on the adjoint and primal system. For the time being, it is not clear if such a procedure may actually work, since even for non-moving Navier-Stokes problem, the convergence of the penalized adjoint is not established.

3. A third choice is to include the divergence free condition directly into the Lagrangian functional thanks to a multiplier that may play the role of the adjoint variable associated to the primal pressure variable. This leads in a certain sense to a saddle point formulation or mixed formulation of the Navier-Stokes system. It is well known that the well-posedness of such formulations is only established for the Stokes system, and that the Navier-Stokes suffers from a lack of convexity while taken into account in the Lagrangian functional. But still, it seems to be the easiest way, at least on the mathematical computation point of view, to deal with divergence free conditions in a sensitivity analysis of the moving system. In the sequel, we adopt such a strategy, keeping in minds, its lack of rigorous mathematical justification.

**Non-homogeneous boundary conditions**

The Navier-Stokes system (Eq. (5.2)) involves an essential non-homogeneous Dirichlet boundary condition,

$$u = V, \quad \text{on } \Gamma_t(V) \qquad (5.98)$$

Again, there exists different methods to take into account this boundary condition in a Min-Max formulation,

1. We can use a lifting of the boundary conditions inside the fluid domain and define a change of variable inside the coupled system, as done in Section (5.5). It has the drawback to put additional terms inside the Lagrangian functionals and to impose more regularity on the boundary conditions.

2. We can use a very weak formulation of the state equation, consisting in totally transposing the Laplacian operator,

$$\int_{\Omega_t} -\nu \Delta u \cdot \phi = \int_{\Omega_t} -\nu \Delta \phi \cdot u + \int_{\Gamma_t} \nu \left[ u \cdot \partial_n \phi - \phi \cdot \partial_n u \right] d\Gamma \quad (5.99)$$

Then we shall substitute inside this identity the desirable boundary conditions. We recover the boundary constraints in performing an integration by parts in the optimality conditions corresponding to the sensitivity with respect to the multipliers. This procedure has been already used in [135] to perform shape optimization problems for elliptic equations using Min-Max principles.

**REMARK 5.7**    This method has been popularized in [99] as a systematic way to study non-homogeneous linear partial differential equations. These formulations are usually called very weak formulations or transposed formulations. We shall notice that these methods are still valid in the non-linear case to obtain regularity or existence results. We refer to [4] for recent applications to the Navier-Stokes system. ⧠

## Fluid state operator

In this section we shall summarize the different options that we have chosen for the Lagrangian framework and define the variational state operator constraint. In the sequel, we will need to define precise state and multiplier spaces in order to endow our problem with a Lagrangian functional framework.

Following the existence result stated previously, we introduce the fluid state spaces:

$$X(\Omega_t) \stackrel{\text{def}}{=} \left\{ u \in H^2(0, T; (H^2(\Omega_t))^d \cap (H^1(\Omega_t))^d) \right\}$$

$$Z \stackrel{\text{def}}{=} \left\{ p \in H^1(0, T; (H^1(D))^d) \right\}$$

We also need test function spaces that will be useful to define Lagrange multipliers:

$$Y(\Omega_t) \stackrel{\text{def}}{=} \left\{ v \in L^2(0, T; (H^2(\Omega_t))^2 \cap (H_0^1(\Omega_t))^d) \right\}$$
$$Q \stackrel{\text{def}}{=} \left\{ q \in H^1(0, T; (H^1(D))^2) \right\}$$

We define the fluid weak state operator,

$$e_V : X \times Z \longrightarrow (Y \times Q)^*$$

whose action is defined by :

$$\langle e_V(u,p),(v,q)\rangle = \int_0^T \int_{\Omega_t(V)} \left[ -u \cdot \partial_t v + (\mathrm{D}\,u \cdot u) \cdot v - \nu u \cdot \Delta v \right.$$

$$\left. + u \cdot \nabla q - p \,\mathrm{div}\,v \right] + \int_0^T \int_{\Gamma_t(V)} V \cdot (\sigma(v,q) \cdot n)$$

$$+ \int_{\Omega_T} u(T) \cdot v(T) - \int_{\Omega_0} u_0 \cdot v(t=0), \quad \forall (v,q) \in Y \times Q$$

**Min-Max problem**

In this section, we introduce the Lagrangian functional associated with Eq. (5.2) and Eq. (5.3) :

$$\mathcal{L}_V(u,p;v,q) \stackrel{\text{def}}{=} J_V(u,p) - \langle e_V(u,p),(v,q)\rangle \tag{5.100}$$

with

$$J_V(u,p) = \frac{\alpha}{2} \int_0^T \int_{\Omega_t(V)} |u|^2 + \frac{\gamma}{2} \int_0^T \int_{\Gamma_t(V)} |V|^2 \tag{5.101}$$

Using this functional, the optimal control problem Eq. (5.3) can be put in the following form:

$$\min_{V \in \mathcal{U}} \quad \min_{(u,p) \in X(\Omega_t(V)) \times Z} \quad \max_{(v,q) \in Y(\Omega_t(V)) \times Q} \quad \mathcal{L}_V(u,p;v,q) \tag{5.102}$$

By using the Min-Max framework, we avoid the computation of the state derivative with respect to $V$. First-order optimality conditions will furnish the gradient of the original cost functional using the solution of an adjoint problem.

Let us first study the saddle point problem,

$$\min_{(u,p) \in X \times Z} \quad \max_{(v,q) \in Y \times Q} \quad \mathcal{L}_V(u,p;v,q) \tag{5.103}$$

**Optimality Conditions**

In this section, we are interested in establishing the first order optimality condition for problem Eq. (5.103), better known as Karusch-Kuhn-Tucker optimality conditions. This step is crucial, because it leads to the formulation of the adjoint problem satisfied by the Lagrange multipliers $(\varphi(V), \pi(V))$. The KKT system will have the following structure :

$$\partial_{(v,q)} \mathcal{L}_V(u,p;v,q) \cdot (\delta v, \delta q) = 0,$$
$$\forall (\delta v, \delta q) \in Y \times Q \rightarrow \quad \text{State Equations}$$
$$\partial_{(u,p)} \mathcal{L}_V(u,p;v,q) \cdot (\delta u, \delta p) = 0,$$
$$\forall (\delta u, \delta p) \in X \times Z \rightarrow \quad \text{Adjoint Equations}$$

**LEMMA 5.17**
For $V \in \mathcal{U}$, $(p, v, q) \in Z \times Y \times Q$, $\mathcal{L}_V(u, p; v, q)$ is differentiable with respect to $u \in X$ and we have

$$\langle \partial_u \mathcal{L}_V(u, p; v, q), \delta u \rangle =$$

$$\int_0^T \int_{\Omega_t(V)} [\alpha\, u \cdot \delta u + \delta u \cdot \partial_t v - [\mathrm{D}\, \delta u \cdot u + \mathrm{D}\, u \cdot \delta u] \cdot v + \nu \delta u \cdot \Delta v - \delta u \cdot \nabla q]$$

$$+ \int_{\Omega_T} \delta u(T) \cdot v(T), \quad \forall \delta u \in X$$

In order to obtain a strong formulation of the fluid adjoint problem, we perform some integration by parts :

**LEMMA 5.18**

$$\int_{Q(V)} (\mathrm{D}\, \delta u \cdot u) \cdot v = - \int_{Q(V)} [\mathrm{D}\, v \cdot u + \mathrm{div}(u) \cdot v] \cdot \delta u + \int_{\Sigma(V)} (\delta u \cdot v)(u \cdot n)$$

It leads to the following identity :

$$\langle \partial_{\hat{u}} \mathcal{L}_V(u, p; \varphi, \pi), \delta u \rangle =$$

$$- \int_{Q(V)} [-\partial_t \varphi + (^*Du) \cdot \varphi - (D\varphi) \cdot u - \mathrm{div}(u) \cdot \varphi - \nu \Delta \varphi + \nabla \pi - \alpha u] \cdot \delta u$$

$$- \nu \int_{\Sigma(V)} (\partial_n \delta u) \cdot \varphi - \int_{\Omega_T} \varphi(T) \cdot \delta u(T)$$

**LEMMA 5.19**
For $V \in \mathcal{U}$, $(u, v, q) \in X \times Y \times Q$, $\mathcal{L}_V(u, p; v, q)$ is differentiable with respect to $p \in Z$ and we have

$$\langle \partial_p \mathcal{L}_V(u, p; \varphi, \pi), \delta p \rangle = \int_0^T \int_{\Omega_t} (\delta p)\, \mathrm{div}\, \varphi, \quad \forall \delta p \in Z \qquad (5.104)$$

This leads to the following fluid adjoint strong formulation,

$$\begin{cases} -\partial_t \varphi - \mathrm{D}\, \varphi \cdot u + (^*Du) \cdot \varphi - \nu \Delta \varphi + \nabla q = \alpha u, & Q(V) \\ \mathrm{div}(\varphi) = 0, & Q(V) \\ \varphi = 0, & \Sigma(V) \\ \varphi(T) = 0, & \Omega_T \end{cases} \qquad (5.105)$$

**REMARK 5.8**  Existence and regularity results for the linearized Navier-Stokes adjoint problem can be found in [1, 86] for the 2D case. These results can be easily adapted for the moving domain case. There is a lack of results for the 3D case. ⌷

### 5.6.2 Function space parametrization

To compute the first-order derivative of $j(V)$, we perturb the moving domain $\Omega_t(V)$ by a velocity field $W$ which generates the family of transformation $T_t^\rho \stackrel{\text{def}}{=} T_t(V+\rho W)$, with $\rho \geq 0$ and the family of domains and their boundaries,

$$\Omega_t^\rho \stackrel{\text{def}}{=} T_t(V+\rho W)(\Omega_0)$$
$$\Gamma_t^\rho \stackrel{\text{def}}{=} T_t(V+\rho W)(\Gamma_0)$$

We set

$$g(\rho) = j(V+\rho W) = \min_{(u,p)\in X(\Omega_t^\rho)\times Z}\ \max_{(v,q)\in Y(\Omega_t^\rho)\times Q}\ \mathcal{L}_{(V+\rho W)}(u,p;v,q) \tag{5.106}$$

The objective of this section is to compute the following derivative :

$$\lim_{\rho\searrow 0}\frac{1}{\rho}(g(\rho)-g(0)) \tag{5.107}$$

We need a theorem that would give the derivative of a Min-Max function with respect to a real parameter $\rho \geq 0$. In our case, it is not trivial since the state and multiplier spaces $X(\Omega_t^\rho) \times Y(\Omega_t^\rho)$ depend on the perturbation parameter $\rho$. This point can be solved using particular parametrization of the functional spaces. To this aim, we use the transverse map introduced in Section (5.3),

$$\mathcal{T}_\rho^t : \overline{\Omega_t} \longrightarrow \overline{\Omega_t^\rho}$$
$$x \mapsto T_t(V+\rho W)\circ T_t(V)^{-1}$$

and we define the following parametrization,

$$X(\Omega_t^\rho) = \{u\circ(T_t^\rho)^{-1},\quad u\in X(\Omega_t(V))\} \tag{5.108}$$
$$Y(\Omega_t^\rho) = \{v\circ(T_t^\rho)^{-1},\quad u\in Y(\Omega_t(V))\} \tag{5.109}$$

This parametrization does not affect the value of the saddle point functional $g(\rho)$, but changes the parametrization of the Lagrangian functional,

$$g(\rho) = j(V+\rho W) =$$
$$\min_{(u,p)\,\in\,X(\Omega_t)\,\times\,Z}\ \max_{(v,q)\,\in\,Y(\Omega_t)\,\times\,Q}\ \mathcal{L}_{(V+\rho W)}(u\circ R_\rho^t,p;v\circ R_\rho^t,q) \tag{5.110}$$

with $R_\rho^t \stackrel{\text{def}}{=} (T_t^\rho)^{-1}$.

We set

$$\mathcal{L}_{V,W}^\rho (u, p; v, q) = J_{V+\rho W}(u \circ R_\rho^t, p) =$$

$$- \int_0^T \int_{\Omega_t^\rho} \left[ - u \circ R_\rho^t \cdot \partial_t(v \circ R_\rho^t) + (D u \circ R_\rho^t \cdot u \circ R_\rho^t) \cdot v \circ R_\rho^t \right.$$

$$\left. - \nu u \circ R_\rho^t \cdot \Delta(v \circ R_\rho^t) + u \circ R_\rho^t \cdot \nabla q - p \, \mathrm{div}(v \circ R_\rho^t) \right]$$

$$- \int_0^T \int_{\Gamma_t^\rho} (V + \rho W) \cdot (\sigma(v \circ R_\rho^t, q) \cdot n^\rho)$$

$$- \int_{\Omega_T} u(T) \cdot v(T) + \int_{\Omega_0} u_0 \cdot v(t = 0)$$

$$\forall (v, q) \in Y(\Omega_t(V)) \times Q$$

where $n^\rho$ stands for unit exterior normal of the perturbed boundary $\Gamma_t^\rho$.

### 5.6.3 Differentiability of the saddle point problem

In this section, we first state a general theorem concerning the differentiability of a Min-Max problem with respect to a scalar parameter. Then we apply it to our case of study. Finally, using a fundamental identity, we are able to express the gradient $\nabla j(V)$ as stated in the main theorem of this chapter.

**General theorem**

We a consider a functional,

$$G : [0, \rho_0] \times X \times Y \to \mathcal{R} \tag{5.111}$$

with $\rho_0 \geq 0$ and two topological spaces $(X, Y)$. For each $\rho \in I \stackrel{\text{def}}{=} [0, \rho_0]$, we define

$$g(\rho) = \inf_{x \in X} \sup_{y \in Y} G(\rho, x, y) \tag{5.112}$$

and the sets

$$X(\rho) = \left\{ x^\rho \in X, \quad \sup_{y \in Y} G(\rho, x^\rho, y) = g(\rho) \right\} \tag{5.113}$$

$$Y(\rho, x) = \left\{ y^\rho \in Y, \quad G(\rho, x, y^\rho) = \sup_{y \in Y} G(\rho, x, y) \right\} \tag{5.114}$$

In a similar way, we define dual functions and sets,

$$h(\rho) = \sup_{y \in Y} \inf_{x \in X} G(\rho, x, y) \tag{5.115}$$

and the sets

$$Y(\rho) = \left\{ y^\rho \in Y, \quad \inf_{x \in X} \ G(\rho, x, y^\rho) = h(\rho) \right\} \tag{5.116}$$

$$X(\rho, y) = \left\{ x^\rho \in X, \quad G(\rho, x^\rho, y) = \inf_{x \in X} \ G(\rho, x, y) \right\} \tag{5.117}$$

Finally we define the sets of saddle points,

$$S(\rho) = \{(x, y) \in X \times Y, \quad g(\rho) = G(\rho, x, y) = h(\rho)\} \tag{5.118}$$

**THEOREM 5.14 [40]**
*Assume that the following hypothesis hold,*

*(H1) The set $S(\rho) \neq \emptyset$, $\rho \in I$.*

*(H2) The partial derivative $\partial_\rho G(\rho, x, y)$ exists in $I$ for all*

$$(x, y) \in \left[ \bigcup_{\rho \in I} X(\rho) \times Y(0) \right] \bigcup \left[ X(0) \times \bigcup_{\rho \in I} Y(\rho) \right]$$

*(H3) There exists a topology $\mathcal{T}_X$ on $X$ such that,*
*for any sequence $(\rho_n)_{n \geq 0} \in I$ with $\lim_{n \nearrow \infty} \rho_n = 0$, there exists $x^0 \in X(0)$*
*and a subsequence $\rho_{n_k}$ and for each $k \geq 1$, there exists $x_{n_k} \in X(\rho_{n_k})$*
*such that,*

*i) $\lim_{n \nearrow \infty} x_{n_k} = x^0$ for the $\mathcal{T}_X$ topology,*

*ii)*

$$\liminf_{(\rho, k) \searrow \nearrow (0, \infty)} \partial_\rho G(\rho, x_{n_k}, y) \geq \partial_\rho G(0, x^0, y)$$

*$\forall y \in Y(0)$.*

*(H4) There exists a topology $\mathcal{T}_Y$ on $Y$ such that,*
*for any sequence $(\rho_n)_{n \geq 0} \in I$ with $\lim_{n \nearrow \infty} \rho_n = 0$, there exists $y^0 \in Y(0)$*
*and a subsequence $\rho_{n_k}$ and for each $k \geq 1$, there exists $y_{n_k} \in Y(\rho_{n_k})$*
*such that,*

*i) $\lim_{n \nearrow \infty} y_{n_k} = y^0$ for the $\mathcal{T}_Y$ topology,*

*ii)*

$$\liminf_{(\rho, k) \searrow \nearrow (0, \infty)} \partial_\rho G(\rho, x, y_{n_k}) \leq \partial_\rho G(0, x, y^0)$$

*$\forall x \in X(0)$.*

*Then there exists $(x^0, y^0) \in X(0) \times Y(0)$ such that*

$$dg(0) = \lim_{\rho \searrow 0} \frac{g(\rho) - g(0)}{\rho} = \inf_{x \in X(0)} \sup_{y \in Y(0)} \partial_\rho G(0, x, y)$$

$$= \partial_\rho G(0, x^0, y^0) = \sup_{y \in Y(0)} \inf_{x \in X(0)} \partial_\rho G(0, x, y)$$

$$(5.119)$$

*This means that $(x^0, y^0) \in X(0) \times Y(0)$ is a saddle point of $\partial_\rho G(0, x, y)$.*

## Derivative of the perturbed Lagrangian

Following Th. (5.14), we need to differentiate the perturbed Lagrangian functional $\mathcal{L}(\rho)$. We shall successively differentiate the distributed and the boundary integrals involved in the perturbed Lagrangian:

a) Distributed terms:
  We set

$$G(\rho, .) = \left[ -u \circ \mathcal{R}_\rho^t \cdot \partial_t (v \circ \mathcal{R}_\rho^t) + D(u \circ \mathcal{R}_\rho^t) \cdot (u \circ \mathcal{R}_\rho^t) \cdot v \circ \mathcal{R}_\rho^t \right.$$
$$\left. -\nu(u \circ \mathcal{R}_\rho^t) \cdot \Delta(v \circ \mathcal{R}_\rho^t) + (u \circ \mathcal{R}_\rho^t) \cdot \nabla q - p \operatorname{div}(v \circ \mathcal{R}_\rho^t) \right]$$

with $\mathcal{R}_\rho^t \stackrel{\text{def}}{=} (\mathcal{T}_\rho^t)^{-1}$.
We shall need the following lemmas in order to derivate $G(\rho, .)$ with respect to $\rho$,

### LEMMA 5.20

$$\left( \frac{d\mathcal{T}_\rho^t}{d\rho} \right) \bigg|_{\rho=0} = Z_t$$

$$\left( \frac{d\mathcal{R}_\rho^t}{d\rho} \right) \bigg|_{\rho=0} = -Z_t$$

### LEMMA 5.21

$$\left( \frac{d \left( u \circ \mathcal{R}_\rho^t \right)}{d\rho} \right) \bigg|_{\rho=0} = -D u \cdot Z_t$$

**PROOF**    Using the chain rule we get

$$
\left. \left( \frac{\mathrm{D}\left(u \circ \mathcal{R}_\rho^t\right)}{\mathrm{D}\rho} \right) \right|_{\rho=0} = \left(\mathrm{D}\,u \circ \mathcal{R}_\rho^t\right) \cdot \left. \left( \frac{\mathrm{D}\,\mathcal{R}_\rho^t}{\mathrm{D}\rho} \right) \right|_{\rho=0}
$$

$$
= - \left(\mathrm{D}\,u \circ \mathcal{R}_\rho^t\right) \cdot \mathcal{Z}^t(\rho,.)\big|_{\rho=0}
$$

$$
= - \mathrm{D}\,u \cdot Z_t
$$

☐

**LEMMA 5.22**

*Then, we have the following result*

$$
\partial_\rho G(\rho,.)|_{\rho=0} = [(\mathrm{D}\,u \cdot Z_t) \cdot \partial_t v + u \cdot (\partial_t(\mathrm{D}\,v \cdot Z_t))
$$
$$
- [(\mathrm{D}(\mathrm{D}\,u \cdot Z_t)) \cdot u + \mathrm{D}\,u \cdot (\mathrm{D}\,u \cdot Z_t)] \cdot v - (\mathrm{D}\,u \cdot u) \cdot (\mathrm{D}\,v \cdot Z_t)
$$
$$
+ \nu(\mathrm{D}\,u \cdot Z_t) \cdot \Delta v + \nu u \cdot (\Delta(\mathrm{D}\,v \cdot Z_t)) + p\,\mathrm{div}(\mathrm{D}\,v \cdot Z_t) - (\mathrm{D}\,u \cdot Z_t) \cdot \nabla q]
$$

**PROOF**    It comes easily using definition of $G(\rho,.)$ and Lem. (5.20)-(5.21). ☐

Then we have an expression of the derivative of distributed terms coming from the Lagrangian with respect to $\rho$,

$$
\frac{d}{d\rho}\left(\int_{\Omega_t^\rho} G(\rho,x)dx\right)\bigg|_{\rho=0} = \int_{\Omega_t} [(\mathrm{D}\,u \cdot Z_t) \cdot \partial_t v + u \cdot (\partial_t(\mathrm{D}\,v \cdot Z_t))
$$
$$
- [(\mathrm{D}(\mathrm{D}\,u \cdot Z_t)) \cdot u + \mathrm{D}\,u \cdot (\mathrm{D}\,u \cdot Z_t)] \cdot v - (\mathrm{D}\,u \cdot u) \cdot (\mathrm{D}\,v \cdot Z_t)
$$
$$
+ \nu(\mathrm{D}\,u \cdot Z_t) \cdot \Delta v + \nu u \cdot (\Delta(\mathrm{D}\,v \cdot Z_t)) + p\,\mathrm{div}(\mathrm{D}\,v \cdot Z_t) - (\mathrm{D}\,u \cdot Z_t) \cdot \nabla q]
$$
$$
+ \int_{\Gamma_t} [-u \cdot \partial_t v + (Du \cdot u) \cdot v - \nu u \cdot \Delta v + u \cdot \nabla q - p\,\mathrm{div}(v)] \langle Z_t, n \rangle
$$

b) Boundary terms :
We must now take into account the terms coming from the moving boundary $\Gamma_t^\rho$. Then we set

$$
\phi(\rho,.) = (V + \rho W) \cdot \left[-q\,\mathrm{I} + \nu\,\mathrm{D}(v \circ \mathcal{R}_\rho^t)\right] \cdot n^\rho
$$
$$
= E(\rho) \cdot n^\rho \tag{5.120}
$$

Since $\phi(\rho,.)$ is defined on the boundary $\Gamma_t^\rho$, we need some extra identities corresponding to boundary shape derivates of terms involved in $\phi(\rho,.)$.

**LEMMA 5.23 [53]**

$$\partial_\rho n^\rho|_{\rho=0} = n'_\Gamma = -\nabla_\Gamma(Z_t \cdot n)$$

**LEMMA 5.24**

$$\frac{d}{d\rho}\left(\int_{\Gamma_t^\rho} \langle E(\rho), n^\rho \rangle d\Gamma\right)\Bigg|_{\rho=0} = \int_{\Gamma_t} \langle E'|_{\Gamma_t}, n \rangle + (\operatorname{div} E)\langle Z_t, n \rangle$$

$$= \int_{\Gamma_t} \langle E'_{\Gamma_t}, n \rangle + (\operatorname{div}_\Gamma E)\langle Z_t, n \rangle$$

$$(5.121)$$

**PROOF**   First, we use that

$$\int_{\Gamma_t^\rho} \langle E(\rho), n^\rho \rangle = \int_{\Omega_t^\rho} \operatorname{div} E(\rho)$$

Then we derive this quantity using Th. (5.4),

$$\frac{d}{d\rho}\left(\int_{\Omega_t^\rho} \operatorname{div} E(\rho)\right)\Bigg|_{\rho=0} = \int_{\Omega_t} \operatorname{div} E' + \int_{\Gamma_t} (\operatorname{div} E)\langle Z_t, n \rangle$$

We conclude using $\int_{\Omega_t} \operatorname{div} E' = \int_{\Gamma_t} \langle E', n \rangle$. For the second identity, using the Th. (5.5), we have

$$\frac{d}{d\rho}\left(\int_{\Gamma_t^\rho} \langle E(\rho), (n \circ \mathcal{R}_\rho^t) \rangle d\Gamma\right)\Bigg|_{\rho=0} = \int_{\Gamma_t} \langle E'_\Gamma, n \rangle + \langle E, n'_\Gamma \rangle + H\langle E, n \rangle\langle Z_t, n \rangle$$

Using Lem. (5.23), we get

$$\frac{d}{d\rho}\left(\int_{\Gamma_t^\rho} \langle E(\rho), (n \circ \mathcal{R}_\rho^t) \rangle d\Gamma\right)\Bigg|_{\rho=0} = \int_{\Gamma_t} \langle E'_\Gamma, n \rangle - \langle E, \nabla_\Gamma(Z_t \cdot n) \rangle$$

$$+ H\langle E, n \rangle\langle Z_t, n \rangle$$

Then using the tangential Stokes identity from Lem. (5.5), we obtain the correct result.   ☐

Hence, we only need to compute the quantity $E'_\Gamma$. To this end, we need the following identities,

**LEMMA 5.25**

$$\left(v \circ \mathcal{R}_\rho^t\right)'_\Gamma\Big|_{\rho=0} = -\,\mathrm{D}_\Gamma\, v \cdot Z_t$$

**PROOF**  Since $\left(v \circ \mathcal{R}_\rho^t\right)^{\cdot}\big|_{\rho=0} = \partial_\rho \left(v \circ \mathcal{R}_\rho^t \circ T_\rho^t\right)\big|_{\rho=0} = \partial_\rho v|_{\rho=0} = 0.$

∎

**LEMMA 5.26**

$$\left(\mathrm{D}\left(v \circ \mathcal{R}_\rho^t\right)\right)'_\Gamma\Big|_{\rho=0} = -\,\mathrm{D}\,v \cdot \mathrm{D}\,Z_t - \left(\mathrm{D}_\Gamma(\mathrm{D}\,v)\right) \cdot Z_t$$

**PROOF**  By definition we have

$$
\begin{aligned}
\left(\mathrm{D}\left(v \circ \mathcal{R}_\rho^t\right)\right)'_\Gamma\Big|_{\rho=0} &= \left(\mathrm{D}\left(v \circ \mathcal{R}_\rho^t\right)\right)^{\cdot}_\Gamma\Big|_{\rho=0} - \mathrm{D}_\Gamma\left(\mathrm{D}(v \circ \mathcal{R}_\rho^t)\right) \cdot \mathcal{Z}_\rho^t\Big|_{\rho=0} \\
&= \partial_\rho\left(\mathrm{D}\left(v \circ \mathcal{R}_\rho^t\right) \circ T_\rho^t\right)\Big|_{\rho=0} - \left(\mathrm{D}_\Gamma(\mathrm{D}\,v)\right) \cdot Z_t \\
&= \partial_\rho\left[\left((\mathrm{D}\,v) \circ \mathcal{R}_\rho^t \cdot \mathrm{D}\,\mathcal{R}_\rho^t\right) \circ T_\rho^t\right]\Big|_{\rho=0} - \left(\mathrm{D}_\Gamma(\mathrm{D}\,v)\right) \cdot Z_t \\
&= \partial_\rho\left[(\mathrm{D}\,v) \cdot \mathrm{D}\,\mathcal{R}_\rho^t \circ T_\rho^t\right]\Big|_{\rho=0} - \left(\mathrm{D}_\Gamma(\mathrm{D}\,v)\right) \cdot Z_t \\
&= -\,\mathrm{D}\,v \cdot \mathrm{D}\,Z_t + \left[\mathrm{D}\,v \cdot \mathrm{D}(\mathrm{D}\,\mathcal{R}_\rho^t) \cdot \partial_\rho(T_\rho^t)\right]\Big|_{\rho=0} \\
&\quad - \left(\mathrm{D}_\Gamma(\mathrm{D}\,v)\right) \cdot Z_t \\
&= -\,\mathrm{D}\,v \cdot \mathrm{D}\,Z_t - \left(\mathrm{D}_\Gamma(\mathrm{D}\,v)\right) \cdot Z_t
\end{aligned}
$$

∎

Using these results, we can state the following :

**LEMMA 5.27**

$$E'_\Gamma = W \cdot [-q\,\mathrm{I} + \nu\,\mathrm{D}\,v] + \nu\,V \cdot [-\,\mathrm{D}\,v \cdot \mathrm{D}\,Z_t - \mathrm{D}_\Gamma(\mathrm{D}\,v) \cdot Z_t] \quad (5.122)$$

This means that we have

$$\frac{d}{d\rho}\left(\int_{\Gamma_t^\rho} \phi(\rho, x)d\Gamma\right)\Big|_{\rho=0} = \int_{\Gamma_t(V)} W \cdot [-q\,n + \nu\,\mathrm{D}\,v \cdot n]$$

$$+ \nu\,V \cdot [-(\mathrm{D}\,v \cdot \mathrm{D}\,Z_t) \cdot n - (\mathrm{D}_\Gamma(\mathrm{D}\,v) \cdot Z_t) \cdot n] + \mathrm{div}_\Gamma(V \cdot [-q\,\mathrm{I} + \nu\,\mathrm{D}\,v])\langle Z_t, n\rangle$$

We have also,

**LEMMA 5.28**

$$E'|_\Gamma = W \cdot [-q\,\mathrm{I} + \nu\,\mathrm{D}\,v] - \nu\,V \cdot [\mathrm{D}(\mathrm{D}\,v) \cdot Z_t] \quad (5.123)$$

Hence, we have

$$\frac{d}{d\rho}\left(\int_{\Gamma_t^\rho}\phi(\rho,x)d\Gamma\right)\Bigg|_{\rho=0} = \int_{\Gamma_t(V)} W\cdot[-q\,n+\nu\,\mathrm{D}\,v\cdot n]$$

$$-\nu\,V\cdot[\mathrm{D}(\mathrm{D}\,v)\cdot Z_t\cdot n]+\mathrm{div}(V\cdot[-q\,\mathrm{I}+\nu^*\,\mathrm{D}\,v])\langle Z_t,n\rangle$$

**REMARK 5.9**    We recall that

$$\int_{\Gamma_t} V\cdot(\mathrm{D}\,v\cdot n) = \int_{\Omega_t}\mathrm{div}(^*\mathrm{D}\,v\cdot V)$$

$$= \int_{\Omega_t}\mathrm{D}\,v\cdot\cdot\mathrm{D}\,V+V\cdot\Delta v \qquad (5.124)$$

☐

We shall use this expression in the sequel. We recall that the perturbed Lagrangian has the following form,

$$\mathcal{L}_{V,W}^\rho = J_{V,W}^\rho - \int_0^T\int_{\Omega_t^\rho}G(\rho) - \int_0^T\int_{\Gamma_t^\rho}\phi(\rho)$$

$$- \int_{\Omega_T}u(T)\cdot v(T)+\int_{\Omega_0}u_0\cdot v(t=0)$$

$$\forall\,(v,q)\in Y(\Omega_t)\times Q$$

Hence its derivative with respect to $\rho$ at point $\rho=0$ has the following expression,

$$\frac{d}{d\rho}\left(\mathcal{L}_{V,W}^\rho\right)\Bigg|_{\rho=0} = \frac{d}{d\rho}\left(J_{V,W}^\rho\right)\Bigg|_{\rho=0} - \int_0^T\frac{d}{d\rho}\left(\int_{\Omega_t^\rho}G(\rho)\right)\Bigg|_{\rho=0}$$

$$- \int_0^T\frac{d}{d\rho}\left(\int_{\Gamma_t^\rho}\phi(\rho)\right)\Bigg|_{\rho=0} \quad \forall\,(v,q)\in Y(\Omega_t)\times Q$$

Furthermore we have,

**LEMMA 5.29**

$$\frac{d}{d\rho}\left(J_{V,W}^\rho\right)\Bigg|_{\rho=0} = -\alpha\int_0^T\int_{\Omega_t(V)}u\cdot(\mathrm{D}\,u\cdot Z_t)+\int_0^T\int_{\Gamma_t(V)}\gamma\,V\cdot W$$

$$+ \int_0^T\int_{\Gamma_t(V)}\left[\frac{\alpha}{2}|u|^2+\frac{\gamma}{2}H|V|^2\right]\langle Z_t,n\rangle$$

Using the last identities concerning the derivative of the distributed and the boundary terms with respect to $\rho$, we shall get the following expression,

$$\left. \frac{d}{d\rho}\left(\mathcal{L}_{V,W}^{\rho}\right)\right|_{\rho=0} = -A_{Z_t} - B_{Z_t} - C_W \qquad (5.125)$$

with

$$A_{Z_t} = \int_0^T \int_{\Omega_t(V)} \left\{ \left[ \alpha u \cdot (\mathrm{D}\, u \cdot Z_t) + (\mathrm{D}\, u \cdot Z_t) \cdot \partial_t v - \left[ \left( \mathrm{D}(\mathrm{D}\, u \cdot Z_t) \right) \right] \cdot u \right.\right.$$

$$+ \mathrm{D}\, u \cdot \left( \mathrm{D}\, u \cdot Z_t \right) \big] \cdot v + \nu (\mathrm{D}\, u \cdot Z_t) \cdot \Delta v - (\mathrm{D}\, u \cdot Z_t) \cdot \nabla q \big]$$

$$+ \left[ u \cdot (\partial_t (\mathrm{D}\, v \cdot Z_t)) - (\mathrm{D}\, u \cdot u) \cdot (\mathrm{D}\, v \cdot Z_t) + \nu u \cdot (\Delta (\mathrm{D}\, v \cdot Z_t)) + p \operatorname{div}(\mathrm{D}\, v \cdot Z_t) \right] \right\}$$

$$B_{Z_t} = \int_0^T \int_{\Gamma_t(V)} \left\{ \left[ -u \cdot \partial_t v + (Du \cdot u) \cdot v - \nu u \cdot \Delta v + u \cdot \nabla q \right. \right.$$

$$- p \operatorname{div}(v) \big] (Z_t \cdot n) - \nu\, V \cdot [(\mathrm{D}(\mathrm{D}\, v) \cdot Z_t) \cdot n] + \operatorname{div}(V \cdot [-q\,\mathrm{I} + \nu\, \mathrm{D}\, v]) \langle Z_t, n \rangle$$

$$- \left[ \frac{\alpha}{2} |u|^2 + \frac{\gamma}{2} H |V|^2 \right] \langle Z_t, n \rangle \right\}$$

$$C_W = \int_0^T \int_{\Gamma_t(V)} \left[ W \cdot [-q\, n + \nu\, \mathrm{D}\, v \cdot n] - \gamma\, V \cdot W \right]$$

## The shape derivative kernel identity

We shall now assume that $(u, p, v, q) = (u, p, \varphi, \pi)$ is a saddle point of the Lagrangian functional $\mathcal{L}_V$. This will help us to simplify several terms involved in the derivative of $\mathcal{L}_V$ with respect to $V$.

Indeed, we would like to express the distributed term $A_{Z_t}$ as a boundary quantity defined on the moving boundary $\Gamma_t$.

### THEOREM 5.15

*For $(u, p, \varphi, \pi)$ saddle points of the Lagrangian functional (Eq. (5.100)), the* *following identity holds,*

$$\int_0^T \int_{\Omega_t(V)} \left\{ [\alpha\, u \cdot (D\, u \cdot Z_t) + (D\, u \cdot Z_t) \cdot \partial_t v - [(D(D\, u \cdot Z_t)) \cdot u \right.$$
$$+ D\, u \cdot (D\, u \cdot Z_t)] \cdot v + \nu(D\, u \cdot Z_t) \cdot \Delta v - (D\, u \cdot Z_t) \cdot \nabla q] + \left[ u \cdot (\partial_t(D\, v \cdot Z_t)) \right.$$
$$\left. - (D\, u \cdot u) \cdot (D\, v \cdot Z_t) + \nu u \cdot (\Delta(D\, v \cdot Z_t)) + p \operatorname{div}(D\, v \cdot Z_t)] \right\}$$
$$- \int_0^T \int_{\Gamma_t(V)} [\nu V \cdot (D(D\, \varphi \cdot Z_t) \cdot n) - (D\, \varphi \cdot Z_t) \cdot (-p\, n + \nu(D\, u \cdot n))] = 0,$$

$$\forall W \in \mathcal{U}$$

**PROOF**    We shall use extremal conditions associated to variations with respect to $(u, v)$ in the Lagrangian functional where we add a boundary integral since we consider test functions $v$ that do not vanish on the boundary $\Gamma_t(V)$, i.e.,

$$\mathcal{L}_V^2(u, p; v, q) =$$
$$J_V(u, p) - \int_0^T \int_{\Omega_t(V)} [-u \cdot \partial_t v + (D\, u \cdot u) \cdot v - \nu u \cdot \Delta v + u \cdot \nabla q - p \operatorname{div} v]$$
$$- \int_0^T \int_{\Gamma_t(V)} V \cdot (\sigma(v, q) \cdot n) + \int_0^T \int_{\Gamma_t(V)} v \cdot (\sigma(u, p) \cdot n)$$
$$- \int_{\Omega_T} u(T) \cdot v(T) + \int_{\Omega_0} u_0 \cdot v(t = 0), \quad \forall\, (v, q) \in Y \times Q$$

This leads to the following perturbation identity,

$$\partial_{(u,v)} \mathcal{L}_V^2 \cdot (\delta u, \delta v) =$$
$$- \int_{Q(V)} [-\alpha u \cdot \delta u - \delta u \cdot \partial_t v - u \cdot \partial_t \delta v + D(\delta u \cdot u) \cdot v + D(u \cdot \delta u) \cdot v$$
$$+ D(u \cdot u) \cdot \delta v - \nu(\delta u \cdot \Delta v) - \nu(u \cdot \Delta \delta v) + \delta u \cdot \nabla q - p \operatorname{div}(\delta v)]$$
$$+ \int_0^T \int_{\Gamma_t(V)} \left\{ -\nu V \cdot (D\, \delta v \cdot n) + \nu v \cdot (D\, \delta u \cdot n) + \delta v \cdot (-p\, n + \nu(D\, u \cdot n)) \right\}$$
$$- \int_{\Omega_T} [\delta u(T) v(T) + u(T) \delta v(T)], \quad \forall\, (\delta u, \delta v) \in X(\Omega_t) \times Y(\Omega_t)$$

We choose specific perturbation directions, i.e.,

$$\delta u = D\, u \cdot Z_t \quad \delta v = D\, v \cdot Z_t$$

with $\delta u(T) = \delta v(T) = \delta u(0) = \delta v(0) = 0$, where $(u, v)$ are saddle points of the Lagrangian, i.e., solutions of respectively the primal and adjoint fluid problem. We recognize immediately the distributed and boundary terms involved in the shape derivative kernel identity.

☐

**Cost functional gradient**

Now, we set $(u, v) = (u, \varphi)$ and we use the fact that $u = V$, on $\Gamma_t$ and $\varphi = 0$, on $\Gamma_t$ to simplify the remaining terms.

$$A_{Z_t} = \int_0^T \int_{\Gamma_t(V)} [\nu V \cdot (D(D\varphi \cdot Z_t) \cdot n) - \nu(D\varphi \cdot Z_t) \cdot (D u \cdot n)] \quad (5.126)$$

**REMARK 5.10** We have used that $(D\varphi \cdot Z_t) \cdot (p n) = (D\varphi \cdot (n \otimes n) \cdot Z_t) \cdot (p n) = p((D\varphi \cdot n) \cdot n) \cdot \langle Z_t, n \rangle = (p \operatorname{div} \varphi)\langle Z_t, n \rangle = 0.$  ☐

$$B_{Z_t} = \int_0^T \int_{\Gamma_t(V)} [-\nu V \cdot \Delta\varphi + V \cdot \nabla\pi] \langle Z_t, n \rangle - \frac{1}{2}[\alpha + \gamma H] |V|^2 \langle Z_t, n \rangle$$

$$-\nu V \cdot [(D(D\varphi) \cdot Z_t) \cdot n] + [-\pi \operatorname{div} V - V \cdot \nabla\pi + \nu D\varphi \cdot \cdot DV + \nu V \cdot \Delta\varphi])\langle Z_t, n \rangle$$

$$C_W = \int_0^T \int_{\Gamma_t(V)} [W \cdot [-\pi n + \nu D\varphi \cdot n] - \gamma V \cdot W]$$

We need to establish the following identity,

**LEMMA 5.30**
$$\int_{\Gamma_t} (D\varphi \cdot Z_t) \cdot (D u \cdot n) = \int_{\Gamma_t} (D\varphi \cdot n) \cdot (D u \cdot n)\langle Z_t, n \rangle \quad (5.127)$$

Then

$$-\frac{d}{d\rho}\left(\mathcal{L}^\rho_{V,W}\right)\Big|_{\rho=0} =$$

$$\int_{\Sigma(V)} \{\nu V \cdot (D(D\varphi \cdot Z_t) \cdot n) + [-\nu(D\varphi \cdot n) \cdot (D u \cdot n) - \nu V \cdot \Delta\varphi$$

$$+ V \cdot \nabla\pi] \langle Z_t, n \rangle + [-\pi \operatorname{div} V - V \cdot \nabla\pi + \nu D\varphi \cdot \cdot DV + \nu V \cdot \Delta\varphi])\langle Z_t, n \rangle$$

$$- \nu V \cdot [(D(D\varphi) \cdot Z_t) \cdot n] - \frac{1}{2}[\alpha + \gamma H] |V|^2 \langle Z_t, n \rangle$$

$$+ [W \cdot [-\pi n + \nu D\varphi \cdot n] - \gamma V \cdot W] \}$$

This allows us to derive the expression of the cost function directional derivative,

**PROPOSITION 5.8**

$$dg(0) =$$

$$\int_{\Sigma(V)} \left[ -\nu(D\,\varphi \cdot n) \cdot (D\,V \cdot n - D\,u \cdot n) + \pi \operatorname{div} V + \frac{1}{2}(\alpha + \gamma\,H)|V|^2 \right] \langle Z_t, n \rangle$$

$$+ \int_{\Sigma(V)} [-\sigma(\varphi, q) \cdot n + \gamma V] \cdot W$$

$$(5.128)$$

Then we use Th. (5.8) with,

$$E = -\nu(D\,\varphi \cdot n) \cdot (D\,V \cdot n - D\,u \cdot n) + \pi \operatorname{div} V + \frac{1}{2}(\alpha + \gamma\,H)|V|^2$$

and we get the correct result.

---

## 5.7   Min-Max and function space embedding

In the previous section, we have used a function space parametrization in order to get the gradient of a given functional related to the solution of the Navier-Stokes system in moving domain, with respect to the speed of the moving domain. In this section, we use a different method based on function space embedding particulary suited for non-homogeneous Dirichlet boundary problems. It means that the state and multiplier variables are defined in a hold-all domain $D$ that contains the moving domain $\Omega_t(V)$ for $t \in (0, T)$ and $\forall V \in \mathcal{U}$.

### 5.7.1   Saddle point formulation of the fluid state system

We recall that we are dealing with the Navier-Stokes in a moving domain $\Omega_t(V)$ which is driven by an Eulerian velocity field $V \in \mathcal{U}$,

$$\begin{cases} \partial_t u + D\,u \cdot u - \nu\Delta u + \nabla p = 0, & Q(V) \\ \operatorname{div}(u) = 0, & Q(V) \\ u = V, & \Sigma(V) \\ u(t = 0) = u_0, & \Omega_0 \end{cases} \qquad (5.129)$$

and

$$\mathcal{U} = \{ V \in H^1(0, T; (H^m(D))^d), \quad \operatorname{div} V = 0 \text{ in } D, \quad V \cdot n = 0 \text{ on } \partial D \} \tag{5.130}$$

with $m > 5/2$.

We introduce a Lagrange multiplier $\mu$ and a functional,

$$E_V(u, p; v, q, \mu) = \int_0^T \int_{\Omega_t(V)} [\partial_t u + \mathrm{D}\,u \cdot u - \nu\Delta u + \nabla p] \cdot v$$

$$- \int_0^T \int_{\Omega_t(V)} q\,\mathrm{div}\,u - \int_0^T \int_{\Gamma_t(V)} (u - V) \cdot \mu$$

for $(u, p) \in X \times P$, $(v, q) \in Y \times Q$ and $\mu \in M$ with

$$X \stackrel{\mathrm{def}}{=} Y \stackrel{\mathrm{def}}{=} H^1(0, T; H^2(D))$$

$$P \stackrel{\mathrm{def}}{=} Q \stackrel{\mathrm{def}}{=} H^1(0, T; H^1(D))$$

$$M = H^1(0, T; H^{3/2}(\Gamma_t))$$

We are interested in the following Min-Max problem,

$$\min_{(u,p)\in X\times P} \max_{(v,q,\mu)\in Y\times Q\times M} E_V(u, p; v, q, \mu) \tag{5.131}$$

The solution $(y, p, \varphi, \pi, \lambda)$ of this problem is characterized by the following optimality system,

- The primal state $(y, p)$ is a solution of the Navier-Stokes system,

$$\begin{cases} \partial_t y + \mathrm{D}\,y \cdot y - \nu\Delta y + \nabla p = 0, & Q(V) \\ \mathrm{div}(y) = 0, & Q(V) \\ y = V, & \Sigma(V) \\ y(t = 0) = y_0, & \Omega_0 \end{cases} \tag{5.132}$$

- The dual state $(\varphi, \pi)$ is the solution of the fluid adjoint system,

$$\begin{cases} -\partial_t\varphi - \mathrm{D}\,\varphi \cdot u + (^*\mathrm{D}u) \cdot \varphi - \nu\Delta\varphi + \nabla\pi = 0, & Q(V) \\ \mathrm{div}(\varphi) = 0, & Q(V) \\ \varphi = 0, & \Sigma(V) \\ \varphi(t = T) = 0, & \Omega_T \end{cases} \tag{5.133}$$

- The multiplier satisfies the following identity,

$$\mu = -q\,n + \nu(\mathrm{D}\,\varphi \cdot n), \quad \text{on } \Gamma_t(V) \tag{5.134}$$

Then we can choose the above particular representation of the boundary Lagrange multiplier $\mu$. This yields to the following functional,

$$E_V(u, p; v, q) = \int_0^T \int_{\Omega_t(V)} [\partial_t u + \mathrm{D}\,u \cdot u - \nu\Delta u + \nabla p] \cdot v - \int_0^T \int_{\Omega_t(V)} q\,\mathrm{div}\,u$$

$$- \int_0^T \int_{\Gamma_t(V)} (u - V) \cdot \sigma(v, q) \cdot n$$

for $(u, p) \in X \times P$, $(v, q) \in Y \times Q$, with

$$\sigma(v, q) \cdot n = -q\, n + \nu(D\varphi \cdot n), \quad \text{on } \Gamma_t(V)$$

The following identities hold true,

**LEMMA 5.31**

$$\int_{\Gamma_t(V)} (u - V) \cdot (D\, v \cdot n) = \int_{\Omega_t(V)} \operatorname{div} [{}^* D\, v \cdot (u - V)]$$

$$= \int_{\Omega_t(V)} [D(u - V) \cdot\cdot D\, v + (u - V) \cdot \Delta v]$$

*and*

$$\int_{\Gamma_t(V)} (u - V) \cdot q\, n = \int_{\Omega_t(V)} \operatorname{div} [q(u - V)]$$

$$= \int_{\Omega_t(V)} [(u - V) \cdot \nabla q + q \operatorname{div}(u - V)]$$

Using this identity, we may get the final expression of our saddle functional,

$$E_V(u, p; v, q) = \int_0^T \int_{\Omega_t(V)} [\partial_t u + D\, u \cdot u - \nu \Delta u + \nabla p] \cdot v - \int_0^T \int_{\Omega_t(V)} q \operatorname{div} u$$

$$+ \int_0^T \int_{\Omega_t(V)} [(u - V) \cdot \nabla q + q \operatorname{div}(u - V) - \nu D(u - V) \cdot\cdot D\, v - \nu(u - V) \cdot \Delta v]$$

for $(u, p) \in X \times P$, $(v, q) \in Y \times Q$.

**REMARK 5.11**     The above expression of the Lagrange functional has the advantage to include only distributed terms. This will be useful for its differentiation with respect to $V$. ⬚

## 5.7.2   The Lagrange functional

We are interested in the following minimization problem,

$$\min_{V \in \mathcal{U}} j(V) \tag{5.135}$$

where $j(V) = J_V(u(V), p(V))$ with $(u(V), p(V))$ is a weak solution of problem (5.2) and $J_V(u, p)$ is a real functional of the following form :

$$J_V(u, p) = \frac{\alpha}{2} \int_0^T \int_{\Omega_t(V)} |u|^2 + \frac{\gamma}{2} \int_0^T \int_{\Gamma_t(V)} |V|^2 \tag{5.136}$$

We may solve this problem by studying the equivalent Min-Max problem,

$$\min_{V \in \mathcal{U}} \quad \min_{(u,p) \in X \times P} \quad \max_{(v,q) \in Y \times Q} \quad \mathcal{L}_V(u, p; v, q) \tag{5.137}$$

with

$$\mathcal{L}_V(u, p; v, q) = J_V(u, p) - E_V(u, p; v, q) \tag{5.138}$$

Our main concern is the differentiation of the above functional with respect to $V \in \mathcal{U}$. As in the previous section we perturb the tubes using a vector field $W \in \mathcal{U}$ with an increment parameter $\rho \geq 0$. Since the functions are embedded in the hold-all domain $D$, the perturbed Lagrangian has the following form,

$$\mathcal{L}^\rho(u, p; v, q) = J_{V+\rho W}(u, p) - E_{V+\rho W}(u, p; v, q) \tag{5.139}$$

The set of saddle points,

$$S(\rho) = X(\rho) \times P \times Y(\rho) \times Q \in X \times P \times Y \times Q$$

is not a singleton since

$$X(\rho) = \left\{ u \in X, u|_{\Omega_t^\rho} = y(\rho) \right\}$$

$$Y(\rho) = \left\{ v \in Y, v|_{\Omega_t^\rho} = \varphi(\rho) \right\}$$

We make the conjecture that we can bypass the min-max, and state

$$\frac{d}{d\rho} j(V + \rho W)\Big|_{\rho=0} = \min_{(u,p) \in X \times P} \quad \max_{(v,q) \in Y \times Q} \quad \frac{d}{d\rho} \mathcal{L}^\rho(u, p; v, q)\Big|_{\rho=0} \tag{5.140}$$

Using non-cylindrical shape derivative framework, we can state

### LEMMA 5.32

$$\partial_V \mathcal{L}_V(u, p; v, q) \cdot W = - \int_0^T \int_{\Gamma_t(V)} [(\partial_t u + \mathrm{D}\, u \cdot u - \nu \Delta u + \nabla p) \cdot v - q \operatorname{div} u$$

$$+ (u - V) \cdot \nabla q + q \operatorname{div}(u - V) - \nu \mathrm{D}(u - V) \cdot \cdot \mathrm{D}\, v - \nu(u - V) \cdot \Delta v - \frac{\alpha}{2}|u|^2$$

$$- H\frac{\gamma}{2}|V|^2 \Big] \langle Z_t, n \rangle - \int_0^T \int_{\Omega_t(V)} [-W \cdot \nabla q - q \operatorname{div} W + \nu \mathrm{D}\, W \cdot \cdot \mathrm{D}\, v + \nu W \cdot \Delta v]$$

$$+ \int_0^T \int_{\Gamma_t(V)} \gamma V \cdot W$$

Then we set $(u, p) = (y, p)$ and $(v, q) = (\varphi, \pi)$ with

$$\begin{cases} -\partial_t \varphi - D\varphi \cdot u + {}^* D u \cdot \varphi - \nu \Delta \varphi + \nabla \pi = \alpha u, \, Q(V) \\ \operatorname{div}(\varphi) = 0, & Q(V) \\ \varphi = 0, & \Sigma(V) \\ \varphi(T) = 0, & \Omega_T \end{cases} \tag{5.141}$$

and we use that

$$(y, \varphi) = (V, 0) \text{ on } \Gamma_t(V)$$

and

$$\int_{\Omega_t(V)} [-W \cdot \nabla q - q \operatorname{div} W + \nu D W \cdot \cdot D v + \nu W \cdot \Delta v] = \int_{\Gamma_t(V)} W \cdot \sigma(v, q) \cdot n$$

Then,

$$\partial_V j(V) \cdot W = -\int_0^T \int_{\Gamma_t(V)} \left\{ \left[ -\pi \operatorname{div} y + \pi \operatorname{div}(y - V) - \nu D(y - V) \cdot \cdot D\varphi \right. \right.$$
$$\left. \left. -\frac{1}{2}(\alpha + H\gamma)|V|^2 \right] \langle Z_t, n \rangle + (\sigma(\varphi, \pi) \cdot n - \gamma V) \cdot W \right\}$$

Using regularity assumptions on $y$ and the free divergence condition on $y$, we may state that $\operatorname{div} y|_{\Gamma_t} = 0$.

**LEMMA 5.33**

$$D y \cdot \cdot D \varphi|_{\Gamma_t(V)} = (D y \cdot n) \cdot (D \varphi \cdot n) \tag{5.142}$$

**PROOF**    Using that $\varphi = 0$ on $\Gamma_t(V)$ yields to

$$D \varphi|_{\Gamma_t} = D \varphi \cdot (n \otimes n)|_{\Gamma_t}$$

then, we get

$$D y \cdot \cdot D \varphi = D y \cdot \cdot (D \varphi \cdot (n \otimes n))$$
$$= (D y \cdot n) \cdot (D \varphi \cdot n)$$

$\square$

Consequently we get

$$\langle j'(V), W \rangle = \int_0^T \int_{\Gamma_t(V)} \left[ -\nu(D \varphi \cdot n) \cdot (DV \cdot n - D u \cdot n) + \pi \operatorname{div} V \right.$$
$$\left. +\frac{1}{2}(\alpha + \gamma H)|V|^2 \right] \langle Z_t, n \rangle + \int_0^T \int_{\Gamma_t(V)} [-\sigma(\varphi, q) \cdot n + \gamma V] \cdot W$$

$$\tag{5.143}$$

We then use theorem (5.8) with

$$E = -\nu(\mathrm{D}\,\varphi \cdot n) \cdot (\mathrm{D}\,V \cdot n - \mathrm{D}\,u \cdot n) + \pi \operatorname{div} V + \frac{1}{2}(\alpha + \gamma\,H)|V|^2$$

and we get the correct result.
□

---

## 5.8  Conclusion

In this chapter, we have been dealing with a particular shape optimization problem involving the Navier-Stokes equations. Its originality lies in the fact that the domain containing the fluid is moving. We have introduced an open loop control problem based on the velocity of the moving domain with the goal of reaching a given objective related to the behaviour of the fluid. Our main concern was to show how the gradient of the cost functional involved in the optimal control problem can be obtained by using non-cylindrical shape optimization concepts. In addition to the classical method based on the state derivative with respect to shape motions, we have introduced two different methods based on the Min-Max principle. Even if for the time being these methods lack from a rigorous mathematical framework, they allow more flexible computations which can be very useful for practical purposes. On the numerical point of view, an implementation of the open loop control is under study in the 2D case. We believe that the concepts introduced in this chapter will prove large efficiency for coupled problems involving a moving boundary, as it will be shown in the next chapters.

# Chapter 6

## Tube derivative in a Lagrangian setting

## 6.1 Introduction

In this chapter, we state similar results as the previous chapter but with a method based directly on the Lagrangian flows which do not define a linear space, but has the advantage to first avoid the introduction of transverse fields and could prove appropriate to deal with fluid-solid interaction problems. The chapter is organised as follows,

1. In section 6.2, we introduce the general framework of non-cylindrical Lagrangian shape derivative and we state the structure of the gradient for general functionals under or without state constraints. We also establish an equivalence result between Eulerian and Lagrangian derivatives.

2. In section 6.3, we apply the general framework to mechanical systems involving a fluid in a moving domain. The fluid is described by the Navier-Stokes equations. We deal with the state derivative with respect to the Lagrangian flow and an optimal control with a general cost function. We finally recover results obtained in [57] using Eulerian derivative concepts.

## 6.2 Evolution maps

Let us consider a smooth map

$$\hat{\theta} : (0, T) \times \mathbb{R}^d \longrightarrow \mathbb{R}^d$$
$$(t, x) \longmapsto \hat{\theta}(t, x)$$

and the image $\Omega_{\hat{\theta}}$ by $\hat{\theta}$ of a fixed domain $\Omega_0 \subset \mathbb{R}^d$ with a Lipschitzian boundary $\Gamma_0$.

We set $\mathcal{C}^k \overset{\text{def}}{=} (\mathcal{C}^k(\mathbb{R}^d))^d$ and we define the space of evolution map,

$$\Theta \overset{\text{def}}{=} \left\{ \hat{\theta} \mid \ (\hat{\theta}, \hat{\theta}^{-1}) \in (\mathcal{C}^1([0,T]; \mathcal{C}^k))^2 \right\} \tag{6.1}$$

**REMARK 6.1**    For $\hat{\theta} \in \Theta$, the set $\Omega_{\hat{\theta}} = \hat{\theta}(\Omega_0)$ is still an open set and its boundary $\Gamma_{\hat{\theta}}$ is Lipschitzian.    ▯

### 6.2.1    Basic differentiation results

Given a function $f$ defined on $(0,T) \times \mathbb{R}^d$, we would like to perform the differentiation of the function $\hat{\theta} \mapsto f \circ \hat{\theta}$.

*LEMMA 6.1*
*Let us consider the scalar function*

$$f \in L^2(0,T; W^{m,p}(\mathbb{R}^d)), \ m \geq 1, \ 1 \leq p \leq \infty, \ k \geq m-1$$
$$f \in \mathcal{C}^m(\mathbb{R}^d), \qquad\qquad \text{if } p = \infty$$

*Then the application*

$$\Theta \longrightarrow L^2(0,T; W^{m,p}(\mathbb{R}^d))$$
$$\hat{\theta} \mapsto f \circ \hat{\theta}$$

*is differentiable at point $\hat{\theta}$ and its derivative in the direction $\delta\hat{\theta} \in \Theta$ is given by*

$$[\frac{\partial}{\partial\hat{\theta}}(f \circ \hat{\theta})] \cdot \delta\hat{\theta} = \nabla f \cdot \delta\hat{\theta}, \quad \in L^2(0,T; W^{m-1,p}(\mathbb{R}^d)) \tag{6.2}$$

Now, let us consider a function $\phi(\hat{\theta})$ which depends on the evolution map $\hat{\theta}$. We shall define the derivative of this function in the following way,

*LEMMA 6.2*
*Let us consider the scalar function*

$$\phi \in L^2(0,T; W^{m,p}(\mathbb{R}^d)), \ m \geq 1, \ 1 \leq p \leq \infty, \ k \geq m-1$$
$$f \in \mathcal{C}^m(\mathbb{R}^d), \qquad\qquad \text{if } p = \infty$$

*Let us assume that the function*

$$\Theta \longrightarrow L^2(0,T; W^{m,p}(\mathbb{R}^d))$$
$$\hat{\theta} \mapsto \phi(\hat{\theta}) \circ \hat{\theta}$$

*is differentiable, then the application*

$$\Theta \longrightarrow L^2(0, T; W^{m,p}(\mathbb{R}^d))$$
$$\hat{\theta} \mapsto \phi(\hat{\theta})$$

*is differentiable and its derivative in the direction* $\delta\hat{\theta} \in \Theta$ *is given by*

$$[\frac{\partial}{\partial\hat{\theta}}\phi(\hat{\theta})] \cdot \delta\hat{\theta} = \left([\frac{\partial}{\partial\hat{\theta}}(\phi(\hat{\theta}) \circ \hat{\theta})] \cdot \delta\hat{\theta}\right) \circ \hat{\theta}^{-1} - \nabla\phi(\hat{\theta}) \cdot (\delta\hat{\theta} \circ \hat{\theta}^{-1}),$$
$$\in L^2(0, T; W^{m-1,p}(\mathbb{R}^d))$$

In this section, we would like to perform a sensitivity analysis with respect to the shape of integrals defined on evolution sets $Q \in \mathcal{E}^T$ with $\overline{Q} \subset D \times (0, T)$. This analysis will be performed using the Lagrangian horizontal mapping $(t, \hat{\theta})$.

## 6.2.2 Mathematical setting

Generally speaking, we would like to solve the following problem,

$$\min_{Q \in \mathcal{A}} J(Q) \tag{6.3}$$

with

$$\mathcal{A} \equiv \{Q \in \mathcal{E}^T, \quad Q \subset D\} \tag{6.4}$$

and

$$J : \mathcal{A} \longrightarrow \mathcal{R}$$
$$Q \mapsto J(Q) \tag{6.5}$$

In order to work with standard differential calculus, we define the functional $j = J \circ \theta_t$ such that

$$j : \Theta_{ad} \longrightarrow \mathcal{R}$$
$$\hat{\theta} \mapsto j(\hat{\theta}) = J(Q_{\hat{\theta}}) \tag{6.6}$$

with $Q_{\hat{\theta}} = \bigcup_{0 < t < T} (\{t\} \times \Omega_t) = (t, \hat{\theta})(Q_0)$.

Then, derivatives in the sense of Gâteaux or Fréchet could make sense for $j(\hat{\theta})$.

## 6.2.3 Elements of shape calculus

Using extension of classical elements of cylindrical shape calculus using the perturbation of identity ([116], [101],[135]), it is possible to define derivation concepts of functions defined in $Q_{\hat{\theta}}$ or $\Sigma_{\hat{\theta}}$ with respect to $\hat{\theta}$.

## Derivation of moving domain integrals

In this paragraph, we are interested in the differentiability properties of integrals defined on the non-cylindrical set $Q_{\hat{\theta}}$,

$$J_1(Q_{\hat{\theta}}) = \int_{Q_{\hat{\theta}}} f(Q_{\hat{\theta}}) dQ_{\hat{\theta}} \tag{6.7}$$

We are trying to characterize the variations of $J_1(Q_{\hat{\theta}})$ with respect to $\hat{\theta}$. We shall state a differentiability result that uses the notion of non-cylindrical material derivative,

**DEFINITION 6.1** *A function $f(\hat{\theta}) \in H(Q_{\hat{\theta}})$ admits a non-cylindrical material derivative $\dot{f}(\hat{\theta}) \cdot \delta\theta$ defined over $Q_{\hat{\theta}}$ at point $\hat{\theta} \in \Theta_{ad}$ in the direction $\delta\hat{\theta} \stackrel{\text{def}}{=} \delta\theta \circ \hat{\theta} \in \Theta$, if the following composed function,*

$$f^\rho : [0, \rho_0[ \rightarrow H(Q_{\hat{\theta}})$$
$$\rho \mapsto f((I+\rho\delta\theta) \circ \hat{\theta}) \circ (I+\rho\delta\theta)$$

*is differentiable at point $\rho = 0$, a.e. $(x,t) \in Q_{\hat{\theta}}$ and $\dot{f}(\hat{\theta}) \cdot \delta\theta = \frac{d}{d\rho} f^\rho \big|_{\rho=0}$.*

Under regularity conditions, it is possible to define a different derivative that will be useful in the context of Eulerian state equations.

**DEFINITION 6.2** *A function $f(\hat{\theta}) \in H(D)$ admits a non-cylindrical shape derivative $f'(\hat{\theta}) \cdot \delta\theta$ defined over $Q_{\hat{\theta}}$ at point $\hat{\theta} \in \Theta_{ad}$ in the direction $\delta\hat{\theta} \stackrel{\text{def}}{=} \delta\theta \circ \hat{\theta} \in \Theta$, if the following composed function,*

$$f_\rho : [0, \rho_0[ \rightarrow H(D)$$
$$\rho \mapsto f((I+\rho\delta\theta) \circ \hat{\theta})$$

*is differentiable at point $\rho = 0$, a.e. $(x,t) \in Q_{\hat{\theta}}$ and $f'(\hat{\theta}) \cdot \delta\theta = \frac{d}{d\rho} f_\rho \big|_{\rho=0}$.*

**REMARK 6.2** It can be proven that the following identity holds,

$$\dot{f}(\hat{\theta}) \cdot \delta\theta = f'(\hat{\theta}) \cdot \delta\theta + \nabla f(\hat{\theta}) \cdot \delta\theta \tag{6.8}$$

☐

We can now state the differentiability properties of non-cylindrical integrals with respect to their moving supports,

**THEOREM 6.1**

*For a bounded measurable domain $\Omega_0$ with boundary $\Gamma_0$, let us assume that for any direction $\delta\theta \overset{\text{def}}{=} \delta\theta \circ \hat{\theta} \in \Theta$ the following hypothesis holds,*

i) *$f(\hat{\theta})$ admits a non-cylindrical material derivative $\dot{f}(\hat{\theta}) \cdot \delta\theta$ then $J_1(.)$ is Gâteaux differentiable at point $\hat{\theta} \in \Theta_{ad}$ and its derivative is given by the following expression,*

$$\frac{d}{d\hat{\theta}} J_1(\hat{\theta}) \cdot \delta\theta = \int_{Q_{\hat{\theta}}} \left[ \dot{f}(\hat{\theta}) \cdot \delta\theta + f(\hat{\theta}) \operatorname{div} \delta\theta \right] dQ_{\hat{\theta}} \qquad (6.9)$$

*Futhermore, if*

ii) *$f(\hat{\theta})$ admits a non-cylindrical shape derivative $f'(\hat{\theta}) \cdot \delta\theta \in H(Q_{\hat{\theta}})$, then*

$$\frac{d}{d\hat{\theta}} J_1(\hat{\theta}) \cdot \delta\theta = \int_{Q_{\hat{\theta}}} \left[ f'(\hat{\theta}) \cdot \delta\theta + \operatorname{div}(f(\hat{\theta}) \delta\theta) \right] dQ_{\hat{\theta}} \qquad (6.10)$$

*Furthermore, if $Q_0$ is an open domain with a Lipschitzian boundary $\Sigma_0$, then*

$$\frac{d}{d\hat{\theta}} J_1(\hat{\theta}) \cdot \delta\theta = \int_{Q_{\hat{\theta}}} f'(\hat{\theta}) \cdot \delta\theta dQ_{\hat{\theta}} + \int_{\Sigma_{\hat{\theta}}} f(\hat{\theta}) \langle \delta\theta, \nu \rangle d\Sigma_{\hat{\theta}} \qquad (6.11)$$

**REMARK 6.3**  Since we deal with horizontal transformations of the type $(t, \hat{\theta})$, it means that perturbations are of the type $(0, \delta\hat{\theta})$, then

$$\langle \delta\theta, \nu \rangle = (1 + v_\nu^2)^{-1/2} \left( \langle 0, -v_\nu \rangle + \langle \delta\theta, n_t \rangle \right)$$

using that $d\Sigma_{\hat{\theta}} = (1 + v_\nu^2)^{1/2} dt\, d\Gamma_{\hat{\theta}}$, we get

$$\int_{\Sigma_{\hat{\theta}}} f(\hat{\theta}) \langle \delta\theta, \nu \rangle d\Sigma_{\hat{\theta}} = \int_0^T \int_{\Gamma_{\hat{\theta}}} f(\hat{\theta}) \langle \delta\theta, n_{\hat{\theta}} \rangle d\Gamma_{\hat{\theta}}\, dt \qquad (6.12)$$

▯

**Derivation of moving boundary integrals**

It is also possible to establish a similar result for integrals over moving boundaries,

$$J_2(\Sigma_{\hat{\theta}}) = \int_{\Sigma_{\hat{\theta}}} g(\Sigma_{\hat{\theta}}) d\Sigma_{\hat{\theta}} \qquad (6.13)$$

For that purpose, we need to define the non-cylindrical tangential material derivative,

**DEFINITION 6.3**    *A function $g(\hat{\theta}) \in H(\Sigma_{\hat{\theta}})$ admits a non-cylindrical material derivative $\dot{g}(\hat{\theta}) \cdot \delta\theta$ at point $\hat{\theta} \in \Theta_{ad}$ in the direction $\delta\hat{\theta} \overset{def}{=} \delta\theta \circ \hat{\theta} \in \Theta$ if the following composed function,*

$$g^\rho : [0, \rho_0] \to H(\Sigma_{\hat{\theta}})$$
$$\rho \mapsto g((\mathrm{I} + \rho\delta\theta) \circ \hat{\theta}) \circ (\mathrm{I} + \rho\delta\theta)$$

*is differentiable at point $\rho = 0$, a.e. $(t, x) \in \Sigma_{\hat{\theta}}$ and $\dot{g}(\hat{\theta}) \cdot \delta\theta = \frac{d}{d\rho} g^\rho \big|_{\rho=0}$.*

Under regularity conditions, it is possible to define a different derivative that will be useful in the context of Eulerian state equations.

**DEFINITION 6.4**    *A function $g(\hat{\theta}) \in H(\Sigma_{\hat{\theta}})$ admits a non-cylindrical shape derivative $g'(\hat{\theta}) \cdot \delta\theta$ defined over $\Sigma_{\hat{\theta}}$ at point $\hat{\theta} \in \Theta_{ad}$ in the direction $\delta\hat{\theta} \overset{def}{=} \delta\theta \circ \theta \in \Theta$, if the following composed function,*

$$g_\rho : [0, \rho_0[ \to H(D)$$
$$\rho \mapsto g((\mathrm{I} + \rho\delta\hat{\theta}) \circ \theta) \circ p$$

*is differentiable at point $\rho = 0$, a.e. $(x, t) \in Q_{\hat{\theta}}$ and $g'(\hat{\theta}) \cdot \delta = \frac{d}{d\rho} g_\rho \big|_{\rho=0}$. Here $p$ stands for the projection mapping on $\Gamma_{\hat{\theta}}$ ([51]).*

**REMARK 6.4**    It can be proven that the following identity holds,

$$\dot{g}(\hat{\theta}) \cdot \delta\theta = g'(\hat{\theta}) \cdot \delta\theta + \nabla_\Gamma\, g(\hat{\theta}) \cdot \delta\theta \tag{6.14}$$

☐

These concepts are involved in the differentiability property of boundary integrals,

**THEOREM 6.2**
*For a bounded measurable domain $\Omega_0$ with boundary $\Gamma_0$, let us assume that for any direction $\delta\hat{\theta} \overset{def}{=} \delta\theta \circ \hat{\theta} \in \Theta$ the following hypothesis holds,*

  i) *$g(\hat{\theta})$ admits a non-cylindrical material derivative $\dot{g}(\hat{\theta}) \cdot \delta\theta$ then $J_2(.)$ is Gâteaux differentiable at point $\hat{\theta} \in \Theta_{ad}$ and its derivative is given by the following expression,*

$$\frac{d}{d\hat{\theta}} J_2(\hat{\theta}) \cdot \delta\theta = \int_{\Sigma_{\hat{\theta}}} \left[ \dot{g}(\hat{\theta}) \cdot \delta\theta + g(\hat{\theta})\, \mathrm{div}_\Gamma\, \delta\theta \right] d\Sigma_{\hat{\theta}} \tag{6.15}$$

  *and if,*

*ii)* $g(\hat\theta)$ *admits a non-cylindrical shape derivative ,* $g'(\hat\theta)\cdot\delta\theta$, *then*

$$\frac{d}{d\hat\theta}J_2(\hat\theta)\cdot\delta\theta = \int_{\Sigma_{\hat\theta}}\left[g'(\hat\theta)\cdot\delta\theta + H\,g(\hat\theta)\langle\delta\theta,\nu\rangle\right]d\Sigma_{\hat\theta} \qquad (6.16)$$

*Furthermore, if* $g(\hat\theta)=\tilde g(\hat\theta)|_{\Sigma_{\hat\theta}}$ *with* $\tilde g\in H(Q_{\hat\theta})$, *then*

$$\frac{d}{d\hat\theta}J_2(\hat\theta)\cdot\delta\theta = \int_{\Sigma_{\hat\theta}}\left[\tilde g'(\hat\theta)\cdot\delta\theta + (\nabla_\Gamma\,\tilde g(V)\cdot n_{\hat\theta} + H\,g(\hat\theta))\,\langle\delta\theta,\nu\rangle\right]d\Sigma_{\hat\theta}$$
$$(6.17)$$

**REMARK 6.5** Using the fact that the transformations $\hat\theta$ on $Q_{\hat\theta}$ are horizontal, i.e., of the type $(t,\hat\theta)$ on $(0,T)\times\Omega_0$, we get

$$\frac{d}{d\hat\theta}J_2(\hat\theta)\cdot\delta\theta = \int_0^T\int_{\Gamma_{\hat\theta}}\left[g'(\hat\theta)\cdot\delta\theta(1+v_\nu^2)^{1/2} + H\,g(\hat\theta)\langle\delta\theta,n_t\rangle\right]dt\,d\Gamma_{\hat\theta} \quad (6.18)$$

⬚

## An adjoint identity

In the remaining part of this chapter, we will try to obtain the structure of several functionals with respect to the Lagrangian mapping $\hat\theta$. To this end, we will need to use the following fundamental integration by parts formula,

**THEOREM 6.3**
*For any* $E\in L^2(\Sigma_{\hat\theta})$ *and* $\delta\theta\overset{\mathrm{def}}{=}\delta\theta\circ\hat\theta\in\Theta$, *the following identity holds,*

$$\int_0^T\int_{\Gamma_{\hat\theta}}E\partial_t(\delta\theta\circ\hat\theta)\circ\hat\theta^{-1} = \int_0^T\int_{\Gamma_{\hat\theta}}\left[-\partial_t E - (\operatorname{div}_\Gamma V_{\hat\theta})\,E - D\,E\cdot V_{\hat\theta}\right]\cdot\delta\theta$$
$$(6.19)$$

*with* $V_{\hat\theta} = (\partial_t\hat\theta)\circ\hat\theta^{-1}$.

**PROOF** We first need to prove the following lemma,

**LEMMA 6.3**

$$\int_0^T\int_{\Gamma_{\hat\theta}}\partial_t f\cdot g = \left[\int_{\Gamma_{\hat\theta}(\tau)}f\cdot g\right]_0^T - \int_0^T\int_{\Gamma_{\hat\theta}}\left[f\cdot\partial_t g + H(f\cdot g)\,(V_{\hat\theta}\cdot n_{\hat\theta})\right.$$
$$\left. + \nabla(f\cdot g)\cdot n_{\hat\theta}\,(V_{\hat\theta}\cdot n_{\hat\theta})\right]$$
$$(6.20)$$

Using classical integral derivatives formulas, we get

$$\int_0^T \partial_t \left( \int_{\Gamma_{\hat\theta}} f \cdot g \right) = \int_0^T \int_{\Gamma_{\hat\theta}} \left[ (f \cdot g)'_{\Gamma_{\hat\theta}} + H(f \cdot g)\,(V_{\hat\theta} \cdot n_{\hat\theta}) \right]$$

$$= \int_0^T \int_{\Gamma_{\hat\theta}} \left[ \partial_t (f \cdot g)|_{\Gamma_{\hat\theta}} + \nabla(f \cdot g) \cdot n_{\hat\theta} + H(f \cdot g)\,(V_{\hat\theta} \cdot n_{\hat\theta}) \right]$$

$$= \int_0^T \int_{\Gamma_{\hat\theta}} \left[ \partial_t f \cdot g + f \cdot \partial_t g + \nabla(f \cdot g) \cdot n_{\hat\theta} \right.$$

$$\left. + H(f \cdot g)\,(V_{\hat\theta} \cdot n_{\hat\theta}) \right]$$

$\square$

In order to use this lemma, we need to modify the expression of $E\partial_t(\delta\theta \circ \hat\theta) \circ \hat\theta^{-1}$. For that purpose, we use the following identities,

### LEMMA 6.4

$$(\partial_t h) \circ \hat\theta^{-1} = \partial_t(h \circ \hat\theta^{-1}) - (\mathrm{D}\,h) \circ \hat\theta^{-1} \partial_t(\hat\theta^{-1}) \tag{6.21}$$

$$\partial_t(\hat\theta^{-1}) = -\mathrm{D}(\hat\theta^{-1}) \cdot V_{\hat\theta} \tag{6.22}$$

$$(\partial_t h) \circ \hat\theta^{-1} = \partial_t(h \circ \hat\theta^{-1}) + \mathrm{D}(h \circ \hat\theta^{-1}) \cdot V_{\hat\theta} \tag{6.23}$$

We apply the last identity with $h = \delta\theta \circ \hat\theta$ and we obtain

$$\partial_t(\delta\theta \circ \hat\theta) \circ \hat\theta^{-1} = \partial_t \delta\theta + \mathrm{D}\,\delta\theta \cdot V_{\hat\theta} \tag{6.24}$$

This allows us to state the following,

$$\int_0^T \int_{\Gamma_{\hat\theta}} E\partial_t(\delta\theta \circ \hat\theta) \circ \hat\theta^{-1} = \int_0^T \int_{\Gamma_{\hat\theta}} E \cdot (\partial_t \delta\theta) + \int_0^T \int_{\Gamma_{\hat\theta}} E \cdot (\mathrm{D}\,\delta\theta \cdot V_{\hat\theta}) \tag{6.25}$$

Using lemma (6.3) with $f = \delta\theta$ and $g = E$, we obtain

$$\int_0^T \int_{\Gamma_{\hat\theta}} E \cdot (\partial_t \delta\theta) = \int_0^T \int_{\Gamma_{\hat\theta}} \left[ -\delta\theta \cdot \partial_t E - \nabla(E \cdot \delta\theta) \cdot n_{\hat\theta}\,(V_{\hat\theta} \cdot n_{\hat\theta}) \right.$$

$$\left. - H(E \cdot \delta\theta)\,(V_{\hat\theta} \cdot n_{\hat\theta}) \right]$$

For the second right hand side we need to establish the following lemma,

### LEMMA 6.5

$$\int_0^T \int_{\Gamma_{\hat\theta}} E \cdot (\mathrm{D}\,\delta\theta \cdot V_{\hat\theta}) = \int_0^T \int_{\Gamma_{\hat\theta}} \left[ -\delta\theta \cdot (\mathrm{D}\,E \cdot V_{\hat\theta}) - (\mathrm{div}_\Gamma V_{\hat\theta})\,\delta\theta \cdot E \right.$$

$$\left. + \nabla(E \cdot \delta\theta) \cdot n_{\hat\theta}\,(V_{\hat\theta} \cdot n_{\hat\theta}) + H\,(E \cdot \delta\theta)\,(V_{\hat\theta} \cdot n_{\hat\theta}) \right]$$

$$\tag{6.26}$$

Combining these two identities, we obtain the expected result. $\qquad\square$

### 6.2.4 Unconstraint non-cylindrical shape optimization

We would like to minimize over $\theta \in \Theta_{ad}$ the following functional,

$$j(\hat{\theta}) = \int_{Q_{\hat{\theta}}} F(\hat{\theta}, y(\hat{\theta})) dx dt + \int_{\Sigma_{\hat{\theta}}} S(\hat{\theta}, y(\hat{\theta})) d\Sigma_{\hat{\theta}} \qquad (6.27)$$

with $Q_{\hat{\theta}} = \bigcup_{0 < t < T} (\{t\} \times \hat{\theta}(\Omega_0))$ and $y(\hat{\theta}) \in H(Q_{\hat{\theta}})$. Using the shape derivative calculus tools introduced in the previous section, it is possible to obtain the derivative with respect to $\hat{\theta}$ of the functional $j(\hat{\theta})$,

**PROPOSITION 6.1**
*Let us assume that for $\hat{\theta} \in \Theta_{ad}$ and for any direction $\delta\hat{\theta} \stackrel{\text{def}}{=} \delta\theta \circ \hat{\theta} \in \Theta$,*

*1. $y(\hat{\theta})$ admits a material derivative $\dot{y}(\hat{\theta}) \cdot \delta\theta \in H(Q_{\hat{\theta}})$,*

*2. the partial derivative $y'(\hat{\theta})\delta\theta = \dot{y}(\hat{\theta}) \cdot \delta\theta - \nabla y(\hat{\theta}) \cdot \delta\theta$ exists in $H(Q_{\hat{\theta}})$,*

*3. $\tau \mapsto j((I+\tau\delta\theta) \circ \hat{\theta})$ is differentiable at $\tau = 0$,*

*then $j(.)$ is differentiable at point $\hat{\theta}$ and there exists a distribution $G(\hat{\theta}) \in L^2(0, T; D'(\mathcal{R}^d))$ with support inside $\overline{Q_{\hat{\theta}}}$ such that*

$$\frac{d}{d\hat{\theta}} j(\hat{\theta}) \cdot \delta\theta = \int_0^T \langle G(\hat{\theta}), \delta\theta \rangle_{D',D} dt \qquad (6.28)$$

*Furthermore $G(\hat{\theta})$ admits the two following representations,*

- *using material derivative,*

$$\langle G(\hat{\theta}), \delta\theta \rangle =$$
$$\int_{Q_{\hat{\theta}}} \left[ \dot{F}(\hat{\theta}, y(\hat{\theta}))\delta\theta + \nabla_y F(\hat{\theta}, y(\hat{\theta})) \cdot \dot{y}(\hat{\theta})\delta\theta + F(\hat{\theta}, y(\hat{\theta})) \cdot \operatorname{div} \delta\theta \right] dQ_{\hat{\theta}}$$
$$+ \int_{\Sigma_{\hat{\theta}}} \left[ \dot{S}(\hat{\theta}, y(\hat{\theta}))\delta\theta + \nabla_y S(\hat{\theta}, y(\hat{\theta})) \cdot \dot{y}(\hat{\theta})\delta\theta + S(\hat{\theta}, y(\hat{\theta})) \cdot \operatorname{div}_\Gamma \delta\theta \right] d\Sigma_{\hat{\theta}}$$
$$(6.29)$$

- *using shape partial derivative,*

$$\langle G(\hat{\theta}), \delta\hat{\theta} \rangle = \int_{Q_{\hat{\theta}}} \left[ F'(\hat{\theta}, y(\hat{\theta}))\delta\theta + \nabla_y F(\hat{\theta}, y(\hat{\theta})) \cdot y'(\hat{\theta}) \cdot \delta\theta \right] dQ_{\hat{\theta}}$$
$$+ \int_{\Sigma_{\hat{\theta}}} \left[ S'(\hat{\theta}, y(\hat{\theta})) \cdot \delta\theta + \nabla_y S(\hat{\theta}, y(\hat{\theta})) \cdot y'(\hat{\theta})\delta\theta \right] d\Sigma_{\hat{\theta}}$$
$$+ \int_0^T \int_{\Gamma_{\hat{\theta}}} \left[ F(\hat{\theta}, y(\hat{\theta})) + \partial_n S(\hat{\theta}, y(\hat{\theta})) + \nabla_y S(\hat{\theta}, y(\hat{\theta}))\partial_n y(\hat{\theta}) \right.$$
$$\left. + HS(\hat{\theta}, y(\hat{\theta})) \right] \langle \delta\theta, n_{\hat{\theta}} \rangle \, dt \, d\Gamma_{\hat{\theta}} \qquad (6.30)$$

**PROOF**    We perform the same steps as in [135] and we use the non-cylindrical calculus introduced previously.                                        □

From now on, let us assume that the partial shape derivative $y'(\hat{\theta})\delta\theta$ is the unique solution of the following linear tangent problem,

$$\langle Ay'(\hat{\theta})\cdot\delta\theta,\phi\rangle = L(\delta\theta,\phi), \quad \forall\,\phi\in H(Q_{\hat{\theta}}) \tag{6.31}$$

with $A\in\mathcal{L}(H(Q_{\hat{\theta}},H(Q_{\hat{\theta}})^*))$ and $L(\delta\theta,.)\in H(Q_{\hat{\theta}})^*$. We define the element $p(\hat{\theta})\in H(Q_{\hat{\theta}})^*$ as the solution of the following adjoint problem,

$$\langle A^*p,\psi\rangle = \int_{Q_{\hat{\theta}}}\left[\nabla_y F(\hat{\theta},y(\hat{\theta}))\cdot\psi\right]dQ_{\hat{\theta}} + \int_{\Sigma_{\hat{\theta}}}\left[\nabla_y S(\hat{\theta},y(\hat{\theta}))\cdot\psi\right]d\Sigma_{\hat{\theta}},$$

$$\forall\,\psi\in H(Q_{\hat{\theta}})$$

We then use the following identity

$$\langle A^*p(\hat{\theta}),y'(\hat{\theta})\cdot\delta\theta\rangle = \langle Ay'(\hat{\theta})\cdot\delta\theta,p(\hat{\theta})\rangle = L(\delta\theta,p(\hat{\theta})) \tag{6.32}$$

This allows us to conclude that the derivative of $j(\hat{\theta})$ with respect to $\hat{\theta}$ has the following structure,

$$j'(\hat{\theta})\cdot\delta\theta = L(\delta,p(\hat{\theta})) + \int_{Q_{\hat{\theta}}}F'(\hat{\theta},y(\hat{\theta}))\cdot\delta\theta\,dQ_{\hat{\theta}} + \int_{\Sigma_{\hat{\theta}}}F(\hat{\theta},y(\hat{\theta}))\langle\delta\theta,n_{\hat{\theta}}\rangle$$

$$+ \int_{\Sigma_{\hat{\theta}}}\left[S'(\hat{\theta},y(\hat{\theta}))\cdot\delta\theta + \left[\partial_n S(\hat{\theta},y(\hat{\theta})) + \nabla_y S(\hat{\theta},y(\hat{\theta}))\partial_n y(\hat{\theta})\right.\right.$$

$$\left.\left. + HS(\hat{\theta},y(\hat{\theta}))\right]\langle\delta\theta,n_{\hat{\theta}}\rangle\right]$$

**REMARK 6.6**    Particular case :
We assume the following hypothesis:

1.  $L(\delta\theta,p(\hat{\theta})) = \int_{\Sigma_{\hat{\theta}}}\ell(p(\hat{\theta}))\,\langle\delta\theta,n_{\hat{\theta}}\rangle\,dt\,d\Gamma_{\hat{\theta}},$

2.  $\int_{Q_{\hat{\theta}}}F'(\hat{\theta},y(\hat{\theta}))\cdot\delta\theta\,dQ_{\hat{\theta}} = 0,$

3.  $\int_{\Sigma_{\hat{\theta}}}S'(\hat{\theta},y(\hat{\theta}))\cdot\delta\theta\,d\Sigma_{\hat{\theta}} = \int_{\Sigma_{\hat{\theta}}}s(\hat{\theta},y(\hat{\theta}))\,\langle\delta\theta,n_{\hat{\theta}}\rangle\,dt\,d\Gamma_{\hat{\theta}}.$

In this case the gradient of $j(\hat{\theta})$ is only supported on $\Gamma_{\hat{\theta}}$ with a density $g_{\hat{\theta}}$

such that the following identity holds,

$$
\begin{aligned}
j^{'}(\hat{\theta}) \cdot \delta\theta &= \int_0^T \langle g(\hat{\theta}), \delta\theta \cdot n_{\hat{\theta}} \rangle \\
&= \int_0^T \int_{\Gamma_{\hat{\theta}}} \Big[ \ell(p(\hat{\theta})) + s(\hat{\theta}, y(\hat{\theta})) + \partial_n S(\hat{\theta}, y(\hat{\theta})) \\
&\quad + \nabla_y S(\hat{\theta}, y(\hat{\theta})) \partial_n y(\hat{\theta}) + HS(\hat{\theta}, y(\hat{\theta})) \Big] \, \langle \delta\theta, n_{\hat{\theta}} \rangle \, dt \, d\Gamma_{\hat{\theta}}
\end{aligned}
$$

(6.33)

$\square$

## 6.2.5 Shape optimization under state constraints

In the previous section, we have established for a general functional, the structure of its gradient with respect to the moving domain. This general functional involved distributed and boundary terms and a state function $y(\hat{\theta}) \in H(Q_{\hat{\theta}})$ that depends on the moving domain $Q_{\hat{\theta}}$ and its parametrization $\hat{\theta}$. This gradient will be involved in the first-order optimality conditions of the minimization problem for $j(\hat{\theta})$.

In this paragraph, we are interested by optimization problems involving state constraints. The design variables are the Lagrange mapping and a state variable $(\hat{\theta}, y) \in \hat{\theta}_{ad} \times H(D)$ related by an abstract state variable

$$
e(\hat{\theta}, y) = 0, \quad \text{in } H(D)^* \tag{6.34}
$$

This setting allows to avoid the derivation of the state variable $y(\theta)$ solution of the implicit equation (6.34) with respect to the mapping $\hat{\theta} \in \Theta_{ad}$. Hence we would like to solve the following minimization problem,

$$
\begin{aligned}
&\min_{} \quad j(\hat{\theta}, y) \\
&\begin{cases} (\hat{\theta}, y) \in \hat{\theta}_{ad} \times H(D), \\ e(\hat{\theta}, y) = 0, \text{ in } H(D)^* \end{cases}
\end{aligned} \tag{6.35}
$$

with

$$
j(\hat{\theta}, y) = \int_{Q_{\hat{\theta}}} F(\hat{\theta}, y) \, dQ_{\hat{\theta}} + \int_{\Sigma_{\hat{\theta}}} S(\hat{\theta}, y) \, d\Sigma_{\hat{\theta}} \tag{6.36}
$$

and

$$
\langle e(\hat{\theta}, y), \phi \rangle = \int_{Q_{\hat{\theta}}} E_\Omega(\hat{\theta}, y, \phi) \, dQ_{\hat{\theta}} + \int_{\Sigma_{\hat{\theta}}} E_\Gamma(\hat{\theta}, y, \phi) \, d\Sigma_{\hat{\theta}}, \quad \forall \phi \in H(D) \tag{6.37}
$$

where we suppose that the applications $(., ., \phi) \longrightarrow \dfrac{E_\Omega(., ., \phi)}{E_\Gamma(., ., \phi)}$ are linear continuous on $H(D)$.

In order to derive the first-order optimality conditions for problem (6.35), we introduce the Lagrangian functional,

$$\mathcal{L}(\hat{\theta}, y; \phi) = j(\hat{\theta}, y) - \langle e(\hat{\theta}, y), \phi \rangle_{H(D)^*, H(D)} \tag{6.38}$$

Consequently, solving problem (6.35) is equivalent to finding the saddle point $(\bar{\hat{\theta}}, \bar{y}, \bar{\phi}) \in \Theta_{ad} \times H(Q_{\bar{\hat{\theta}}}) \times H(Q_{\bar{\hat{\theta}}})^*$ of the following Min-Max problem,

$$\min_{(\hat{\theta}, y) \in \Theta_{ad} \times H(D)} \quad \max_{\phi \in H(D)^*} \mathcal{L}(\hat{\theta}, y; \phi) \tag{6.39}$$

### PROPOSITION 6.2
*Let us assume the following points,*

1. *Problem (6.35) admits a local solution $(\bar{\hat{\theta}}, \bar{y})$ and the functional $j(\hat{\theta}, y)$ is Fréchet differentiable in a neighborhood of point $(\bar{\hat{\theta}}, \bar{y}) \in \Theta_{ad} \times H(Q_{\bar{\hat{\theta}}})$.*

2. *The local solution $(\bar{\hat{\theta}}, \bar{y})$ is a regular point, i.e., that the linear tangent application $e'((\bar{\hat{\theta}}, \bar{y}))$ is surjective.*

*Then there exists a Lagrange multiplier $\bar{\phi} \in H(Q_{\bar{\hat{\theta}}})^*$, such that the following first-order optimality conditions hold,*

$$e(\bar{\hat{\theta}}, \bar{y}) = 0, \tag{6.40}$$

$$\mathcal{L}'_{(\hat{\theta}, y)}(\bar{\hat{\theta}}, \bar{y}; \bar{\phi}) = 0 \tag{6.41}$$

We shall now derive the explicit form of the optimality conditions for problem (6.39). Using differentiability results of $j(.,.)$, the following identity comes easily,

$$\langle \mathcal{L}'_y(\hat{\theta}, y; \phi), \delta y \rangle = \int_{Q_{\hat{\theta}}} \left[ \nabla_y F(\hat{\theta}, y) \cdot \delta y + \nabla_y E_\Omega(\hat{\theta}, y, \phi) \cdot \delta y \right] dQ_{\hat{\theta}} +$$

$$\int_{\Sigma_{\hat{\theta}}} \left[ \nabla_y S(\hat{\theta}, y) \cdot \delta y + \nabla_y E_\Gamma(\hat{\theta}, y, \phi) \cdot \delta y \right] d\Sigma_{\hat{\theta}} \tag{6.42}$$

and

$$\langle \mathcal{L}'_{\hat{\theta}}, \delta\hat{\theta} \rangle = \int_{Q_{\hat{\theta}}} \left[ F'(\hat{\theta}, y) \cdot \delta\theta + E'_\Omega(\hat{\theta}, y, \phi) \cdot \delta\theta \right] dQ_{\hat{\theta}}$$

$$+ \int_{\Sigma_{\hat{\theta}}} \left[ S'(\hat{\theta}, y) \cdot \delta\theta + E'_\Gamma(\hat{\theta}, y, \phi) \cdot \delta\theta \right]$$

$$+ \int_{\Sigma_{\hat{\theta}}} \left[ F(\hat{\theta}, y) + E_\Omega(y, \hat{\theta}; \phi) + (H \, \mathrm{I} + \partial_n) \cdot (S(\hat{\theta}, y) + E_\Gamma(\hat{\theta}, y, \phi)) \right] \cdot \langle \delta\theta, n_{\hat{\theta}} \rangle \, d\Sigma_{\hat{\theta}}$$

At a regular point $(\bar{\bar{\theta}}, \bar{y})$, there exists a Lagrange multiplier $\bar{\phi}$ solution of the following linear tangent adjoint equation,

$$
\int_{Q_{\bar{\theta}}} \left[ \nabla_y F(\bar{\bar{\theta}}, \bar{y}) \cdot \psi + \nabla_y E_\Omega(\bar{\bar{\theta}}, \bar{y}, \bar{\phi}) \cdot \psi \right] dQ_{\hat{\theta}}
$$

$$
+ \int_{\Sigma_{\bar{\theta}}} \left[ \nabla_y S(\bar{\bar{\theta}}, \bar{y}) \cdot \psi + \nabla_y E_\Gamma(\bar{\bar{\theta}}, \bar{y}, \bar{\phi}) \cdot \psi \right] d\Sigma_{\hat{\theta}} = 0,
$$

$$
\forall \psi \in H(D)
$$

Then the optimality condition reduces to

$$
j'(\bar{\bar{\theta}}; \delta\hat{\theta}) = \int_{Q_{\bar{\theta}}} \left[ F'(\bar{\bar{\theta}}, \bar{y}) \cdot \delta\theta + E'_\Omega(\bar{\bar{\theta}}, \bar{y}, \bar{\phi}) \cdot \delta\theta \right] dQ_{\hat{\theta}}
$$

$$
+ \int_{\Sigma_{\bar{\theta}}} \left[ S'(\bar{\bar{\theta}}, \bar{y}) \cdot \delta\theta + E'_\Gamma(\bar{\bar{\theta}}, \bar{y}, \bar{\phi}) \cdot \delta\theta \right]
$$

$$
+ \int_{\Sigma_{\bar{\theta}}} \left[ F(\bar{\bar{\theta}}, \bar{y}) + E_\Omega(\bar{\bar{\theta}}, \bar{y}; \bar{\phi}) + (H\,I + \partial_n) \cdot (S(\bar{\bar{\theta}}, \bar{y}) \right.
$$

$$
\left. + E_\Gamma(\bar{\bar{\theta}}, \bar{y}, \bar{\phi})) \right] \cdot \langle \delta\theta, n_{\hat{\theta}} \rangle \, d\Sigma_{\hat{\theta}}
$$

$$
= 0 \tag{6.43}
$$

**REMARK 6.7**  Let us add the following hypothesis,

1. $\displaystyle\int_{Q_{\hat{\theta}}} \left[ F'(\hat{\theta}, y) \cdot \delta\theta + E'_\Omega(\hat{\theta}, y, \phi) \cdot \delta\theta \right] dQ_{\hat{\theta}} = 0,$

2. $\displaystyle\int_{\Sigma_{\hat{\theta}}} \left[ S'(\hat{\theta}, y) \cdot \delta\theta + E'_\Gamma(\hat{\theta}, y, \phi) \cdot \delta\theta \right] d\Sigma_{\hat{\theta}} = \int_{\Sigma_{\bar{\theta}}} \left[ s(\hat{\theta}, y) + e_\Gamma(\hat{\theta}, y, \phi) \right] \cdot \langle \delta\theta, n_{\hat{\theta}} \rangle \, d\Sigma_{\hat{\theta}}.$

In this case, the gradient is only supported on $\Sigma_{\bar{\bar{\theta}}}$, and its representation has the following form,

$$
j'(\bar{\bar{\theta}}) \cdot \delta\theta = \int_{\Sigma_{\bar{\theta}}} \left[ s(\bar{\bar{\theta}}, \bar{y}) + e_\Gamma(\bar{\bar{\theta}}, \bar{y}, \bar{\phi}) + F(\bar{\bar{\theta}}, \bar{y}) + E_\Omega(\bar{\bar{\theta}}, \bar{y}, \bar{\phi}) \right.
$$

$$
\left. + (H\,I + \partial_n) \cdot (S(\bar{\bar{\theta}}, \bar{y}) + E_\Gamma(\bar{\bar{\theta}}, \bar{y}, \bar{\phi})) \right] \langle \delta\theta, n_{\hat{\theta}} \rangle \, d\Sigma_{\hat{\theta}} = 0 \tag{6.44}
$$

▯

## 6.2.6  Eulerian vs. Lagrangian non-cylindrical derivative

We shall now ask a fundamental question concerning the link between the derivative using perturbations of the identity and Eulerian derivative introduced in [59].

Let us consider a functional $J$,

$$\begin{aligned} J : \mathcal{A} &\longrightarrow \mathcal{R} \\ Q &\mapsto J(Q) \end{aligned} \tag{6.45}$$

with

$$\mathcal{A} \equiv \{Q \in \mathcal{E}^T, \quad Q \subset D\} \tag{6.46}$$

Let us assume that this functional can be parametrized by either the Lagrangian mapping parametrizing the tubes in $\mathcal{A}$ with $j_\ell = J \circ \hat{\theta}$ such that

$$\begin{aligned} j_\ell : \Theta_{ad} &\longrightarrow \mathcal{R} \\ \hat{\theta} &\mapsto j_\ell(\hat{\theta}) = J(Q_{\hat{\theta}}) \end{aligned} \tag{6.47}$$

or the velocity field $V \in \mathcal{V}$ building the tubes in $\mathcal{A}$ with $j_V = J \circ T_t(V)$ such that

$$\begin{aligned} j_e : \mathcal{V} &\longrightarrow \mathcal{R} \\ V &\mapsto j_e(V) = J(Q(V)) \end{aligned} \tag{6.48}$$

The following result holds true,

### THEOREM 6.4

*The differentiability of the functional $j_\ell$ at point $\hat{\theta} \in \Theta_{ad}$ in the direction $\delta\hat{\theta} \overset{def}{=} \delta\theta \circ \hat{\theta} \in \Theta$ is equivalent to the differentiability of $j_e$ with respect to $V \in \mathcal{V}_{ad}$ in the direction $W \in \mathcal{V}$. Furthermore the respective functional derivatives are related by the following relation,*

$$\langle j'_\ell(\hat{\theta}), \delta\theta \rangle = \langle j'_e(V_{\hat{\theta}}), W_{\hat{\theta}} \rangle, \quad \delta\hat{\theta} \overset{def}{=} \delta\theta \circ \hat{\theta} \in \Theta \tag{6.49}$$

*with $V_{\hat{\theta}} = \partial_t \hat{\theta} \circ \hat{\theta}^{-1}$, $W_{\hat{\theta}} = \partial_t(\delta\theta \circ \hat{\theta}) \circ \hat{\theta}^{-1} - D V_{\hat{\theta}} \cdot \delta\theta$ and the transverse field involved in the Eulerian derivative is given by $Z_t = \delta\theta$.*

**PROOF**   As shown here, in the Lagrangian mapping case, we deal with a first-order perturbation of the identity inside $\Omega_{\hat{\theta}}$.

$$\Omega_0 \quad \overset{\hat{\theta}}{\longrightarrow} \quad \Omega_{\hat{\theta}}$$

$$\hat{\theta} + \rho\hat{\theta} \quad \downarrow \quad \diagup \; I + \rho\delta\theta$$

$$\Omega_{\hat{\theta} + \rho\delta\hat{\theta}}$$

The structure of the perturbation rules is similar in the case of an Eulerian description using velocity fields,

$$\Omega_0 \quad \xrightarrow{T_t(V)} \quad \Omega(V)$$

$$T_t(V + \rho W) \quad \downarrow \quad \nearrow \quad T_\rho^t \overset{\text{def}}{=} T_t(V + \rho W) \circ T_t(V)^{-1}$$

$$\Omega(V + \rho W)$$

As proved in [59], the transverse map $T_\rho^t$ is the flow of a transverse velocity field $\mathcal{Z}_\rho^t$. This velocity field is written $Z_t$ for $\rho = 0$ and it satisfies the following Cauchy problem,

$$\begin{cases} \partial_t Z_t + \mathrm{D}\, Z_t \cdot V - \mathrm{D}\, V \cdot Z_t = W, \ (0, T) \\ Z_{t=0} = 0 \end{cases} \tag{6.50}$$

Using chain rule it is possible to deduce the following identity,

$$j'_e(V) \cdot W = \frac{d}{dV} j_e(T_t(V)) \cdot W = \frac{d}{d\theta} j_e(T_t(V)) \cdot (\frac{d}{dV} T_t(V) \cdot W) \tag{6.51}$$

As recalled above, the derivative of the mapping $T_t(V)$ with respect to $V$ is

$$\frac{d}{dV} T_t(V) \cdot W = Z_t \circ T_t(V) \tag{6.52}$$

We then set $\delta \hat{\theta} \overset{\text{def}}{=} \delta \theta \circ \hat{\theta} = Z_t \circ T_t(V)$, which means that $\delta \theta = Z_t$. For a given Lagrangian map $\hat{\theta} \in \Theta_{ad}$, there exists $V_{\hat{\theta}} = \partial_t \hat{\theta} \circ \hat{\theta}^1 \in V$ such that $\hat{\theta} = T_t(V_{\hat{\theta}})$. Hence if we perturbed the Lagrangian map in the direction $\delta \hat{\theta} \overset{\text{def}}{=} \delta \theta \circ \hat{\theta}$, we generate a velocity field perturbation $W_{\hat{\theta}}$ given by the following expression,

$$W_{\hat{\theta}} = \partial_t(\delta \theta \circ \hat{\theta}) \circ \hat{\theta}^{-1} - \mathrm{D}\, V_{\hat{\theta}} \circ \delta \theta \tag{6.53}$$

Indeed, using the chain rule, we have

$$\partial_t(\delta \theta \circ \hat{\theta}) \circ \hat{\theta}^{-1} = \partial_t \delta \theta + \mathrm{D}\, \delta \theta \cdot \partial_t \hat{\theta} \circ \hat{\theta}^{-1}$$
$$= \partial_t \delta \theta + \mathrm{D}\, \delta \theta \cdot V_{\hat{\theta}}$$

As stated before, we can identify the transverse velocity field $Z_t$ with $\delta \theta$. Using the transverse equation we define $W_{\hat{\theta}}$ such that

$$\partial_t \delta \theta + \mathrm{D}\, \delta \theta \cdot V_{\hat{\theta}} = W_{\hat{\theta}} + \mathrm{D}\, V_{\hat{\theta}} \cdot \delta \theta \tag{6.54}$$

from which we deduce that

$$W_{\hat{\theta}} = \partial_t(\delta \theta \circ \hat{\theta}) \circ \hat{\theta}^{-1} - \mathrm{D}\, V_{\hat{\theta}} \cdot \delta \theta \tag{6.55}$$

$\square$

## 6.3   Navier-Stokes equations in moving domain

In this section, we shall apply the previous non-cylindrical shape derivative concepts, in order to solve two questions related to the Navier-Stokes equations in a moving domain:

1. We will first obtain a differentiability result concerning the sensitivity of the solution of Navier-Stokes equations in a moving domain with respect to the Lagrangian mapping $\hat{\theta} \in \Theta_{ad}$. This result allows us to justify in a rigorous manner transpiration boundary conditions [63] in fixed domain as a first order approximation of the Navier-Stokes system in moving domain. This approximation is valid around a fixed domain configuration, i.e., around $\hat{\theta} = I$.

2. Then, we deal with an optimal control problem for the Navier-Stokes system where the control is the displacement of the moving domain $\hat{\theta}$. Using results obtained in the first section, we derive the structure of the gradient for a general cost functional. We recover early results obtained using Eulerian field derivative methods.

### 6.3.1   Transpiration boundary conditions

We consider a viscous incompressible newtonian fluid inside a moving domain $\Omega_{\hat{\theta}}$. Its evolution is described by its velocity $u$ and its pressure $p$. The couple $(u, p)$ satisfies the classical Navier-Stokes equations written in non-conservative form,

$$\begin{cases} \partial_t u + \mathrm{D}\, u \cdot u - \nu \Delta u + \nabla p = 0, & Q_{\hat{\theta}} \\ \mathrm{div}(u) = 0, & Q_{\hat{\theta}} \\ u = V_{\hat{\theta}}, & \Sigma_{\hat{\theta}} \\ u(t=0) = u_0, & \Omega_0 \end{cases} \qquad (6.56)$$

A kinematic continuity boundary condition is satisfied on $\Gamma_{\hat{\theta}}$; this means that the fluid velocity is equal to the velocity of the moving boundary $V_{\hat{\theta}} = \partial_t \hat{\theta} \circ \hat{\theta}^{-1}$ on $\Gamma_{\hat{\theta}}$. The main result of this paragraph is the following,

**THEOREM 6.5**
*Let us assume that*

- *The mapping $\hat{\theta} \in \Theta$ is such that $V_{\hat{\theta}} = \partial_t \hat{\theta} \circ \hat{\theta}^{-1} \in H^1(0, T; H_0^m(D))$ with $m \geq 5/2$,*

*then the solution $(u_{\hat{\theta}}, p_{\hat{\theta}})$ of the Navier-Stokes system (6.56) admits a material shape derivative in $L^2(0, T; H^2(\Omega_{\hat{\theta}})) \cap L^\infty(0, T; H^1(\Omega_{\hat{\theta}}))$. Furthermore,*

$(u_{\hat\theta}, p_{\hat\theta})$ *admits a partial shape derivative* $(u'(\hat\theta) \cdot \delta\theta, p'(\hat\theta) \cdot \delta\theta)$ *solution of the following linear tangent system,*

$$\begin{cases} \partial_t u' + \mathrm{D}\, u' \cdot u + \mathrm{D}\, u \cdot u' - \nu\Delta u' + \nabla p' = 0, & Q_{\hat\theta} \\ \mathrm{div}(u') = 0, & Q_{\hat\theta} \\ u' = \partial_t(\delta\theta \circ \hat\theta) \circ \hat\theta^{-1} - \mathrm{D}\, u \cdot \delta\theta, & \Sigma_{\hat\theta} \\ u'(0) = 0, & \Omega_0 \end{cases} \quad (6.57)$$

**PROOF**  We follow the proof used in [59],[57]. The only difference is that we do not need to use the transerve field $Z_t$. We refer to [13] for the details in case of homogeneous Dirichlet boundary conditions.

We shall detail the computation of the linearized boundary conditions. We perturbed the Lagrangian mapping with the increment $\rho\delta\hat\theta$ and we get the following perturbed boundary conditions written on the fixed boundary $\Gamma_0$,

$$\left.\frac{d}{d\rho} u(\hat\theta + \rho\delta\hat\theta) \circ (\hat\theta + \rho\delta\hat\theta)\right|_{\rho=0} = \left.\frac{d}{d\rho}(\partial_t(\hat\theta + \rho\delta\hat\theta))\right|_{\rho=0}, \quad \text{on } \Gamma_0$$

$$\left.\frac{d}{d\rho} u(\hat\theta + \rho\delta\hat\theta)\right|_{\rho=0} \circ \hat\theta + (\mathrm{D}\, u(\hat\theta)) \circ \hat\theta \cdot \delta\hat\theta = \partial_t \delta\hat\theta$$

$$(u'(\hat\theta) \cdot \delta\theta) \circ \hat\theta + \left[\mathrm{D}\, u(\hat\theta) \cdot \delta\theta\right] \circ \hat\theta = \partial_t(\delta\theta \circ \hat\theta)$$

then we get

$$u'(\hat\theta) \cdot \delta\theta = \partial_t(\delta\theta \circ \hat\theta) \circ \hat\theta^{-1} - \mathrm{D}\, u(\hat\theta) \cdot \delta\theta, \quad \text{on } \Gamma_{\hat\theta} \quad (6.58)$$

$\square$

Applying this theorem in the particular case, where the reference flow is considered in a fixed domain, i.e., with $\hat\theta = I$, leads to the derivation of transpiration boundary condition for linearized system,

### COROLLARY 6.1

*Considering a reference flow* $(u^0, p^0)$ *defined in a fixed domain* $\Omega_0 \times (0, T)$, *i.e., satisfying the following Navier-Stokes system,*

$$\begin{cases} \partial_t u^0 + \mathrm{D}\, u^0 \cdot u^0 - \nu\Delta u^0 + \nabla p^0 = 0, & \Omega_0 \times (0, T) \\ \mathrm{div}(u^0) = 0, & \Omega_0 \times (0, T) \\ u^0 = 0, & \Gamma_0 \times (0, T) \\ u^0(t = 0) = u_0^0, & \Omega_0 \end{cases} \quad (6.59)$$

*The first order variation* $(\delta u, \delta p) = (\mathrm{D}\, u^0 \cdot \delta\theta, \mathrm{D}\, p^0 \cdot \delta\theta) + (u', p')$ *around this flow of the solution* $(u^0, p^0)$ *of problem (6.59) exists and* $(u', p')$ *satisfies the following linear tangent system written in the reference fixed domain,*

$$\begin{cases} \partial_t u' + \mathrm{D}\, u' \cdot u^0 + \mathrm{D}\, u^0 \cdot u' - \nu \Delta u' + \nabla p' = 0, & \Omega_0 \times (0,T) \\ \mathrm{div}(u') = 0, & \Omega_0 \times (0,T) \\ u' = \partial_t(\delta\theta \circ \hat{\theta}) \circ \hat{\theta}^{-1} - (\mathrm{D}\, u^0 \cdot n)\,\langle \delta\theta, n \rangle, & \Gamma_0 \times (0,T) \\ u'(0) = 0, & \Omega_0 \end{cases} \qquad (6.60)$$

**PROOF**    This a direct consequence of theorem (6.5). We only need to check the boundary conditions. In the general case, we have

$$u'(\hat{\theta}) \cdot \delta\theta = \partial_t(\delta\theta \circ \hat{\theta}) \circ \hat{\theta}^{-1} - \mathrm{D}\, u(\hat{\theta}) \cdot \delta\theta, \quad \text{on } \Gamma_{\hat{\theta}} \qquad (6.61)$$

We set $\hat{\theta} = \mathrm{I}$ and we get

$$u' = \partial_t(\delta\theta \circ \hat{\theta}) \circ \hat{\theta}^{-1} - \mathrm{D}\, u^0 \cdot \delta\theta, \quad \Gamma_0 \times (0,T)$$

We then use

$$\mathrm{D}\, u^0 \cdot \delta\theta = (\mathrm{D}_\Gamma\, u^0 + \mathrm{D}\, u^0 n \otimes n) \cdot \delta\theta$$

But $u^0 = 0$ on $\Gamma_0$, then $\mathrm{D}_\Gamma\, u^0 = 0$ on $\Gamma_0$, from which we deduce

$$u' = \partial_t(\delta\theta \circ \hat{\theta}) \circ \hat{\theta}^{-1} - (\mathrm{D}\, u^0 \cdot n)\,\langle \delta\theta, n \rangle, \quad \Gamma_0 \times (0,T)$$

⬚

### 6.3.2   Optimal control of the Navier-Stokes system by moving the domain

In this paragraph, we shall deal with an optimal control for the Navier-Stokes system, where the control is the displacement of the domain boundary $\Gamma_{\hat{\theta}}$. We choose to minimize a general cost functional based on $(\hat{\theta}, u_{\hat{\theta}}, p_{\hat{\theta}})$. We are interested in solving the following minimization problem :

$$\min_{\hat{\theta} \in \mathcal{U}} j(\hat{\theta}) \qquad (6.62)$$

where $j(\hat{\theta}) = J_{\hat{\theta}}(u(\hat{\theta}), p(\hat{\theta}))$ with $(u(\hat{\theta}), p(\hat{\theta}))$ is a weak solution of problem (6.56) and $J_{\hat{\theta}}(u, p)$ is a real functional of the following form :

$$J_{\hat{\theta}}(u, p) = \frac{\alpha}{2}\|\mathcal{B}\, u\|^2_{L^2(Q_{\hat{\theta}})} + \frac{\gamma}{2}\|\mathcal{K}\, V_{\hat{\theta}}\|^2_{L^2(\Sigma_{\hat{\theta}})} \qquad (6.63)$$

where $\mathcal{B} \in \mathcal{L}(\mathcal{H}, L^2(Q_{\hat{\theta}}))$ is a general linear differential operator satisfying the following identity,

$$\langle \mathcal{B}\, u, v \rangle_{L^2(Q_{\hat{\theta}})} + \langle u, \mathcal{B}^* v \rangle_{L^2(Q_{\hat{\theta}})} = \langle \mathcal{B}_\Sigma\, u, v \rangle_{L^2(\Sigma_{\hat{\theta}})} \qquad (6.64)$$

and $\mathcal{K} \in \mathcal{L}(\mathcal{U}, L^2(\Sigma_{\hat\theta}))$ is a general linear differential operator satisfying the following identity,

$$\langle \mathcal{K}\, u, v\rangle_{L^2(\Sigma_{\hat\theta})} + \langle u, \mathcal{K}^* v\rangle_{L^2(\Sigma_{\hat\theta})} = \langle \mathcal{K}_\Sigma\, u, v\rangle_{L^2(\Sigma_{\hat\theta})} \tag{6.65}$$

The main difficulty in dealing with such a minimization problem is related to the fact that integrals over the domain $\Omega_{\hat\theta}$ depend on the control variable $\hat\theta$.

**THEOREM 6.6**
*For $V_{\hat\theta} \stackrel{\text{def}}{=} \partial_t \hat\theta \circ \hat\theta^{-1} \in \mathcal{U}$ and $\Omega_0$ of class $\mathcal{C}^2$, the functional $j(\hat\theta)$ possesses a gradient $\nabla j(\hat\theta)$ which is supported on the moving boundary $\Gamma_{\hat\theta}$ and can be represented by the following expression,*

$$\nabla j(\hat\theta) = -\,\partial_t E_{\hat\theta} - (\mathrm{div}_\Gamma\, V_{\hat\theta})\, E_{\hat\theta} - \mathrm{D}\, E_{\hat\theta} \cdot V_{\hat\theta}$$
$$-\,(-\sigma(\varphi, \pi) \cdot n_{\hat\theta} + \alpha \mathcal{B}_\Sigma \mathcal{B}\, u) \cdot \mathrm{D}\, u - \gamma(-\mathcal{K}^* \mathcal{K} + \mathcal{K}_\Sigma \mathcal{K})\, V_{\hat\theta} \cdot \mathrm{D}_\Gamma\, V_{\hat\theta}$$
$$+\,\frac{1}{2}\left[ \alpha\, |\mathcal{B} u|^2 + \gamma\, |\mathcal{K} V_{\hat\theta}|^2\right] n_{\hat\theta} \tag{6.66}$$

*where $(\varphi, \pi)$ stands for the adjoint fluid state solution of the following system,*

$$\begin{cases} -\partial_t \varphi - \mathrm{D}\,\varphi \cdot u + {}^* \mathrm{D}\, u \cdot \varphi - \nu \Delta \varphi + \nabla \pi = -\alpha\, \mathcal{B}^* \mathcal{B}\, u, & Q_{\hat\theta} \\ \mathrm{div}(\varphi) = 0, & Q_{\hat\theta} \\ \varphi = 0, & \Sigma_{\hat\theta} \\ \varphi(T) = 0, & \Omega_T \end{cases} \tag{6.67}$$

*and*

$$E_{\hat\theta} = -\sigma(\varphi, \pi) \cdot n_{\hat\theta} + \gamma \left[ -\mathcal{K}^* \mathcal{K} + \mathcal{K}_\Sigma \mathcal{K}\right] V_{\hat\theta} + \alpha \mathcal{B}_\Sigma \mathcal{B}\, V_{\hat\theta} \tag{6.68}$$

**PROOF** Using chain rule, we obtain the differentiability of $j(\hat\theta)$ in the perturbation direction $\delta\hat\theta \stackrel{\text{def}}{=} \delta\theta \circ \hat\theta \in \Theta$. Using theorems (6.1) and (6.2), we can state

$$\langle j'(\hat\theta), \delta\theta\rangle = \alpha\langle \mathcal{B} u, \mathcal{B} u'\rangle_{L^2(Q_{\hat\theta})} + \gamma\langle \mathcal{K} V_{\hat\theta}, \mathcal{K}(\partial_t(\delta\hat\theta) \circ \hat\theta^{-1} - \mathrm{D}_\Gamma\, V_{\hat\theta} \cdot \delta\theta)\rangle_{L^2(\Sigma_{\hat\theta})}$$
$$+\,\langle \frac{1}{2}\left[ \alpha\, |\mathcal{B} u|^2 + \gamma\, |\mathcal{K} V_{\hat\theta}|^2\right] n_{\hat\theta}, \delta\theta\rangle_{L^2(\Sigma_{\hat\theta})}$$

Using the adjoint operators, we perform an integration by parts,

$$\langle j'(\hat\theta), \delta\theta\rangle =$$
$$dis - \alpha\langle \mathcal{B}^* \mathcal{B}\, u, u'\rangle_{L^2(Q_{\hat\theta})} + \gamma\langle (-\mathcal{K}^* \mathcal{K} + \mathcal{K}_\Sigma \mathcal{K})\, V_{\hat\theta}, \partial_t(\delta\hat\theta) \circ \hat\theta^{-1} - \mathrm{D}_\Gamma\, V_{\hat\theta} \cdot \delta\theta\rangle_{L^2(\Sigma_{\hat\theta})}$$
$$+\,\langle \alpha \mathcal{B}_\Sigma \mathcal{B}\, u, u'\rangle_{L^2(\Sigma_{\hat\theta})} + \langle \frac{1}{2}\left[ \alpha\, |\mathcal{B} u|^2 + \gamma\, |\mathcal{K} V_{\hat\theta}|^2\right] n_{\hat\theta}, \delta\theta\rangle_{L^2(\Sigma_{\hat\theta})}$$

We now define the fluid adjoint state $(\varphi, \pi)$ solution of the adjoint problem (5.8) and we deduce the following identity,

$$
\int_{Q_{\hat{\theta}}} [\partial_t u' + \mathrm{D}\, u' \cdot u + \mathrm{D}\, u \cdot u' - \nu \Delta u' + \nabla p'] \, \varphi - \int_{Q_{\hat{\theta}}} \pi \, \mathrm{div}\, u'
$$

$$
= \int_{Q_{\hat{\theta}}} [-\partial_t \varphi - \mathrm{D}\, \varphi \cdot u + {}^* \mathrm{D}\, u \cdot \varphi - \nu \Delta \varphi + \nabla \pi] \, u' - \int_{Q_{\hat{\theta}}} p' \, \mathrm{div}\, \varphi
$$

$$
+ \int_{\Sigma_{\hat{\theta}}} [p'\, n \cdot \varphi - \nu \varphi \cdot \partial_n u' + \nu u' \cdot \partial_n \varphi - u' \cdot \pi n] \quad (6.69)
$$

Hence we get

$$
-\alpha \langle \mathcal{B}^* \mathcal{B}\, u, u' \rangle_{L^2(Q_{\hat{\theta}})} = -\langle \sigma(\varphi, \pi) \cdot n_{\hat{\theta}}, u' \rangle_{L^2(\Sigma_{\hat{\theta}})} \qquad (6.70)
$$

Using the linearized system (5.83) satisfied by $u'$, we deduce that

$$
u' = \partial_t(\delta\theta \circ \hat{\theta}) \circ \hat{\theta}^{-1} - \mathrm{D}\, u \cdot \delta\theta, \quad \text{on } \Sigma_{\hat{\theta}}
$$

then,

$$
\langle j'(\hat{\theta}), \delta\theta \rangle = \langle -\sigma(\varphi, \pi) \cdot n_{\hat{\theta}} + \alpha \mathcal{B}_{\Sigma} \mathcal{B}\, u + \gamma(-\mathcal{K}^* \mathcal{K} + \mathcal{K}_{\Sigma} \mathcal{K}) \, V_{\hat{\theta}}, \partial_t(\delta\hat{\theta}) \circ \hat{\theta}^{-1} \rangle_{L^2(\Sigma_{\hat{\theta}})}
$$

$$
- \langle -\sigma(\varphi, \pi) \cdot n_{\hat{\theta}} + \alpha \mathcal{B}_{\Sigma} \mathcal{B}\, u, \mathrm{D}\, u \cdot \delta\theta \rangle_{L^2(\Sigma_{\hat{\theta}})}
$$

$$
- \langle \gamma(-\mathcal{K}^* \mathcal{K} + \mathcal{K}_{\Sigma} \mathcal{K}) \, V_{\hat{\theta}}, \mathrm{D}_\Gamma \, V_{\hat{\theta}} \cdot \delta\theta \rangle_{L^2(\Sigma_{\hat{\theta}})}
$$

$$
+ \langle \frac{1}{2} \left[ \alpha \,|\mathcal{B}u|^2 + \gamma \,|\mathcal{K}V_{\hat{\theta}}|^2 \right] n_{\hat{\theta}}, \delta\theta \rangle_{L^2(\Sigma_{\hat{\theta}})}
$$

We now set $E_{\hat{\theta}} = -\sigma(\varphi, \pi) \cdot n_{\hat{\theta}} + \alpha \mathcal{B}_{\Sigma} \mathcal{B}\, u + \gamma\,(-\mathcal{K}^* \mathcal{K} + \mathcal{K}_{\Sigma} \mathcal{K}) \, V_{\hat{\theta}}$ and we use theorem (6.3) and we obtain

$$
\langle j'(\hat{\theta}), \delta\theta \rangle = \langle \left[ -\partial_t E_{\hat{\theta}} - (\mathrm{div}_\Gamma \, V_{\hat{\theta}}) \, E_{\hat{\theta}} - \mathrm{D}\, E_{\hat{\theta}} \cdot V_{\hat{\theta}} \right]
$$

$$
- (-\sigma(\varphi, \pi) \cdot n_{\hat{\theta}} + \alpha \mathcal{B}_{\Sigma} \mathcal{B}\, u) \cdot \mathrm{D}\, u - \gamma(-\mathcal{K}^* \mathcal{K} + \mathcal{K}_{\Sigma} \mathcal{K}) \, V_{\hat{\theta}} \cdot \mathrm{D}_\Gamma \, V_{\hat{\theta}}
$$

$$
+ \frac{1}{2} \left[ \alpha \,|\mathcal{B}u|^2 + \gamma \,|\mathcal{K}V_{\hat{\theta}}|^2 \right] n_{\hat{\theta}}, \, \delta\theta \rangle_{L^2(\Sigma_{\hat{\theta}})}
$$

from which we deduce the expression of the gradient $\nabla\, j(\hat{\theta})$. $\qquad\square$

### 6.3.3   Comparison with the Eulerian derivative version

Using theorem (6.4), we can reciprocally deduce the expression of one of the directional derivatives from another. Let us apply this equivalence on the optimal control treaten previously.

We have established the structure of the Lagrangian derivative of the functional $j(\hat{\theta})$,

$$\langle j'(\hat{\theta}), \delta\theta\rangle = \langle -\sigma(\varphi, \pi) \cdot n_{\hat{\theta}} + \alpha \mathcal{B}_\Sigma \mathcal{B} u + \gamma(-\mathcal{K}^*\mathcal{K} + \mathcal{K}_\Sigma \mathcal{K}) V_{\hat{\theta}}, \partial_t(\delta\hat{\theta}) \circ \hat{\theta}^{-1}\rangle_{L^2(\Sigma_{\hat{\theta}})}$$
$$- \langle -\sigma(\varphi, \pi) \cdot n_{\hat{\theta}} + \alpha \mathcal{B}_\Sigma \mathcal{B} u, D\, u \cdot \delta\theta\rangle_{L^2(\Sigma_{\hat{\theta}})}$$
$$- \langle \gamma(-\mathcal{K}^*\mathcal{K} + \mathcal{K}_\Sigma \mathcal{K}) V_{\hat{\theta}}, D_\Gamma V_{\hat{\theta}} \cdot \delta\theta\rangle_{L^2(\Sigma_{\hat{\theta}})}$$
$$+ \langle \frac{1}{2}\left[\alpha\, |\mathcal{B}u|^2 + \gamma\, |\mathcal{K}V_{\hat{\theta}}|^2\right] n_{\hat{\theta}}, \delta\theta\rangle_{L^2(\Sigma_{\hat{\theta}})}$$

Let us set $\hat{\theta} = T_t(V_{\hat{\theta}})$, $\delta\theta = Z_t$ and $\partial_t(\delta\theta \circ \hat{\theta}) \circ \hat{\theta}^{-1} = W_{\hat{\theta}} + D\, V_{\hat{\theta}} \cdot \delta Z_t$; we get

$$\langle j'(T_t(V_{\hat{\theta}})), Z_t\rangle = \langle -\sigma(\varphi, \pi) \cdot n_{\hat{\theta}} + \alpha \mathcal{B}_\Sigma \mathcal{B} u + \gamma(-\mathcal{K}^*\mathcal{K} + \mathcal{K}_\Sigma \mathcal{K}) V_{\hat{\theta}}, W_{\hat{\theta}}$$
$$+ D\, V_{\hat{\theta}} \cdot Z_t\rangle_{L^2(\Sigma(V_{\hat{\theta}}))}$$
$$- \langle -\sigma(\varphi, \pi) \cdot n_{\hat{\theta}} + \alpha \mathcal{B}_\Sigma \mathcal{B} u, D\, u \cdot Z_t\rangle_{L^2(\Sigma_{\hat{\theta}})}$$
$$- \langle \gamma(-\mathcal{K}^*\mathcal{K} + \mathcal{K}_\Sigma \mathcal{K}) V_{\hat{\theta}}, D_\Gamma V_{\hat{\theta}} \cdot Z_t\rangle_{L^2(\Sigma(V_{\hat{\theta}}))}$$
$$+ \langle \frac{1}{2}\left[\alpha\, |\mathcal{B}u|^2 + \gamma\, |\mathcal{K}V_{\hat{\theta}}|^2\right], \langle Z_t, n_{\hat{\theta}}\rangle\rangle_{L^2(\Sigma(V_{\hat{\theta}}))}$$

We recall that

$$D\, V_{\hat{\theta}} \cdot Z_t = D_\Gamma V_{\hat{\theta}} \cdot Z_t + (D\, V_{\hat{\theta}} \cdot n_{\hat{\theta}}) \langle Z_t, n_{\hat{\theta}}\rangle$$
$$D\, u \cdot Z_t = D_\Gamma u \cdot Z_t + (D\, u \cdot n_{\hat{\theta}}) \langle Z_t, n_{\hat{\theta}}\rangle$$

Using $u = V_{\hat{\theta}}$, on $\Gamma_{\hat{\theta}}$ we deduce that $D\, u \cdot Z_t = D_\Gamma V_{\hat{\theta}} \cdot Z_t + (D\, u \cdot n_{\hat{\theta}}) \langle Z_t, n_{\hat{\theta}}\rangle$, then we deduce

$$\langle j'(T_t(V_{\hat{\theta}})), Z_t\rangle = \langle -\sigma(\varphi, \pi) \cdot n_{\hat{\theta}} + \alpha \mathcal{B}_\Sigma \mathcal{B} u + \gamma(-\mathcal{K}^*\mathcal{K} + \mathcal{K}_\Sigma \mathcal{K}) V_{\hat{\theta}}, W_{\hat{\theta}}\rangle_{L^2(\Sigma(V_{\hat{\theta}}))}$$
$$\langle -\sigma(\varphi, \pi) \cdot n_{\hat{\theta}} + \alpha \mathcal{B}_\Sigma \mathcal{B} u, D(V_{\hat{\theta}} - u) \cdot n_{\hat{\theta}} \langle Z_t, n_{\hat{\theta}}\rangle\rangle_{L^2(\Sigma_{\hat{\theta}})}$$
$$+ \langle \gamma(-\mathcal{K}^*\mathcal{K} + \mathcal{K}_\Sigma \mathcal{K}) V_{\hat{\theta}}, D\, V_{\hat{\theta}} \cdot n_{\hat{\theta}} \langle Z_t, n_{\hat{\theta}}\rangle\rangle_{L^2(\Sigma(V_{\hat{\theta}}))}$$
$$+ \langle \frac{1}{2}\left[\alpha\, |\mathcal{B}u|^2 + \gamma\, |\mathcal{K}V_{\hat{\theta}}|^2\right], \langle Z_t, n_{\hat{\theta}}\rangle\rangle_{L^2(\Sigma(V_{\hat{\theta}}))}$$

We choose a canonical extension $V_{\hat{\theta}} \circ p$. This means that $D\, V_{\hat{\theta}} \cdot n_{\hat{\theta}} = 0$, on $\Gamma_{\hat{\theta}}$, then

$$\langle j'_e(V_{\hat{\theta}}), W_{\hat{\theta}}\rangle = \langle j'(T_t(V_{\hat{\theta}})), Z_t\rangle = \langle -\sigma(\varphi, \pi) \cdot n_{\hat{\theta}} + \alpha \mathcal{B}_\Sigma \mathcal{B} u$$
$$+ \gamma(-\mathcal{K}^*\mathcal{K} + \mathcal{K}_\Sigma \mathcal{K}) V_{\hat{\theta}}, W_{\hat{\theta}}\rangle_{L^2(\Sigma(V_{\hat{\theta}}))}$$
$$+ \langle \left[\sigma(\varphi, \pi) \cdot n_{\hat{\theta}} - \alpha \mathcal{B}_\Sigma \mathcal{B} u\right] \cdot D\, u \cdot n_{\hat{\theta}}$$
$$+ \frac{1}{2}\left[\alpha\, |\mathcal{B}u|^2 + \gamma\, |\mathcal{K}V_{\hat{\theta}}|^2\right], \langle Z_t, n_{\hat{\theta}}\rangle\rangle_{L^2(\Sigma_{\hat{\theta}})}$$

This is exactly the expression obtained in [57], using Eulerian methods. Let us define the adjoint transverse boundary field $\lambda_{\hat{\theta}}$ solution of the following tangential dynamical system,

$$\begin{cases} -\partial_t \lambda_{\hat{\theta}} - \nabla_\Gamma \lambda_{\hat{\theta}} \cdot V_{\hat{\theta}} - \mathrm{div}(V_{\hat{\theta}})\,\lambda_{\hat{\theta}} = f, & (0,T) \\ \lambda_{\hat{\theta}}(T) = 0, & \Gamma_T(V) \end{cases} \qquad (6.71)$$

with

$$f = \left[ \sigma(\varphi,\pi)\cdot n_{\hat{\theta}} - \alpha\mathcal{B}_\Sigma\mathcal{B}\,u \right] \cdot \mathrm{D}\,u\cdot n_{\hat{\theta}} + \frac{1}{2}\left[ \alpha\,|\mathcal{B}u|^2 + \gamma\,|\mathcal{K}V_{\hat{\theta}}|^2 \right] \qquad (6.72)$$

then

$$\nabla j_e(V_{\hat{\theta}}) = -\lambda_{\hat{\theta}}\,n_{\hat{\theta}} - \sigma(\varphi,\pi)\cdot n_{\hat{\theta}} + \alpha\,\mathcal{B}_\Sigma\mathcal{B}\,u + \gamma\left[ -\mathcal{K}^*\mathcal{K}\,V_{\hat{\theta}} + \mathcal{K}_\Sigma\mathcal{K}\,V_{\hat{\theta}} \right] \quad (6.73)$$

where $(\varphi,\pi)$ stands for the adjoint fluid state.

# Chapter 7

## Sensitivity analysis for a simple fluid-solid interaction system

### 7.1 Introduction

This chapter deals with the analysis of an inverse boundary problem arising in the study of bridge deck aeroelastic stability. The aeroelastic system consists of an elastically supported rigid solid moving inside an incompressible fluid flow in 2-D. As described in the introduction of this lecture note, the keystone of the sensitivity analysis of coupled fluid-structure model is based on the moving shape analysis introduced in the previous chapters.

The chapter is organized in four parts :

- In section 7.2, we introduce the notations and the mechanical system we shall deal with and we state the main result of this chapter, namely the structure of a cost functional gradient with respect to inflow boundary conditions pertubations. The proof is given in the remaining part of the chapter.

- In section 7.3, we recall early results concerning the well-posedness of the coupled fluid-structure system.

- In section 7.4, we introduce the minimization problem and its associated Lagrangian functional.

- In section 7.5, we perform derivation of the Lagrangian with respect to state variables. This allows us to obtain the structure of the adjoint variables involved in the cost functional gradient.

### 7.2 Mathematical settings

We consider a two dimensionnal flexible structure in rigid motion. For the sake of simplicity, we only consider one degree of freedom for the structural motion : the vertical displacement $d(t)e_2$ where $e_2$ is the element of cartesian

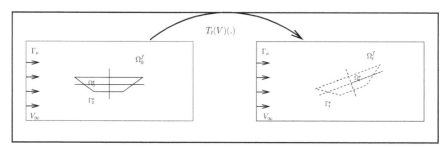

**FIGURE 7.1**:   Arbitrary Euler-Lagrange map

basis $(e_1, e_2, e_3)$ in $\mathbb{R}^3$.

The structure is surrounded by a viscous fluid in the plane $(e_1, e_2)$. We consider a control volume $\Omega \subset \mathbb{R}^2$ containing the solid for every time $t \geq 0$. Hence, the analysis of the coupled problem is set in $\Omega \times (0, T)$ where $T > 0$ is an arbitrary time.

In order to deal with such a coupled system, we introduce a diffeomorphic map sending a fix reference domain $\Omega_0$ into the physical configuration $\Omega$ at time $t \geq 0$.

Without loss of generality, we choose the reference configuration to be the physical configuration at initial time $\Omega(t = 0)$.

Hence, we define a map $T_t \in \mathcal{C}^1(\overline{\Omega_0})$ such that

$$\overline{\Omega_t^f} = T_t(\overline{\Omega_0^f}),$$
$$\overline{\Omega_t^s} = T_t(\overline{\Omega_0^s})$$

Since we only consider one degree of freedom motion, we write

$$\overline{\Omega_t^s} = \overline{\Omega_0^s} + d(t)e_2$$

We set $\Sigma^s \equiv \bigcup_{0<t<T} (\{t\} \times \Gamma_t^s)$, $Q^f \equiv \bigcup_{0<t<T} (\{t\} \times \Omega_t^f)$ and $\Sigma_\infty^f \equiv \Gamma_\infty^f \times (0, T)$

The map $T_t$ can be actually defined as the flow of a particular vector field, as described in the following lemma :

**LEMMA 7.1 [147]**

*Assuming that $d(.)$ is smooth enough, there exists a vector field $V$ that builds $Q$ , i.e,*

$$\overline{\Omega_t^f} = T_t(V)(\overline{\Omega_0^f}),$$
$$\overline{\Omega_t^s} = T_t(V)(\overline{\Omega_0^s})$$

*where $T_t(V)$ is the solution of the following dynamical system :*

$$T_t(V) : \Omega_0 \longrightarrow \Omega$$
$$x_0 \longmapsto x(t, x_0) \equiv T_t(V)(x_0)$$

*with*

$$\frac{\mathrm{d}\,x}{\mathrm{d}\,\tau} = V(\tau, x(\tau)), \ \tau \in [0, T]$$
$$x(\tau = 0) = x_0, \quad in \ \Omega_0 \tag{7.1}$$

In our simple case, we can give an example of an appropriate flow vector field :

$$\begin{cases} V(x, t) = \dot{d}(t)e_2, & x \in \overline{\Omega_t^s} \\ V(x, t) = \mathrm{Ext}(\dot{d}(t)e_2), & x \in \Omega_t^f \\ V(x, t) \cdot n = 0, & x \in \Gamma_\infty^f \end{cases} \tag{7.2}$$

where Ext is an arbitrary extension operator from $\Gamma_0^s$ into $\Omega_0^f$. The map $T_t$ is usually referred to as the Arbitrary Euler-Lagrange map.
The solid is described by the evolution of its displacement and its velocity and the couple $(d, \dot{d})$ is the solution of the following ordinary second order differential equation :

$$\begin{cases} m\ddot{d} + kd = F_f, \\ [d, \dot{d}] (t = 0) = [d_0, d_1] \end{cases} \tag{7.3}$$

where $(m, k)$ stand for the structural mass and stiffness. $F_f$ is the projection of the fluid loads on $\Gamma_t^s$ along the direction of motion $e_2$.
The fluid is assumed to be a viscous incompressible newtonian fluid. Its evolution is described by its velocity $u$ and its pressure $p$. The couple $(u, p)$ satisfies the classical Navier-Stokes equations written in non-conservative form:

$$\begin{cases} \partial_t u + \mathrm{D}\,u \cdot u - \nu \Delta u + \nabla p = 0, & Q^f(V) \\ \mathrm{div}(u) = 0, & Q^f(V) \\ u = u_\infty, & \Sigma_\infty^f \\ u(t = 0) = u_0, & \Omega_0^f \end{cases} \tag{7.4}$$

where $\nu$ stands for the kinematic viscosity and $u_\infty$ is the farfield velocity field. Hence, the projected fluid loads $F_f$ have the following expression :

$$F_f = - \left( \int_{\Gamma_t^s} \sigma(u, p) \cdot n \right) \cdot e_2 \tag{7.5}$$

where $\sigma(u, p) = -p\mathrm{I} + \nu(\mathrm{D}\,u + {}^*\mathrm{D}\,u)$ stands for the fluid stress tensor inside $\Omega_t^f$, with $(\mathrm{D}\,u)_{ij} = \partial_j u_i = u_{i,j}$.
We complete the whole system with kinematic continuity conditions at the fluid-structure interface $\Gamma_t^s$ :

$$u = V = \dot{d}e_2, \ \mathrm{sur} \ \Sigma^s(V) \tag{7.6}$$

To summarize, we get the following coupled system :

$$\begin{cases} \partial_t u + \mathrm{D}\,u \cdot u - \nu \Delta u + \nabla p = 0, & Q^f(V) \\ \mathrm{div}(u) = 0, & Q^f(V) \\ u = u_\infty, & \Sigma^f_\infty \\ u = \dot{d}\,e_2, & \Sigma^s(V) \\ m\,\ddot{d} + k\,d = - \left( \int_{\Gamma^s_t} \sigma(u,p) \cdot n \right) \cdot e_2, & (0,T) \\ \left[ u, d, \dot{d} \right](t=0) = \left[ u_0, d_0, d_1 \right], & \Omega^f_0 \times \mathbb{R}^2 \end{cases} \tag{7.7}$$

**Main Result:** For $u_\infty \in \mathcal{U}_c$ the following minimization problem,

$$\min_{u_\infty \in \mathcal{U}_c} j(u_\infty) \tag{7.8}$$

where $j(u_\infty) = J_{u_\infty}\left( \left[ u, p, d, \dot{d} \right](u_\infty) \right)$ with $\left[ u, p, d, \dot{d} \right](u_\infty)$ is a weak solution of problem (7.7) and $J_{u_\infty}$ is a real functional of the following form :

$$J_{u_\infty}\left( \left[ u, p, d, \dot{d} \right] \right) = \frac{\alpha}{2} \int_0^T \left[ |d - d^1_g|^2 + |\dot{d} - d^2_g|^2 \right] + \frac{\gamma}{2} \|u_\infty\|^2_{\mathcal{U}_c} \tag{7.9}$$

admits at least a solution $u^*_\infty$ which satisfies the following necessary first-order optimality conditions,

$$\nabla j(u^*_\infty) = (\sigma(\varphi_{u^*_\infty}, \pi_{u^*_\infty}) \cdot n)|_{\Sigma^f_\infty} + \gamma\, u^*_\infty = 0 \tag{7.10}$$

with $(\varphi, \pi, b_1, b_2)$ are solutions of the following adjoint system,

$$\begin{cases} -\partial_t \varphi - \mathrm{D}\,\varphi \cdot u + (^*Du) \cdot \varphi - \nu \Delta \varphi + \nabla q = 0, & Q^f(V) \\ \mathrm{div}(\varphi) = 0, & Q^f(V) \\ \varphi = b_2\,e_2, & \Sigma^s(V) \\ \varphi = 0, & \Sigma^f_\infty \\ \varphi(T) = 0, & \Omega^f_T \end{cases} \tag{7.11}$$

$$\begin{cases} -\dot{b}_1 + k\,b_2 = \alpha(d - d^1_g), & (0,T) \\ b_1(T) = 0, \end{cases} \tag{7.12}$$

$$\begin{cases} -\partial_t \lambda - \nabla_\Gamma \lambda \cdot V = f, & \Sigma^s(\dot{d}\,e_2) \\ \lambda(T) = 0, & \Gamma_T(\dot{d}\,e_2) \end{cases} \tag{7.13}$$

with $f = \left[ -\dot{d}\,b_2 + \nu\,(\mathrm{D}\varphi \cdot n) \cdot (\mathrm{D}\,u \cdot n) - |\dot{d}|^2\,(\mathrm{D}\varphi \cdot e_2) \cdot e_2 \right]$ and

$$\int_{\Gamma^s_t(V)} \lambda\,n = \left[ -b_1 - m\,\dot{b}_2 - \alpha(\dot{d} - d^2_g) \right] e_2 + \int_{\Gamma^s_t(V)} \sigma(\phi, \pi) \cdot n \tag{7.14}$$

**REMARK 7.1**     We can eliminate the auxiliary adjoint variables $(\lambda, b_1)$, in order to get a system only involving the adjoint variables $(\varphi, \pi, b_2)$ . We then replace equations (7.2), (7.13) , (7.14) by the following second order ODE,

$$
\begin{cases}
m\ddot{b}_2 + k\, b_2 = \alpha(d - d_g^1) - \alpha(\dot{d} - \dot{d}_g^2) + \partial_t \left( \int_{\Gamma_t^s(V)} \sigma(\phi, \pi) \cdot n \right) \cdot e_2 \\[4mm]
+ \int_{\Gamma_t^s(V)} \left[ |\dot{d}|^2 \, (\mathrm{D}\,\varphi \cdot e_2) \cdot e_2 - \nu \, (\mathrm{D}\,\varphi \cdot n) \cdot (\mathrm{D}\,u \cdot n) \right] \cdot n & (0, T) \\[4mm]
\left[ b_2, \dot{b}_2 \right] (T) = [0, 0]
\end{cases}
$$

$$(7.15)$$

□

---

## 7.3     Well-posedness of the coupled system

We are interested in recalling global existence results for weak solutions to the initial boundary value problem (7.7). One should expect some restrictions on the existence time for the solutions since depending on the data, the solid may vanish outside the control volume $\Omega$.

This problem was recently investigated by several authors [131], [39], [54], [68], [81], [133]. We refer to [80] and [81] for a complete review. In our case, we only have one degree of freedom for the solid motion and we are dealing with two dimensional Navier-Stokes equations. Nevertheless, we deal with non-homogeneous Dirichlet boundary conditions at the farfield fluid boundary.

### *THEOREM 7.1*

*Assume the following hypothesis :*

*i)* $\Omega_0^s$, $\Omega_0$ *are of class* $\mathcal{C}^2$,

*ii)* $a = dist(\Gamma_\infty^f, \Omega_0^s) > 0$,

*iii)*

$$
\begin{aligned}
u_0 &\in (L^2(\Omega_0^f))^2 \quad u_\infty \in (H^m(\Sigma_\infty^f))^2 \; m > \tfrac{3}{4} \\
\mathrm{div}(u_0) &= 0, \qquad dans \; \Omega_0^f \\
u_0 &= d_1 e_2, \qquad sur \; \Gamma_0^s
\end{aligned}
$$

$$(7.16)$$

*then there exists a positive real time* $T_0 = T_0(u_0, a, \Omega_0^s, \Omega_0)$ *such that for any* $T \in (0, T_0)$, *there exists at least one weak solution to the initial-boundary value problem (7.7), with*

$$
d \in W^{1,\infty}(0, T; \mathbb{R}) \cap \mathcal{C}^0([0, T]; \mathbb{R})
$$

$$
(u, \dot{d}) \in L^2(0, T; \mathcal{V}_{d(.)}) \cap L^\infty(0, T; \mathcal{H}_{d(.)})
$$

*with*

$$\mathcal{H}_{d(t)} \equiv \Big\{ (v, \ell) \in (L^2(\Omega_0))^2 \times \mathbb{R}, \quad div(v) = 0,$$

$$v \cdot n = 0, \quad on \ \ \Gamma^f_\infty, \quad v|_{\Omega^s_t} = \ell \cdot e_2, \ supp(v) \subset \Omega^f_t \Big\}$$

*and*

$$\mathcal{V}_{d(t)} \equiv \Big\{ (v, \ell) \in (H^1(\Omega_0))^2 \times \mathbb{R}, \quad div(v) = 0,$$

$$v = 0, \quad on \ \ \Gamma^f_\infty, \quad v|_{\Omega^s_t} = \ell \cdot e_2, \ supp(v) \subset \Omega^f_t \Big\}$$

*and satisfies the following identity :*

$$- \int_0^T \left[ \int_{\Omega^f_t} u \cdot \partial_t v + m \dot{d}\, \dot{\ell} - k\, d\, \ell \right]$$

$$+ \int_0^T \int_{\Omega^f_t} [(D\, u \cdot u) \cdot v + \nu\, D\, u \cdot\cdot\, D\, v] = m d_1\, \ell(0) + \int_{\Omega^f_0} u_0 \cdot v(0) \tag{7.17}$$

$$\forall\, (v, \ell) \in \mathcal{C}^1([0, T];\ \mathcal{V}_{d(.)})\ with\ v(T) = \dot{\ell}(T) = 0$$

**REMARK 7.2**   Using results from Fursikov *et al* [72], we can relax the regularity assumption for $u_\infty$ and ask :

iii)

$$u_\infty \cdot \tau \in\ L^2(0, T;\ (H^{\frac{1}{2}}(\Gamma^f_\infty))^2) \cap H^{\frac{1}{4}}(0, T;\ (L^2(\Gamma^f_\infty))^2)$$

$$u_\infty \cdot n \in\ L^2(0, T;\ (H^{\frac{1}{2}}(\Gamma^f_\infty))^2) \cap H^{\frac{3}{4}}(0, T;\ (H^{-1}(\Gamma^f_\infty))^2)$$

$\Box$

## 7.4   Inverse problem settings

We are interested in solving the following minimization problem :

$$\min_{u_\infty \in \mathcal{U}_c} j(u_\infty) \tag{7.18}$$

where $j(u_\infty) = J_{u_\infty}\left( \left[ u, p, d, \dot{d} \right] (u_\infty) \right)$ with $\left[ u, p, d, \dot{d} \right] (u_\infty)$ is a weak solution of problem (7.7) and $J_{u_\infty}$ is a real functional of the following form :

$$J_{u_\infty}\left( \left[ u, p, d, \dot{d} \right] \right) = \frac{\alpha}{2} \int_0^T \left[ |d - d^1_g|^2 + |\dot{d} - d^2_g|^2 \right] + \frac{\gamma}{2} \|u_\infty\|^2_{\mathcal{U}_c} \tag{7.19}$$

The main difficulty in dealing with such a minimization problem is related to the dependence of integrals on the unknown domain $\Omega_t^f$ which depends also on the control variable $u_\infty$. This point will be solved by using the ALE map $T_t$ introduced previously.

### 7.4.1  Analysis strategy

We shall focus our efforts on the derivation of first-order optimality conditions for problem (7.18). This involves the computation of the gradient with respect to the inflow condition $u_\infty$ of the cost function $J_{u_\infty}$.
There exist several methods to handle such a question,

- Following [58], it is possible to handle the derivative $\frac{D}{Du_\infty}(u, p, d) \cdot \delta u_\infty$ using a back transport map into a fix domain and use the weak implicit function theorem to justify and obtain the linearized system. Once the full linear tangent system is defined, it is possible to define an adjoint system whose solution may be involved in the computation of the objective function gradient.

- An other possible choice is to try to pass through the computation of a linear tangent system and directly get the adjoint system. This may be realized using a Min-Max formulation of the minimization problem (7.18) taking into account the coupled system constraint through Lagrange multipliers.

In this chapter, we shall use the latter choice coupled with the introduction of a transverse map in order to handle the sensitivity analysis of the Lagrangian functional with respect to variation of the fluid domain.

### 7.4.2  Free divergence and non-homogeneous Dirichlet boundary condition constraints

We shall now describe the way to take into account inside the Lagrangian functional several constraints associated to the coupled system.
The divergence free condition coming from the fact that the fluid has an homogeneous density and evolves as an incompressible flow is difficult to impose on the mathematical and numerical point of view. We choose to include the divergence free condition directly into the Lagrangian functional thanks to a multiplier that may play the role of the adjoint variable associated to the primal pressure variable. This leads in a certain sense to a saddle point formulation or mixed formulation of the Navier-Stokes subsystem. We shall use the following identity,

$$-\int_{\Omega_t^f} q \operatorname{div} u = \int_{\Omega_t^f} u \cdot \nabla q - \int_{\Gamma_\infty^f} u \cdot (q\,n) - \int_{\Gamma_t^s} u \cdot (q\,n) \qquad (7.20)$$

The coupled system (7.7) involves two essential non-homogeneous Dirichlet boundary conditions,

$$u = u_\infty, \ \Sigma_\infty^f \tag{7.21}$$

$$u = \dot{d}\, e_2, \ \Sigma_0^s \tag{7.22}$$

We use a very weak formulation of the state equation, consisting in totally transposing the Laplacian operator,

$$\int_{\Omega_t^f} -\nu \Delta u \cdot \phi = \int_{\Omega_t^f} -\nu \Delta \phi \cdot u + \int_{\Gamma_\infty^f \cup \Gamma_t^s} \nu \left[ u \cdot \partial_n \phi - \phi \cdot \partial_n u \right] d\Gamma \tag{7.23}$$

We shall substitute inside this identity the desirable boundary conditions and we will recover the boundary constraints in performing an integration by parts inside the optimality conditions corresponding to the sensitivity with respect to the multipliers. This procedure has been already used in [135] and [45, 46] to perform shape optimization problems for elliptic equations using Min-Max principles.

We shall also choose to transpose the time operator inside the weak formulation. This has to be performed very carefully, since we are dealing with a moving domain,

$$\int_0^T \int_{\Omega_t^f} \partial_t u \cdot v = - \int_0^T \int_{\Omega_t^f} u \cdot \partial_t v - \int_0^T \int_{\partial \Omega_t^f} (u \cdot v) \langle V, n \rangle \tag{7.24}$$

$$+ \int_{\Omega_T^f} u(T) \cdot v(T) - \int_{\Omega_0^f} u(0) \cdot v(0) \tag{7.25}$$

**REMARK 7.3**    This kind of technique has been popularized in [99] as a systematic way to study non-homogeneous linear partial differential equations. These formulations are usually called very weak formulations or transposition formulations. We shall notice that these methods are still valid in the non-linear case to obtain regularity or existence results. We refer to [4] for recent applications on the Navier-Stokes system. □

## 7.4.3    Solid reduced order and solid weak state operator

For the sake of simplicity, we reduce the order of the solid governing equation by defining the global solid variable,

$$\tilde{d} = \begin{pmatrix} d_1 \\ d_2 \end{pmatrix} = \begin{pmatrix} d \\ \dot{d} \end{pmatrix} \tag{7.26}$$

leading to the first order ordinary differential equation,

$$M\,\dot{\tilde{d}} + K\,\tilde{d} = F \tag{7.27}$$

with

$$M = \begin{bmatrix} 1 & 0 \\ 0 & m \end{bmatrix}$$

$$K = \begin{bmatrix} 0 & -1 \\ k & 0 \end{bmatrix}$$

and

$$F = \begin{pmatrix} 0 \\ F_f \end{pmatrix}$$

In the case of the coupled fluid-solid system, the loads depend on the fluid state variable,

$$F_f(u, p) = - \left( \int_{\Gamma_t^s} \sigma(u, p) \cdot n \right) \cdot e_2$$

We introduce the state and multiplier spaces in order to define a proper solid weak state operator,

$$X_1^s = \{ d_1 \in \mathcal{C}^1([0, T]) \} \tag{7.28}$$

$$X_2^s = \{ d_2 \in \mathcal{C}^1([0, T]) \} \tag{7.29}$$

$$Y_1^s = \{ b_1 \in \mathcal{C}^1([0, T]) \} \tag{7.30}$$

$$Y_2^s = \{ b_2 \in \mathcal{C}^1([0, T]) \} \tag{7.31}$$

and the load space,

$$L = \{ F \in \mathcal{C}^1([0, T]) \}$$

This allows us to define a solid state operator,

$$e^s : X_1^s \times X_2^s \times L \longrightarrow (Y_1^s \times Y_2^s)^*$$

whose action is defined by the following identity,

$$\langle e^s(d_1, d_2, F), (b_1, b_2) \rangle = \int_0^T \left[ -d_1 \, \dot{b}_1 - d_2 \, b_1 \right] - d_1^0 b_1(0) + d_1(T) b_1(T)$$

$$+ \int_0^T \left[ -m \, d_2 \, \dot{b}_2 + k \, d_1 \, b_2 \right] - m \, d_2^0 \, b_2(0) + m \, d_2(T) \, b_2(T) - \int_0^T F \, b_2$$

## 7.4.4   Fluid state operator

In this section, we summarize the different options that we have chosen for the Lagrangian framework and define the variational state operator constraint.

In order to deal with rigid displacement vector fields, we introduce the following spaces :

Rigid displacement spaces:

$$\mathcal{H}^1 \stackrel{\text{def}}{=} \{ \phi \in (H^1(\Omega_0))^2; \quad \nabla \phi = 0, \quad \text{in } \Omega_0^s \}$$

$$\mathcal{H}_0^1 \stackrel{\text{def}}{=} \{ \phi \in (H^1(\Omega_0))^2; \quad \nabla \phi = 0, \quad \text{in } \Omega_0^s, \quad \phi = 0, \quad \text{on } \Gamma_\infty^f \}$$

**LEMMA 7.2**
*For $\varphi \in \mathcal{H}^1$, assuming that $\Omega_0^s$ is connected, there exists $C \in \mathbb{R}^2$ such that*

$$\varphi|_{\Omega_0^s} = C$$

In the sequel, we will need to define precise state and multiplier spaces in order to endow our problem with a Lagrangian functional framework.
Following the existence result stated previously, we introduce the fluid state space :

$$X^f \overset{\text{def}}{=} \left\{ u \in H^2(0, T; (H^2(\Omega_t^f))^2 \cap \mathcal{H}^1) \right\}$$

$$Z^f \overset{\text{def}}{=} \left\{ p \in H^1(0, T; (H^1(D))^2) \right\}$$

We also need test function spaces that will be useful to define Lagrange multipliers :

$$Y^f \overset{\text{def}}{=} \left\{ v \in L^2(0, T; (H^2(\Omega_t^f))^2 \cap \mathcal{H}_0^1) \right\}$$

$$V^f \overset{\text{def}}{=} \left\{ q \in H^1(0, T; (H^1(D))^2) \right\}$$

$$W_s^f \overset{\text{def}}{=} \left\{ (v, b_2) \in Y^f \times Y_2^s, \ v|_{\Gamma_t^s} = b_2 \cdot e_2 \right\}$$

We define the fluid weak state operator,

$$e_{u_\infty}^f : X^f \times Z^f \times U^s \longrightarrow (Y^f \times V^f)^*$$

whose action is defined by :

$$\langle e_{u_\infty}^f(u, p, u^s), (v, q) \rangle =$$

$$\int_0^T \int_{\Omega_t^f} [-u \cdot \partial_t v + (\mathrm{D}\, u \cdot u) \cdot v - \nu u \cdot \Delta v + u \cdot \nabla q - p \,\mathrm{div}\, v]$$

$$+ \int_0^T \int_{\Gamma_\infty^f} u_\infty \cdot (\sigma(v, q) \cdot n) + \int_0^T \int_{\Gamma_t^s} [u^s \cdot (\sigma(v, q) \cdot n) - (u \cdot v)\langle u^s, n \rangle]$$

$$- \int_0^T \int_{\Gamma_t^s} v \cdot (\sigma(u, p) \cdot n) + \int_{\Omega_T} u(T) \cdot v(T) - \int_{\Omega_0} u_0 \cdot v(t = 0)$$

$$\forall\, (v, q) \in Y^f \times V^f$$

## 7.4.5   Coupled system operator

Our mechanical system consists of a solid part and a fluid part. These subsystems have been represented thanks to a solid and a fluid state operator. It is now possible to couple these two operators in order to build an ad-hoc coupled system operator.

The major point here is to notice that the fluid load $F_f$ appears in the fluid state operator and then can be coupled with the solid part thanks to the input load $F$ of the solid operator. To achieve this coupling, we need to decide whether or not the fluid and the solid multipliers match at the fluid-solid interface. If not, we have to work with the fluid constraint tensor at the fluid-solid boundary, which may be not convenient due to a regularity requirement. Hence, we choose to work with continuous test functions on $\Gamma_t^s$. This means that we shall choose the fluid and second solid multiplier spaces to be the space $W_s^f$. We define the coupled system weak state operator as follows,

$$e_{u_\infty} : Y^f \times Z^f \times X_1^s \times X_2^s \longrightarrow (W_s^f \times V^f \times Y_1^s)^*$$

whose action is defined by the following identity,

$$\langle e_{u_\infty}(u,p,d_1,d_2),(v,q,b_1,b_2)\rangle =$$
$$\langle e^s(d_1,d_2,F_f),(b_1,b_2)\rangle + \langle e_{u_\infty}^f(u,p,d_2\,e_2),(v,q,d_1\cdot e_2)\rangle$$
$$= \int_0^T \int_{\Omega_t^f} [-u\cdot\partial_t v + (D\,u\cdot u)\cdot v - \nu u\cdot\Delta v + u\cdot\nabla q - p\,\mathrm{div}\,v]$$
$$+ \int_0^T \int_{\Gamma_\infty^f} u_\infty\cdot(\sigma(v,q)\cdot n) + \int_0^T \int_{\Gamma_t^s} [(d_2\,e_2)\cdot(\sigma(v,q)\cdot n) - (u\cdot v)\langle d_2\,e_2,n\rangle]$$
$$+ \int_0^T \left[-d_1\,\dot{b}_1 - d_2\,b_1\right] + \int_0^T \left[-m\,d_2\,\dot{b}_2 + k\,d_1\,b_2\right]$$
$$+ \int_{\Omega_T} u(T)\cdot v(T) - \int_{\Omega_0} u_0\cdot v(t=0)$$
$$- d_1^0\,b_1(0) + d_1(T)\,b_1(T) - md_2^0\,b_2(0) + m\,d_2(T)\,b_2(T),$$
$$\forall\,(v,b_2,q,b_1)\in W_s^f \times V^f \times Y_1^s$$

## 7.4.6   Min-Max problem

In this section, we introduce the Lagrangian functional associated with problem (7.7) and problem (7.18) :

$$\mathcal{L}_{u_\infty}(u,p,d_1,d_2;v,q,b_1,b_2) \overset{\text{def}}{=} J_{u_\infty}(u,p,d_1,d_2) - \langle e_{u_\infty}(u,p,d_1,d_2),(v,q,b_1,b_2)\rangle \tag{7.32}$$

with

$$J_{u_\infty}(u,p,d_1,d_2) = \frac{\alpha}{2}\left[\int_0^T |d_1 - d_g^1|^2 + |d_2 - d_g^2|^2\right] + \frac{\gamma}{2}\|u_\infty\|_{u_c}^2$$

Using this functional, the optimal control problem (7.18) can be put in the following form :

$$\min_{u_\infty \in \mathcal{U}_c} \quad \min_{\substack{(u,p,d_1,d_2) \in \\ \mathcal{X}^f \times Z^f \times X_1^s \times X_2^s}} \quad \max_{\substack{(v,b_2,q,b_1) \in \\ W_s^f \times V^f \times Y_1^s}} \quad \mathcal{L}_{u_\infty}(u,p,d_1,d_2;v,q,b_1,b_2)$$

(7.33)

By using the Min-Max framework, we avoid the computation of the state derivative with respect to $u_\infty$. First-order optimality conditions will furnish the gradient of the original cost function using the solution of an adjoint problem.

We would like to apply min-max differentiability results to problem (7.33). The main issue is to prove that the min-max subproblem

$$\min_{(u,p,d_1,d_2) \in X^f \times Z^f \times X_1^s \times X_2^s} \quad \max_{(v,b_2,q,b_1) \in W_s^f \times V^f \times Y_1^s} \quad \mathcal{L}_{u_\infty}(u,p,d_1,d_2;v,q,b_1,b_2)$$

(7.34)

admits at least one saddle point for $u_\infty \in \mathcal{U}_c$.

### Reduced Gradient

We assume that the conditions to apply the Min-Max principle [40] are fulfilled so we can bypass the derivation with respect to the control variable $u_\infty$ through the min-max subproblem (7.34). It leads to the following result :

**THEOREM 7.2**
*For $u_\infty \in \mathcal{U}_c$, and $(u_{u_\infty}, p_{u_\infty}, \tilde{d}_{u_\infty}, \varphi_{u_\infty}, \pi_{u_\infty}, \tilde{b}_{u_\infty})$ the unique saddle point of problem (7.34), the gradient of the cost function $j$ at point $u_\infty \in \mathcal{U}_c$ is given by the following expression :*

$$\nabla j(u_\infty) = (\sigma(\varphi_{u_\infty}, \pi_{u_\infty}) \cdot n)|_{\Sigma_\infty^f} + \gamma\, u_\infty \tag{7.35}$$

**PROOF**    Using theorem (3) from [45, 46], we bypass the derivation with respect to $u_\infty$ inside the min-max subproblem (7.34) :

$$< j'(u_\infty), \delta u_\infty >=$$

$$\langle \frac{D}{Du_\infty} \mathcal{L}_{u_\infty}(u_{u_\infty}, p_{u_\infty}, d_{u_\infty}^1, d_{u_\infty}^2; \phi_{u_\infty}, \pi_{u_\infty}, b_{u_\infty}^1, b_{u_\infty}^2), \delta u_\infty \rangle$$

$$= \min_{\substack{(u,p,d_1,d_2) \in \\ X^f \times Z^f \times X_1^s \times X_2^s}} \quad \max_{\substack{(v,b_2,q,b_1) \in \\ W_s^f \times V^f \times Y_1^s}} \quad \langle \frac{D}{Du_\infty} \mathcal{L}_{u_\infty}(u,p,d_1,d_2;v,q,b_1,b_2), \delta u_\infty \rangle$$

$$= \int_0^T \int_{\Gamma_\infty^f} (\nu\, \partial_n \phi_{u_\infty} - \pi_{u_\infty} n) \cdot \delta u_\infty + \gamma \int_0^T \int_{\Gamma_\infty^f} u_\infty \cdot \delta u_\infty$$

$\square$

## 7.5  KKT Optimality Conditions

In this section, we are interested in establishing the first order optimality condition for problem (7.34), better known as Karusch-Kuhn-Tucker optimality conditions. This step is crucial, because it leads to the formulation of the adjoint problem satisfied by the Lagrange multipliers $(\varphi_{u_\infty}, \pi_{u_\infty}, b^1_{u_\infty}, b^2_{u_\infty})$. We recall the expression of the Lagrangian,

$$\mathcal{L}_{u_\infty}(u, p, d_1, d_2; v, q, b_1, b_2) \overset{\text{def}}{=} J_{u_\infty}(u, p, d_1, d_2) - \langle e_{u_\infty}(u, p, d_1, d_2), (v, q, b_1, b_2) \rangle \tag{7.36}$$

The KKT system will have the following structure :

$$\partial_{(v,q,b_1,b_2)}\mathcal{L}_{u_\infty}(u, p, d_1, d_2; v, q, b_1, b_2) \cdot (\delta v, \delta q, \delta b_1, \delta b_2) = 0,$$
$$\forall (\delta v, \delta b_2, \delta q, \delta b_1) \in W^f_s \times V^f \times Y^s_1 \rightarrow \quad \text{State Equations}$$
$$\partial_{(u,p,d_1,d_2)}\mathcal{L}_{u_\infty}(u, p, d_1, d_2; v, q, b_1, b_2) \cdot (\delta u, \delta p, \delta d_1, \delta d_2) = 0,$$
$$\forall (\delta u, \delta p, \delta d_1, \delta d_2) \in X^f \times Z^f \times X^s_1 \times X^s_2 \rightarrow \quad \text{Adjoint Equations}$$

### 7.5.1  Derivatives with respect to state variables

**Fluid adjoint system**

**LEMMA 7.3**
*For $u_\infty \in \mathcal{U}_c$, $(p, b_1, b_2, v, b_2, q, b_1) \in Z^f \times X^s_1 \times X^s_2 \times W^f_s \times V^f \times Y^s_1$, $\mathcal{L}_{u_\infty}(u, p, d_1, d_2; v, q, b_1, b_2)$ is differentiable with respect to $u \in Y^f$ and we have*

$$\langle \partial_u \mathcal{L}_{u_\infty}(u, p, d_1, d_2; v, q, b_1, b_2), \delta u \rangle =$$
$$- \int_0^T \int_{\Omega^f_t} [-\delta u \cdot \partial_t v + [\mathrm{D}\, \delta u \cdot u + \mathrm{D}\, u \cdot \delta u] \cdot v - \nu \delta u \cdot \Delta v + \delta u \cdot \nabla q]$$
$$+ \int_0^T \int_{\Gamma^s_t} (\delta u \cdot v) \langle d_2\, e_2, n \rangle - \int_{\Omega_T} \delta u(T) \cdot v(T) \quad \forall \delta u \in X^f$$

In order to obtain a strong formulation of the fluid adjoint problem, we perform some integration by parts :

**LEMMA 7.4**

$$\int_{\Omega^f_t} (\mathrm{D}\, \delta u \cdot u) \cdot v = - \int_{\Omega^f_t} [\mathrm{D}\, v \cdot u + \mathrm{div}(u)\, v] \cdot \delta u + \int_{\Gamma^f_\infty \cup \Gamma^s_t} (\delta u \cdot v) \langle u, n \rangle$$

It leads to the following identity :

$$\langle \partial_{\hat{u}} \mathcal{L}_{u_\infty}(u, p, d_1, d_2; \varphi, \pi, b_1, b_2), \delta u \rangle =$$

$$- \int_{Q^f} [-\partial_t \varphi + (^*Du) \cdot \varphi - (D\varphi) \cdot u - \mathrm{div}(u)\,\varphi - \nu \Delta \varphi + \nabla \pi] \cdot \delta u$$

$$- \int_{\Omega_T^f} \varphi(T) \cdot \delta u(T)$$

**LEMMA 7.5**

*For $u_\infty \in \mathcal{U}_c$, $(u, b_1, b_2, v, b_2, q, b_1) \in X^f \times X_1^s \times X_2^s \times W_s^f \times V^f \times Y_1^s$, $\mathcal{L}_{u_\infty}(u, p, d_1, d_2; v, q, b_1, b_2)$ is differentiable with respect to $p \in Z^f$ and we have*

$$\langle \partial_p \mathcal{L}_{u_\infty}(\hat{u}, p, d_1, d_2; \varphi, q, b_1, b_2), \delta p \rangle = \int_0^T \int_{\Omega_t^f} \delta p \,\mathrm{div}\,\varphi, \quad \forall \delta p \in Z^f$$

This leads to the following fluid adjoint strong formulation,

$$\begin{cases} -\partial_t \varphi - \mathrm{D}\,\varphi \cdot u + (^*Du) \cdot \varphi - \nu \Delta \varphi + \nabla q = 0, \, Q^f \\ \mathrm{div}(\varphi) = 0, & Q^f \\ \varphi = b_2 \cdot e_2, & \Sigma^s \\ \varphi = 0, & \Sigma_\infty^f \\ \varphi(T) = 0, & \Omega_T^f \end{cases} \quad (7.37)$$

## 7.5.2  Solid adjoint system

We recall that the ALE map is built as the flow of a vector field $V$ that matches the solid velocity at the fluid-solid interface and is arbitrary inside the fluid domain, i.e., using the reduced oder model,

$$\begin{cases} V(x,t) = d_2 \cdot e_2, & x \in \overline{\Omega_t^s} \\ V(x,t) = \mathrm{Ext}(d_2 \cdot e_2), & x \in \Omega_t^f \\ V(x,t) \cdot n = 0, & x \in \Gamma_\infty^f, \end{cases} \quad (7.38)$$

Hence, the ALE map depends on $d_2$ through $V$. Furthermore, the derivative with respect to $d_1$ may be simpler since it does not involve a derivative with respect to the geometry. Then, we have the following result,

**LEMMA 7.6**

*For $u_\infty \in \mathcal{U}_c$ and $(u, p, d_1, d_2, v, b_2, q, b_1) \in X^f \times Z^f \times X_1^s \times X_2^s \times W_s^f \times V^f \times Y_1^s$, the Lagrangian $\mathcal{L}_{u_\infty}(u, p, d_1, d_2; v, q, b_1, b_2)$ is differentiable with respect to $d_1 \in$*

$X_1^s$ and we have

$$\langle \partial_{d_1} \mathcal{L}_{u_\infty}(u, p, d_1, d_2; \varphi, \pi, b_1, b_2), \delta d_1 \rangle =$$

$$\int_0^T \left[ \alpha\,(d_1 - d_g^1)\,\delta d_1 - k\,\delta d_1\,b_2 + \delta d_1\,\dot{b}_1 \right] - \delta d_1(T)\,b_1(T)$$

From which we deduce the following adjoint ODE,

$$\begin{cases} -\dot{b}_1 + k\,b_2 = \alpha(d_1 - d_g^1), \ (0, T) \\ b_1(T) = 0 \end{cases} \tag{7.39}$$

The derivative of the Lagrangian with respect to the solid velocity variable $d_2$ involves shape derivatives of domain integrals.

This point has been investigated previously by Zolésio in [157], [156] and in [59], [57]. Then we need to introduce the concept of Transverse Field associated to a perturbation of the solid velocity.

We introduce a perturbation flow $W$ associated to the perturbation $\delta d_2 e_2$. For example,

$$W = \mathrm{Ext}(\delta d_2 \cdot e_2)$$

This flow generates new fluid and solid domains through the ALE map, $T_t(V + \rho W)$, with $\rho \geq 0$. We set

$$\Omega_t^{f,\rho} \overset{\text{def}}{=} T_t(V + \rho W)(\Omega_0^f)$$

$$\Omega_t^{s,\rho} \overset{\text{def}}{=} T_t(V + \rho W)(\Omega_0^s)$$

For the sake of simplicity, we set

$$T_t^\rho \overset{\text{def}}{=} T_t(V + \rho W)$$

The objective of this paragraph is to compute the following derivative :

$$\left. \left( \frac{d}{d\rho} \mathcal{L}(\hat{u}, p, d_1, d_2 + \rho \delta d_2; \hat{v}, q, b_1, b_2) \right) \right|_{\rho=0}$$

**Transverse map and vector field**

Since we would like to differentiate the Lagrangian functional with respect to $\rho$ at point $\rho = 0$, it is convenient to work with functions already defined in $\Omega_t^{f,\rho=0} \overset{\text{def}}{=} \Omega_t^f$ as we proceed to the limit $\rho \to 0$.

To this end, we introduce a new map as in [157] :

$$T_\rho^t \overset{\text{def}}{=} T_t^\rho \circ T_t^{-1} : \begin{array}{c} \Omega_t^f \longrightarrow \Omega_t^{f,\rho} \\ \Omega_t^s \longrightarrow \Omega_t^{s,\rho} \end{array} \tag{7.40}$$

This map is actually the flow of a vector field following for $\rho \in (0, \rho_0)$,

**THEOREM 7.3**
*The Transverse map $T_t^\rho$ is the flow of a transverse field $\mathcal{Z}_\rho^t$ defined as follows:*

$$\mathcal{Z}_\rho^t \stackrel{\text{def}}{=} \mathcal{Z}^t(\rho, .) = \left( \frac{\partial T_\rho^t}{\partial \rho} \right) \circ (T_\rho^t)^{-1} \tag{7.41}$$

*i.e., is the solution of the following dynamical system:*

$$T_t^\rho(\mathcal{Z}_\rho^t) : \Omega \longrightarrow \Omega$$
$$x \longmapsto x(\rho, x) \equiv T_t^\rho(\mathcal{Z}_\rho^t)(x)$$

*with*

$$\frac{\mathrm{d}\, x(\rho)}{\mathrm{d}\, \rho} = (\rho, x(\rho)), \; \rho \geq 0$$
$$x(\rho = 0) = x, \qquad in \; \Omega \tag{7.42}$$

Since, we only consider derivatives at point $\rho = 0$, we set $Z_t \stackrel{\text{def}}{=} \mathcal{Z}_{\rho=0}^t$. We recall a result from [58] which might be useful for the sequel,

**THEOREM 7.4**
*The mapping,*

$$[0, \rho_0] \longrightarrow \mathcal{C}^0([0, T]; W^{k-1,\infty}(\Omega))$$
$$\rho \longmapsto T_t(V + \rho W)$$

*is continuously differentiable and*

$$S^t(\rho, .) \stackrel{\text{def}}{=} \partial_\rho [T_t(V + \rho W)] = \int_0^t [D(V + \rho W) \circ T_\tau(V + \rho W) \cdot S^\tau(\rho, .)$$
$$+ W \circ T_\tau(V + \rho W) d\tau]$$

**COROLLARY 7.1**
$S_t(.) \stackrel{\text{def}}{=} S^t(\rho = 0, .)$ *is the unique solution of the following Cauchy problem,*

$$\begin{cases} \partial_t S_t - (D\, V \circ T_t) \cdot S_t = W \circ T_t, \; \Omega_0 \times (0, T) \\ S_{t=0} = 0, \qquad\qquad\qquad\quad \Omega_0 \end{cases} \tag{7.43}$$

A fundamental result furnishes the dynamical system satisfied by the vector field $Z_t$ related to the vector fields $(V, W)$,

**THEOREM 7.5 [59]**
*The vector field $Z_t$ is the unique solution of the following Cauchy problem,*

$$\begin{cases} \partial_t Z_t + [Z_t, V] = W, & \Omega_0 \times (0, T) \\ Z_{t=0} = 0, & \Omega_0 \end{cases} \tag{7.44}$$

*where $[Z_t, V] \overset{def}{=} DZ_t \cdot V - DV \cdot Z_t$ stands for the Lie bracket of the pair $(Z_t, V)$.*

**Shape derivatives**

In the sequel, we will need general results concerning shape derivatives of integrals over domains or boundaries. We will use the framework developed in Sokolowski-Zolésio [135].

**LEMMA 7.7**

$$\frac{d}{d\rho}\left(\int_{\Omega_t^{f,\rho}} G(\rho)\, d\Omega\right)\bigg|_{\rho=0} = \int_{\Omega_t^f} \partial_\rho G(\rho)\, d\Omega + \int_{\Gamma_t^s} G(\rho = 0)\langle Z_t, n\rangle\, d\Gamma \tag{7.45}$$

**LEMMA 7.8**

$$\frac{d}{d\rho}\left(\int_{\Gamma_t^{s,\rho}} \phi(\rho)\, d\Gamma\right)\bigg|_{\rho=0} = \int_{\Gamma_t^s} \left[\phi'_\Gamma + H\phi\,\langle Z_t, n\rangle\right] d\Gamma \tag{7.46}$$

*where $\phi'_\Gamma$ stands for the tangential shape derivative of $\phi(\rho, .) \in L^1(\Gamma_t^s)$.*

We recall classical definitions of shape derivative functions :

**DEFINITION 7.1** *For $\phi(\rho, x) \in \mathcal{C}^0(0, \rho_0; \mathcal{C}^1(\Gamma_t^{s,\rho}))$, the material derivative is given by the following expression :*

$$\dot{\phi} = \frac{d}{d\rho}\left(\phi(\rho, .) \circ T_\rho^t\right)\bigg|_{\rho=0}$$

*then the tangential shape derivative of $\phi$ is given by the following expression,*

$$\phi'_\Gamma \overset{def}{=} \dot{\phi} - \nabla_\Gamma \phi \cdot Z_t$$

**REMARK 7.4** *If $\phi$ is the trace of a vector field $\tilde{\phi}$ defined over $\Omega$, then we have*

$$\phi'_\Gamma \overset{def}{=} \tilde{\phi}'|_{\Gamma_t^s} + \partial_n \tilde{\phi}\,\langle Z_t, n\rangle$$

*with $\tilde{\phi}'|_{\Gamma_t^s} \overset{def}{=} \frac{d}{d\rho}\left(\tilde{\phi}(\rho, .)\right)\bigg|_{\rho=0}\bigg|_{\Gamma_t^s}$.*     ⊓

Following this remark, we have

**LEMMA 7.9**

$$\frac{d}{d\rho}\left(\int_{\Gamma_t^{s,\rho}}\tilde{\phi}(\rho,x)da\right)\bigg|_{\rho=0} = \int_{\Gamma_t^s}\left[\tilde{\phi}' + \left[H\tilde{\phi} + \partial_n\tilde{\phi}\right]\langle Z_t, n\rangle\right]d\Gamma \qquad (7.47)$$

**Derivation of the perturbed Lagrangian**

Thanks to the introduction of the transverse map, it is now possible to work with functions $(u,v)$ that are defined on $\Omega_t^f$. This leads to the following perturbed Lagrangian :

$$\mathcal{L}_{u_\infty}^\rho \stackrel{\text{def}}{=} \mathcal{L}(u,p,d_1,d_2+\rho\delta d_2;v,q,b_1,b_2)$$
$$= J_{u_\infty}(u,p,d_1,d_2+\rho\delta d_2)$$
$$-\left[\int_0^T\int_{\Omega_t^{f,\rho}}\left[-(u\circ\mathcal{R}_\rho^t)\cdot\partial_t(v\circ\mathcal{R}_\rho^t) + (D\,u\circ\mathcal{R}_\rho^t\cdot u\circ\mathcal{R}_\rho^t)\cdot v\circ\mathcal{R}_\rho^t\right.\right.$$
$$-\nu u\circ\mathcal{R}_\rho^t\cdot\Delta v\circ\mathcal{R}_\rho^t + u\circ\mathcal{R}_\rho^t\cdot\nabla q - p\,\text{div}\,v\circ\mathcal{R}_\rho^t\right] + \int_0^T\int_{\Gamma_\infty^f}u_\infty\cdot(\sigma(v,q)\cdot n)$$
$$+\int_0^T\int_{\Gamma_t^{s,\rho}}\left[((d_2+\rho\delta d_2)\,e_2)\cdot(\sigma(v\circ\mathcal{R}_\rho^t,q)\cdot n^\rho)\right.$$
$$\left.-u\circ\mathcal{R}_\rho^t\cdot v\circ\mathcal{R}_\rho^t\cdot((d_2+\rho\delta d_2)\,e_2)\cdot n^\rho\right]$$
$$+\int_0^T\left[-d_1\,\dot{b}_1 - (d_2+\rho\delta d_2)\,b_1\right] + \int_0^T\left[-m\,(d_2+\rho\delta d_2)\,\dot{b}_2 + k\,d_1\,b_2\right]$$
$$+\int_{\Omega_T}u(T)\cdot v(T) - \int_{\Omega_0}u_0\cdot v(t=0) - d_1^0\,b_1(0) + d_1(T)\,b_1(T) - m\,d_2^0\,b_2(0)$$
$$+m\,(d_2+\rho\delta d_2)(T)\,b_2(T), \quad \forall\,(v,b_2,q,b_1)\in W_s^f\times V^f\times Y_1^s$$

with $\mathcal{R}_\rho^t \stackrel{\text{def}}{=} (\mathcal{T}_\rho^t)^{-1}$. We apply the previous results to the perturbed Lagrangian functional $\mathcal{L}$.

a) **Distributed terms:**
   We set

$$G(\rho,.) = \left[-u\circ\mathcal{R}_\rho^t\cdot\partial_t(v\circ\mathcal{R}_\rho^t) + D(u\circ\mathcal{R}_\rho^t)\cdot(u\circ\mathcal{R}_\rho^t)\cdot v\circ\mathcal{R}_\rho^t\right.$$
$$\left.-\nu(u\circ\mathcal{R}_\rho^t)\cdot\Delta(v\circ\mathcal{R}_\rho^t) + (u\circ\mathcal{R}_\rho^t)\cdot\nabla q - p\,\text{div}(v\circ\mathcal{R}_\rho^t)\right]$$

with $\mathcal{R}_\rho^t \stackrel{\text{def}}{=} (\mathcal{T}_\rho^t)^{-1}$.
We need the following lemmas in order to derivate $G(\rho,.)$ with respect to $\rho$,

**LEMMA 7.10**

$$\left(\frac{dT_\rho^t}{d\rho}\right)\Bigg|_{\rho=0} = Z_t$$

$$\left(\frac{d\mathcal{R}_\rho^t}{d\rho}\right)\Bigg|_{\rho=0} = -Z_t$$

**LEMMA 7.11**

$$\left(\frac{d\left(u \circ \mathcal{R}_\rho^t\right)}{d\rho}\right)\Bigg|_{\rho=0} = -\mathrm{D}\,u \cdot Z_t$$

**PROOF**  Using the chain rule we get

$$\begin{aligned}
\left(\frac{d}{d\rho}\left(u \circ \mathcal{R}_\rho^t\right)\right)\Bigg|_{\rho=0} &= \left(\mathrm{D}\,u \circ \mathcal{R}_\rho^t\right) \cdot \left(\frac{\mathrm{D}\,\mathcal{R}_\rho^t}{\mathrm{D}\,\rho}\right)\Bigg|_{\rho=0}\\
&= -\left(\mathrm{D}\,u \circ \mathcal{R}_\rho^t\right) \cdot \mathcal{Z}^t(\rho,.)\big|_{\rho=0}\\
&= -\mathrm{D}\,u \cdot Z_t
\end{aligned}$$

▯

**LEMMA 7.12**

*Then, we have the following result*

$$\begin{aligned}
\partial_\rho G(\rho,.)|_{\rho=0} = &[(\mathrm{D}\,u \cdot Z_t) \cdot \partial_t v + u \cdot (\partial_t(\mathrm{D}\,v \cdot Z_t))\\
&- [(\mathrm{D}(\mathrm{D}\,u \cdot Z_t)) \cdot u + \mathrm{D}\,u \cdot (\mathrm{D}\,u \cdot Z_t)] \cdot v - (\mathrm{D}\,u \cdot u) \cdot (\mathrm{D}\,v \cdot Z_t)\\
&+ \nu(\mathrm{D}\,u \cdot Z_t) \cdot \Delta v + \nu u \cdot (\Delta(\mathrm{D}\,v \cdot Z_t)) + p\,\mathrm{div}(\mathrm{D}\,v \cdot Z_t) - (\mathrm{D}\,u \cdot Z_t) \cdot \nabla q]
\end{aligned}$$

**PROOF**  It comes easily using definition of $G(\rho,.)$ and lemma (7.10)-(7.11).
▯

Then we have an expression of the derivative of distributed terms coming from the Lagrangian with respect to $\rho$,

$$\frac{d}{d\rho}\left(\int_{\Omega_t^{f,\rho}} G(\rho)d\Omega\right)\bigg|_{\rho=0} = \int_{\Omega_t^f} [(\mathrm{D}\,u \cdot Z_t) \cdot \partial_t v + u \cdot (\partial_t(\mathrm{D}\,v \cdot Z_t))$$
$$- [(\mathrm{D}(\mathrm{D}\,u \cdot Z_t)) \cdot u + \mathrm{D}\,u \cdot (\mathrm{D}\,u \cdot Z_t)] \cdot v - (\mathrm{D}\,u \cdot u) \cdot (\mathrm{D}\,v \cdot Z_t)$$
$$+\nu(\mathrm{D}\,u \cdot Z_t) \cdot \Delta v + \nu u \cdot (\Delta(\mathrm{D}\,v \cdot Z_t)) + p\,\mathrm{div}(\mathrm{D}\,v \cdot Z_t) - (\mathrm{D}\,u \cdot Z_t) \cdot \nabla q]$$
$$+ \int_{\Gamma_t} [-u \cdot \partial_t v + (Du \cdot u) \cdot v - \nu u \cdot \Delta v + u \cdot \nabla q - p\,\mathrm{div}(v)]\,\langle Z_t, n\rangle$$

b) Boundary terms :
   We must now take into account the terms coming from the moving boundary $\Gamma_t^{s,\rho}$. Then we set

$$\phi(\rho,.) =$$
$$\left[(d_2 + \rho\delta d_2)\,e_2\right) \cdot (\sigma(v \circ \mathcal{R}_\rho^t, q) - u \circ \mathcal{R}_\rho^t \cdot v \circ \mathcal{R}_\rho^t \cdot ((d_2 + \rho\delta d_2)\,e_2)\right] \cdot n^\rho$$

We set $V = d_2\,e_2$ and $W = \delta d_2\,e_2$, then

$$\phi(\rho,.) = (V + \rho W) \cdot \left[\sigma(v \circ \mathcal{R}_\rho^t, q) - u \circ \mathcal{R}_\rho^t \cdot v \circ \mathcal{R}_\rho^t\right] \cdot n^\rho$$

Since $\phi(\rho,.)$ is defined on the boundary $\Gamma_t^{s,\rho}$, we need some extra identities corresponding to boundary shape derivates of terms involved in $\phi(\rho,.)$.

**LEMMA 7.13 [53]**

$$\partial_\rho n^\rho|_{\rho=0} = n'_\Gamma = -\nabla_\Gamma(Z_t \cdot n)$$

**LEMMA 7.14**

$$\frac{d}{d\rho}\left(\int_{\Gamma_t^{s,\rho}} \langle E(\rho), n^\rho\rangle d\Gamma\right)\bigg|_{\rho=0} = \int_{\Gamma_t^s} \langle E'|_{\Gamma_t}, n\rangle + (\mathrm{div}\,E)\langle Z_t, n\rangle$$

**PROOF**   First, we use that

$$\int_{\Gamma_t^{s,\rho}} \langle E(\rho), n^\rho\rangle = \int_{\Omega_t^{f,\rho}} \mathrm{div}\,E(\rho)$$

then we derive this quantity using lemma (7.7),

$$\frac{d}{d\rho}\left(\int_{\Omega_t^{f,\rho}} \mathrm{div}\,E(\rho)\right)\bigg|_{\rho=0} = \int_{\Omega_t^f} \mathrm{div}\,E' + \int_{\Gamma_t^s} (\mathrm{div}\,E)\langle Z_t, n\rangle$$

We conclude using $\int_{\Omega_t^f} \operatorname{div} E' = \int_{\Gamma_t^s} \langle E', n \rangle$. ▯

Then we use

**LEMMA 7.15**

$$E'|_{\Gamma_t} =$$
$$W \cdot [-q\,\mathrm{I} + \nu\,\mathrm{D}\,v - u \cdot v] + V \cdot [-\nu\,\mathrm{D}(\mathrm{D}\,v \cdot Z_t) + (\mathrm{D}\,u \cdot Z_t) \cdot v + u \cdot (\mathrm{D}\,v \cdot Z_t)]$$

Hence, we have

$$\frac{d}{d\rho} \left( \int_{\Gamma_t^{s,\rho}} \phi(\rho)\, d\Gamma \right) \bigg|_{\rho=0} = \int_{\Gamma_t^s(V)} W \cdot [-q\,\mathrm{I} + \nu\,\mathrm{D}\,v - u \cdot v] \cdot n$$

$$+ \int_{\Gamma_t^s(V)} V \cdot [-\nu\,\mathrm{D}(\mathrm{D}\,v \cdot Z_t) + (\mathrm{D}\,u \cdot Z_t) \cdot v + u \cdot (\mathrm{D}\,v \cdot Z_t)] \cdot n$$

$$+ \int_{\Gamma_t^s(V)} \operatorname{div}(V \cdot [-q\,\mathrm{I} + \nu^*\,\mathrm{D}\,v - u \cdot v]) \langle Z_t, n \rangle$$

**REMARK 7.5**   We recall that

$$\int_{\Gamma_t} V \cdot (\mathrm{D}\,v \cdot n) = \int_{\Omega_t^f} \operatorname{div}({}^*\mathrm{D}\,v \cdot V)$$

$$= \int_{\Omega_t^f} \mathrm{D}\,v \cdot \cdot\, \mathrm{D}\,V + V \cdot \Delta v \qquad (7.48)$$

▯

We use this expression in the sequel. We recall that the perturbed Lagrangian has the following form,

$$\mathcal{L}_{V,W}^\rho = J_{V,W}^\rho - \int_0^T \int_{\Omega_t^{f,\rho}} G(\rho) - \int_0^T \int_{\Gamma_t^{s,\rho}} \phi(\rho)$$

$$- \int_{\Omega_T} u(T) \cdot v(T) + \int_{\Omega_0} u_0 \cdot v(t=0)$$

$$\forall\, (v, q) \in Y(\Omega_t^f) \times V^f$$

Hence, its derivative with respect to $\rho$ at point $\rho = 0$ has the following expression,

$$\frac{d}{d\rho}\left(\mathcal{L}_{V,W}^\rho\right)\bigg|_{\rho=0} = \frac{d}{d\rho}\left(J_{V,W}^\rho\right)\bigg|_{\rho=0} - \int_0^T \frac{d}{d\rho}\left(\int_{\Omega_t^{f,\rho}} G(\rho)\right)\bigg|_{\rho=0}$$

$$- \int_0^T \frac{d}{d\rho}\left(\int_{\Gamma_t^{s,\rho}} \phi(\rho)\right)\bigg|_{\rho=0}, \quad \forall\, (v,q) \in Y(\Omega_t^f) \times V^f$$

Furthermore we have,

**LEMMA 7.16**

$$\frac{d}{d\rho}\left(J_{V,W}^\rho\right)\bigg|_{\rho=0} = \int_0^T \alpha(d_2 - d_g^2)\, e_2 \cdot W \tag{7.49}$$

Using the last identities concerning the derivative of the distributed and the boundary terms with respect to $\rho$, we get the following expression,

$$\frac{d}{d\rho}\left(\mathcal{L}_{V,W}^\rho\right)\bigg|_{\rho=0} = -A_{Z_t} - B_{Z_t} - C_W \tag{7.50}$$

with

$$A_{Z_t} = \int_0^T \int_{\Omega_t^f(V)} \left\{ [(\mathrm{D}\,u \cdot Z_t) \cdot \partial_t v - [(\mathrm{D}(\mathrm{D}\,u \cdot Z_t)) \cdot u \right.$$

$$+ \mathrm{D}\,u \cdot (\mathrm{D}\,u \cdot Z_t)] \cdot v + \nu(\mathrm{D}\,u \cdot Z_t) \cdot \Delta v - (\mathrm{D}\,u \cdot Z_t) \cdot \nabla q]$$
$$+ [u \cdot (\partial_t(\mathrm{D}\,v \cdot Z_t)) - (\mathrm{D}\,u \cdot u) \cdot (\mathrm{D}\,v \cdot Z_t) + \nu u \cdot (\Delta(\mathrm{D}\,v \cdot Z_t))$$

$$\left. + p\,\mathrm{div}(\mathrm{D}\,v \cdot Z_t)] \right\}$$

$$B_{Z_t} =$$
$$\int_0^T \int_{\Gamma_t^s(V)} [-u \cdot \partial_t v + (Du \cdot u) \cdot v - \nu u \cdot \Delta v + u \cdot \nabla q - p\,\mathrm{div}(v)]\,(Z_t \cdot n)$$

$$+ V \cdot [-\nu\,\mathrm{D}(\mathrm{D}\,v \cdot Z_t) + (\mathrm{D}\,u \cdot Z_t) \cdot v + u \cdot (\mathrm{D}\,v \cdot Z_t)] \cdot n$$

$$+ \mathrm{div}(V \cdot [-q\,\mathrm{I} + \nu^*\,\mathrm{D}\,v - u \cdot v])\langle Z_t, n\rangle$$

$$- \int_0^T \int_{\Gamma_\infty^f} \nu\, u_\infty \cdot \mathrm{D}(\mathrm{D}\,v \cdot Z_t) \cdot n$$

$$C_W = \int_0^T \int_{\Gamma_t^s(V)} W \cdot \left[ -b_1 \, e_2 - m \, \dot{b}_2 \, e_2 + \sigma(v,q) \cdot n - (u \cdot v) \, n - \alpha(d_2 - d_g^2) \, e_2 \right]$$

### The shape derivative kernel identity

We shall now assume that $(u, p, \varphi, \pi)$ is a saddle point of the Lagrangian functional $\mathcal{L}$. This will help us to simplify several terms involved in the derivative of $\mathcal{L}$ with respect to $V$.

Indeed, we would like to express the distributed term $A_{Z_t}$ as a boundary quantity defined on the fluid moving boundary $\Gamma_t^s$ and the fixed boundary $\Gamma_\infty^f$.

### *THEOREM 7.6*

*For $(u, p, \varphi, \pi)$ saddle points of the Lagrangian functional (7.32), the following identity holds,*

$$\int_0^T \int_{\Omega_t^f(V)} \big\{ [(D\,u \cdot Z_t) \cdot \partial_t \varphi - [(D(D\,u \cdot Z_t)) \cdot u$$

$$+ D\,u \cdot (D\,u \cdot Z_t)] \cdot \varphi + \nu(D\,u \cdot Z_t) \cdot \Delta\varphi - (D\,u \cdot Z_t) \cdot \nabla\pi] + \big[ u \cdot (\partial_t(D\,\varphi \cdot Z_t))$$

$$- (D\,u \cdot u) \cdot (D\,\varphi \cdot Z_t) + \nu u \cdot (\Delta(D\,\varphi \cdot Z_t)) + p \, \mathrm{div}(D\,\varphi \cdot Z_t)] \big\}$$

$$- \int_0^T \int_{\Gamma_t^s(V)} V \cdot [\nu \, D(D\,\varphi \cdot Z_t) - (D\,u \cdot Z_t) \cdot \varphi - u \cdot (D\,\varphi \cdot Z_t)] \cdot n$$

$$+ \int_0^T \int_{\Gamma_t^s(V)} [\nu \, (\varphi - b_2 \, e_2) \cdot (D(D\,u \cdot Z_t) \cdot n) + (D\,\varphi \cdot Z_t) \cdot (-p \, n + \nu(D\,u \cdot n))]$$

$$- \int_0^T \int_{\Gamma_\infty^f} \nu \, u_\infty \cdot (D(D\,\varphi \cdot Z_t) \cdot n) = 0, \quad \forall W \overset{\text{def}}{=} \delta d_2 \, e_2$$

**PROOF**    We use extremal conditions associated to variations with respect to $(u, v)$ in the Lagrangian functional where we have added a boundary since we consider test functions $v$ that do not vanish on $\Gamma_\infty^f$ and do not match the solid test functions at the fluid-solid interface $\Gamma_t^s(V)$ , i.e.,

$$\mathcal{L}^2(u,p,d_1,d_2;v,q,b_1,b_2) = J_{u_\infty}(u,p,d_1,d_2)$$

$$-\int_0^T \int_{\Omega_t^f(V)} [-u \cdot \partial_t v + (\mathrm{D}\,u \cdot u) \cdot v - \nu u \cdot \Delta v + u \cdot \nabla q - p \operatorname{div} v]$$

$$-\int_0^T \int_{\Gamma_\infty^f} u_\infty \cdot \sigma(v,q) \cdot n - \int_0^T \int_{\Gamma_t^s(V)} V \cdot [\sigma(v,q) \cdot n - (u \cdot v)\,n]$$

$$+\int_0^T \int_{\Gamma_t^s(V)} (v - b_2\,e_2) \cdot (\sigma(u,p) \cdot n)$$

$$-\int_0^T \left[-d_1\,\dot{b}_1 - d_2\,b_1\right] - \int_0^T \left[-md_2\,\dot{b}_2 + kd_1\,b_2\right]$$

$$-\int_{\Omega_T} u(T)v(T) + \int_{\Omega_0} u_0 v(t=0) + d_1^0 b_1(0) - d_1(T)b_1(T)$$

$$+ md_2^0 b_2(0) - md_2(T)b_2(T)\forall\,(v,q,b_1,b_2) \in\; Y^f \times V^f \times Y_1^s \times Y_2^s$$

This leads to the following perturbation identity,

$$\partial_{(u,v)}\mathcal{L}^2 \cdot (\delta u, \delta v) =$$

$$-\int_0^T \int_{\Omega_t^f} [-\delta u \cdot \partial_t v - u \cdot \partial_t \delta v + D(\delta u \cdot u) \cdot v + D(u \cdot \delta u) \cdot v$$

$$+D(u \cdot u) \cdot \delta v - \nu(\delta u \cdot \Delta v) - \nu(u \cdot \Delta \delta v) + \delta u \cdot \nabla q - p \operatorname{div}(\delta v)]$$

$$+\int_0^T \int_{\Gamma_t^s(V)} [\nu\,(v - b_2\,e_2) \cdot (\mathrm{D}\,\delta u \cdot n) + \delta v \cdot (-p\,n + \nu(\mathrm{D}\,u \cdot n))]$$

$$-\int_0^T \int_{\Gamma_t^s(V)} V \cdot [\nu\,\mathrm{D}(\delta v) - \delta u \cdot v - u \cdot \delta v] \cdot n - \int_0^T \int_{\Gamma_\infty^f} \nu\,u_\infty \cdot (\mathrm{D}\,\delta v \cdot n)$$

$$-\int_{\Omega_T} [\delta u(T)v(T) + u(T)\delta v(T)], \quad \forall\,(\delta u, \delta v) \in\; X(\Omega_t^f) \times Y(\Omega_t^f)$$

We choose specific perturbation directions, i.e.,

$$\delta u = \mathrm{D}\,u \cdot Z_t \quad \delta v = \mathrm{D}\,v \cdot Z_t$$

with $\delta u(T) = \delta v(T) = \delta u(0) = \delta v(0) = 0$, where $(u,v)$ are saddle points of the Lagrangian, i.e., solutions of respectively the primal and adjoint fluid problem. We recognize immediately the distributed and boundary terms involved in the shape derivative kernel identity. $\quad\square$

### Solid and ALE adjoint problem

Using the shape derivative kernel identity, we simplify the Lagrangian derivative at the saddle point $(u,p,d_1,d_2,\varphi,b_1,b_2)$. We set $(u,v) = (u,\varphi)$

and we get

$$A_{Z_t} = \int_0^T \int_{\Gamma_t^s(V)} V \cdot [\nu \, D(D\varphi \cdot Z_t) - (D u \cdot Z_t) \cdot \varphi - u \cdot (D\varphi \cdot Z_t)] \cdot n$$

$$- \int_0^T \int_{\Gamma_t^s(V)} [\nu \,(\varphi - b_2 \, e_2) \cdot (D(D u \cdot Z_t) \cdot n) + (D\varphi \cdot Z_t) \cdot (-p \, n + \nu(D u \cdot n))]$$

$$+ \int_0^T \int_{\Gamma_\infty^f} \nu \, u_\infty \cdot (D(D\varphi \cdot Z_t) \cdot n)$$

We use that $\varphi = b_2 \, e_2$ on $\Gamma_t^s(V)$ and the following identities,

$$(D\varphi \cdot Z_t) \cdot (p \, n) = (p \, \text{div} \, \varphi)\langle Z_t, n \rangle \tag{7.51}$$

$$(D\varphi \cdot Z_t) \cdot (D u \cdot n) = (D\varphi \cdot n) \cdot (D u \cdot n)\langle Z_t, n \rangle \tag{7.52}$$

then,

$$A_{Z_t} = \int_0^T \int_{\Gamma_t^s(V)} V \cdot [\nu \, D(D\varphi \cdot Z_t) - (D u \cdot Z_t) \cdot \varphi - u \cdot (D\varphi \cdot Z_t)] \cdot n$$

$$- \int_0^T \int_{\Gamma_t^s(V)} [-p \, \text{div} \, \varphi + \nu \, (D\varphi \cdot n) \cdot (D u \cdot n)] \, \langle Z_t, n \rangle$$

$$+ \int_0^T \int_{\Gamma_\infty^f} \nu \, u_\infty \cdot (D(D\varphi \cdot Z_t) \cdot n)$$

Using the following identity,

$$\text{div}(V \cdot [-\pi \, I + \nu^* \, D\phi - u \cdot \varphi]) = -\pi \, \text{div} \, V - V \cdot \nabla \pi + \nu \, D\varphi \cdot \cdot \, D V$$
$$+ \nu \, V \Delta\varphi - (\text{div} \, V) \, u \cdot \varphi - V \cdot \nabla(u \cdot \varphi) \tag{7.53}$$

we get

$$B_{Z_t} =$$

$$\int_0^T \int_{\Gamma_t^s(V)} \big\{ [-u \cdot \partial_t \varphi + (D u \cdot u) \cdot \varphi - \nu u \cdot \Delta\varphi + u \cdot \nabla \pi - p \, \text{div}(\varphi)] \, \langle Z_t, n \rangle$$

$$+ V \cdot [-\nu \, D(D\varphi \cdot Z_t) + (D u \cdot Z_t) \cdot \varphi + u \cdot (D\varphi \cdot Z_t)] \cdot n$$

$$+ \big[ - \pi \, \text{div} \, V - V \cdot \nabla \pi + \nu \, D\varphi \cdot \cdot \, D V + \nu \, V \Delta\varphi$$

$$- (\text{div} \, V) \, u \cdot \varphi - V \cdot \nabla(u \cdot \varphi)]\langle Z_t, n \rangle - \int_0^T \int_{\Gamma_\infty^f} \nu \, u_\infty \cdot D(D\varphi \cdot Z_t) \cdot n$$

then we get,

$$A_{Z_t} + B_{Z_t} = \int_0^T \int_{\Gamma_t^s(V)} [\nu \, (D\varphi \cdot n) \cdot (D(V - u) \cdot n) - u \cdot \partial_t \varphi + (D u \cdot u) \cdot \varphi$$

$$- \pi \, \text{div} \, V - (\text{div} \, V) \, u \cdot \varphi - V \cdot \nabla(u \cdot \varphi)] \, \langle Z_t, n \rangle$$

We use the following identity,

$$V \cdot \nabla(u \cdot \varphi) = \varphi \cdot \mathrm{D}\, u \cdot V + V \cdot \mathrm{D}\, \varphi \cdot u$$

and the boundary conditions, $u = d_2\, e_2$ on $\Gamma_t^s$, $\varphi = b_2\, e_2$ on $\Gamma_t^s$. This leads to

$$A_{Z_t} + B_{Z_t} = \int_0^T \int_{\Gamma_t^s(V)} \left[ \nu\, (\mathrm{D}\,\varphi \cdot n) \cdot (\mathrm{D}(V - u) \cdot n) - d_2\, \dot{b}_2 - \pi \, \mathrm{div}\, V \right.$$
$$\left. - (\mathrm{div}\, V)\, u \cdot \varphi - d_2\, e_2 \cdot \mathrm{D}\,\varphi \cdot d_2\, e_2 \right] \langle Z_t, n \rangle$$

We choose the velocity field $V = \mathrm{Ext}(d_2\, e_2) \circ p$, then

$$(\mathrm{D}\, V \cdot n) \cdot n|_{\Gamma_t^s} = 0$$

and

$$\mathrm{div}\, V|_{\Gamma_t^s} = \mathrm{div}_\Gamma V + (\mathrm{D}\, V \cdot n) \cdot n$$
$$= 0$$

Finally, we have

$$A_{Z_t} + B_{Z_t} =$$
$$\int_0^T \int_{\Gamma_t^s(V)} \left[ -d_2\, \dot{b}_2 + \nu\, (\mathrm{D}\,\varphi \cdot n) \cdot (\mathrm{D}\, u \cdot n) - |d_2|^2\, (\mathrm{D}\,\varphi \cdot e_2) \cdot e_2 \right] \langle Z_t, n \rangle$$

and

$$C_W = \int_0^T \int_{\Gamma_t^s(V)} W \cdot \left[ -b_1\, e_2 - m\, \dot{b}_2\, e_2 + \sigma(\phi, \pi) \cdot n - \alpha(d_2 - d_g^2)\, e_2 \right]$$

where we have used that

$$\int_{\Gamma_t^s(V)} d_2\, e_2 \cdot b_2\, e_2 \cdot n = 0$$

We introduce the adjoint field $\lambda$ solution of the following system,

$$\begin{cases} -\partial_t \lambda - \nabla_\Gamma \lambda \cdot V - \lambda\, \mathrm{div}_\Gamma V = f, & (0, T) \\ \lambda(T) = 0, & \Gamma_T(V) \end{cases} \tag{7.54}$$

with

$$f = \left[ -d_2\, \dot{b}_2 + \nu\, (\mathrm{D}\,\varphi \cdot n) \cdot (\mathrm{D}\, u \cdot n) - |d_2|^2\, (\mathrm{D}\,\varphi \cdot e_2) \cdot e_2 \right] \tag{7.55}$$

**REMARK 7.6**   In our case $\mathrm{div}_\Gamma V = 0$. ▯

We recall the following property,

**LEMMA 7.17 [59],[57]**
*For any $E(V) \in L^2(\Sigma^s(V))$ and $(V, W) \in \mathcal{U}$, the following identity holds,*

$$\int_0^T \int_{\Gamma_t^s(V)} E(V) \langle Z_t, n \rangle = -\int_0^T \int_{\Gamma_t^s(V)} \lambda \langle W, n \rangle \qquad (7.56)$$

*where $\lambda \in C^0([0, T]; H^1(\Gamma_t^s))$ is the unique solution of problem (7.54) with $f = E$.*

Then we have

$$A_{Z_t} + B_{Z_t} = \int_0^T \int_{\Gamma_t^s(V)} f \langle Z_t, n \rangle$$

$$= -\int_0^T \int_{\Gamma_t^s(V)} \langle \lambda n, W \rangle$$

However, using the optimality condition for the Lagrangian functional, we obtain

$$-(A_{Z_t} + B_{Z_t}) = C_W.$$

This leads to

$$\int_0^T \int_{\Gamma_t^s(V)} \lambda n \cdot W =$$

$$dis \int_0^T \int_{\Gamma_t^s(V)} \left[ -b_1 e_2 - m \dot{b_2} e_2 + \sigma(\phi, \pi) \cdot n - \alpha(d_2 - d_g^2) e_2 \right] \cdot W,$$

$$\forall W \stackrel{\text{def}}{=} \delta d_2 e_2$$

from which we deduce the following identity,

$$\int_{\Gamma_t^s(V)} \lambda n = \left[ -b_1 - m \dot{b_2} - \alpha(d_2 - d_g^2) \right] e_2 + \int_{\Gamma_t^s(V)} \sigma(\phi, \pi) \cdot n \qquad (7.57)$$

We now use the following lemma,

**LEMMA 7.18**

$$\partial_t \left( \int_{\Gamma_t^s(V)} \lambda n \right) = \int_{\Gamma_t^s(V)} \left[ \partial_t \lambda + \nabla_\Gamma \lambda \cdot V \right] n \qquad (7.58)$$

We get

$$\partial_t \left( \left[ -b_1 - m\,\dot{b}_2 - \alpha(d_2 - d_g^2) \right] e_2 + \int_{\Gamma_t^s(V)} \sigma(\phi, \pi) \cdot n \right) =$$

$$\int_{\Gamma_t^s(V)} \left[ -d_2\,\dot{b}_2 + \nu\,(D\,\varphi \cdot n) \cdot (D\,u \cdot n) - |d_2|^2\,(D\,\varphi \cdot e_2) \cdot e_2 \right] n$$

This can be written as

$$\dot{b}_1 + m\,\ddot{b}_2 = -\alpha(\dot{d}_2 - \dot{d}_g^2) + \partial_t \left( \int_{\Gamma_t^s(V)} \sigma(\phi, \pi) \cdot n \right) \cdot e_2 +$$

$$\int_{\Gamma_t^s(V)} \left[ |d_2|^2\,(D\,\varphi \cdot e_2) \cdot e_2 - \nu\,(D\,\varphi \cdot n) \cdot (D\,u \cdot n) \right] \cdot n$$

and we recall the other solid adjoint equation,

$$\begin{cases} -\dot{b}_1 + k\,b_2 = \alpha(d - d_g^1),\ (0, T) \\ b_1(T) = 0 \end{cases} \tag{7.59}$$

Let us inject $\dot{b}_1$ inside the first one,

$$m\,\ddot{b}_2 + k\,b_2 = \alpha(d - d_g^1) - \alpha(\dot{d}_2 - \dot{d}_g^2) + \partial_t \left( \int_{\Gamma_t^s(V)} \sigma(\phi, \pi) \cdot n \right) \cdot e_2 +$$

$$\int_{\Gamma_t^s(V)} \left[ |d_2|^2\,(D\,\varphi \cdot e_2) \cdot e_2 - \nu\,(D\,\varphi \cdot n) \cdot (D\,u \cdot n) \right] \cdot n$$

This concludes the proof of the main result.

---

## 7.6   Conclusion

In this chapter, we have investigated sensitivity analysis for a simple 2-D coupled fluid-structure system. This analysis was performed using a Lagrangian functional and non-cylindrical shape derivative tools to handle perturbation with respect to the velocity of the solid. This led to the derivation of first-order optimality conditions for an optimal control problem related to a tracking functional. This optimality system can be numerically approximated and included inside a gradient based optimization procedure. This point is under investigation following the strategy adopted for Navier-Stokes optimal control problems as in [69].

The methodology used in this chapter can be generalized to more complex fluid-structure interaction problems. We can either change the fluid model and handle compressibility as in [37], or change the solid equations and use a real 3D non-linear elastic model as described in the next chapter.

# Chapter 8

## Sensitivity analysis for a general fluid-structure interaction system

### 8.1 Introduction

In the previous chapter, we have performed a sensitivity analysis for a coupled fluid-solid model, where the solid was driven by a basic second order differential equation. This has allowed us to simply reduce the differential order by expanding the solid state and use its velocity as an explicit state variable. Then we have been able to use the framework developed in Chapter 5 concerning the derivative of the Navier-Stokes system with respect to the velocity of the moving domain. Here, the situation is quite different since we are considering a more general solid model with a non-linear constitutive law. In this case, the parametrization of the solid by its velocity is more challenging and the straightforward use of the results obtained in Chapter 7 is not obvious.

As a consequence, we choose to work with the solid displacement state variable. Then, we need to use the results obtained in Chapter 6 concerning the derivative of the Navier-Stokes using the non-cylindrical identity perturbations.

The chapter is organized as follows: first of all, we introduce the mechanical problem and its mathematical description. We use the classical arbitrary Euler-Lagrange (ALE) formulation, particularly suited for problems involving moving boundaries. Then, in the second part, we state the main result of this chapter, namely the cost function gradient computation involving the solution of a linear adjoint problem. Finally, its proof is fully developed.

### 8.2 Mechanical problem and main result

In this section we introduce a fluid-structure interaction model. We consider a solid located at time $t \geq 0$ in a domain $\Omega^s(t) \subset \mathbb{R}^3$ with boundary $\Gamma^s(t)$. It is surrounded by a fluid in $\mathbb{R}^3$. We introduce a control volume $\Omega \subset \mathbb{R}^3$ containing the solid at each time $t \geq 0$. The notation $\partial\Omega$ stands for the boundary of $\Omega$. Hence, the fluid evolution is restricted to the domain $\Omega^f(t) = \Omega - \overline{\Omega}^s(t)$. In the sequel we set $\Gamma^f = \partial\Omega$ and

$$\partial\Omega^f(t) = \Gamma^f \cup \Gamma^s(t)$$

stands for the fluid domain boundary; see figure 8.1. We assume the fluid

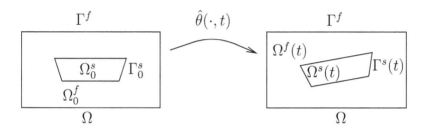

**FIGURE 8.1**:   Geometric description

to be Newtonian viscous, homogeneous and incompressible. Its behavior is described by its velocity and pressure. The elastic solid under large displacements is described by its velocity and its stress tensor. The classical conservation laws of the continuum mechanics drive the evolution of these unknowns. The fluid state satisfies the following incompressible Navier-Stokes equations written in Eulerian conservative formulation:

$$\partial_t u + \mathrm{D}\, u \cdot u - \nu \Delta u + \nabla p = 0, \text{ in } \Omega^f(t),$$

$$\mathrm{div}\, u = 0, \text{ in } \Omega^f(t) \tag{8.1}$$

where $(u, p)$ stand, respectively, for the fluid velocity and pressure. In addition, the fluid stress tensor is given by

$$\sigma(u, p) = -p\,\mathrm{I} + \nu\,[\mathrm{D}\, u + {}^*\mathrm{D}\, u]$$

with $\nu$ the kinematic viscosity of the fluid.

In a fluid-structure interaction framework the evolution of the fluid domain $\Omega^f(t)$ is induced by the structural deformation through the fluid-structure interface $\Gamma^s(t)$. Indeed, by definition $\Omega^f(t) = \Omega - \overline{\Omega}^s(t)$. It leads us to describe $\Omega^f(t)$ according to a map acting in a fixed reference domain. This approach is usually used for the solid domain $\Omega^s(t)$, by means of the Lagrangian formulation [94].

Given a material reference configuration for the solid $\Omega_0^s \subset \Omega$ with boundary

$\Gamma_0^s$, we take $\Omega_0^f = \Omega - \overline{\Omega}_0^s$ as the reference fluid configuration. Then, the control volume $\Omega = \Omega^f(t) \cup \overline{\Omega}^s(t)$ is described by a smooth and injective map:

$$\hat{\theta} : \overline{\Omega} \times \mathbb{R}^+ \longrightarrow \overline{\Omega}$$
$$(x_0, t) \longmapsto \hat{\theta} = \hat{\theta}(x_0, t)$$

We set $\hat{\theta}^f = \hat{\theta}_{|\Omega_0^f}$ and $\hat{\theta}^s = \hat{\theta}_{|\Omega_0^s}$, such that [95]:

- for $x_0 \in \Omega_0^s$, $\hat{\theta}^s(x_0, t)$ represents the position at time $t \geq 0$ of the material point $x_0$. This corresponds to the classical Lagrangian flow,

- the map $\hat{\theta}^f$ is defined from $\hat{\theta}^s_{|\Gamma_0^s}$, as an arbitrary extension over domain $\overline{\Omega}_0^f$, which preserves $\Gamma_0^f = \Gamma^f = \partial\Omega$.

In short, the ALE map $\hat{\theta}$ is given by

$$\hat{\theta}(x_0, t) = \mathrm{Ext}(\hat{\theta}^s_{|\Gamma_0^s})(x_0, t), \quad \forall x_0 \in \overline{\Omega}_0^f,$$
$$\hat{\theta}(x_0, t) = \hat{\theta}^s(x_0, t), \quad \forall x_0 \in \overline{\Omega}_0^s.$$

Here, Ext represents an extension operator from $\Gamma_0^s$ to $\overline{\Omega}_0^f$ such that

$$\mathrm{Ext}(\hat{\theta}^s_{|\Gamma_0^s})_{|\Gamma_0^f} = I_{\Gamma_0^f}$$

**REMARK 8.1**    We set $\Omega_0 = \Omega_0^f \cup \overline{\Omega}_0^s$ as the reference domain, and $\Omega = \Omega^f(t) \cup \overline{\Omega}^s(t)$ stands for the actual configuration at time $t > 0$.    ⬜

**REMARK 8.2**    The operator Ext is arbitrary defined inside $\Omega_0^f$.    ⬜

We set $\Sigma_{\hat{\theta}} \equiv \bigcup_{0<t<T} (\{t\} \times \Gamma_{\hat{\theta}}^s)$, $Q \equiv \bigcup_{0<t<T} (\{t\} \times \Omega_{\hat{\theta}}^f)$ and $\Sigma_\infty^f \equiv \Gamma_\infty^f \times (0, T)$

We set

$$J(\hat{\theta}) = \det(\mathrm{D}\,\hat{\theta}) > 0$$

In the sequel, we shall only use the notation $\hat{\theta}$, keeping in mind that it corresponds to the solid displacement in $\Omega_0^s$.

The solid evolution is given by its motion $\hat{\theta}$ and the stress tensor field $S(\hat{\theta})$ (second Piola-Kirchoff tensor [36]). The field $S(\hat{\theta})$ is related to $\hat{\theta}$ through an appropriate constitutive law. Then, the pair $(\hat{\theta}, S(\hat{\theta}))$ satisfies the non-linear elastodynamic equations [36]:

$$\partial_{tt}\,\hat{\theta} - \mathrm{div}_0[\mathrm{D}\,\hat{\theta} \cdot S(\hat{\theta})] = f, \quad \text{in} \quad \Omega_0^s \qquad (8.2)$$

**REMARK 8.3**   The constitutive law for a St. Venant-Kirchhoff material is given by the following expression,

$$S(\hat{\theta}) = \lambda \left( \operatorname{Tr} E(\hat{\theta}) \right) \mathrm{I} + 2\,\mu\, E(\hat{\theta})$$

where $\lambda, \mu > 0$ stands for the Lamé coefficients and $E(\hat{\theta})$ is the strain tensor,

$$E(\hat{\theta}) = \frac{1}{2} \left[ {}^{*}\mathrm{D}\,\hat{\theta} \cdot \mathrm{D}\,\hat{\theta} - \mathrm{I} \right] \tag{8.3}$$

This kind of constitutive law is often referred to as " large displacement - small strain " model. ◻

We shall also assume that there exists two operators $T_A$ defined on $\Omega_0^s$ and $T_\Gamma$ defined on $\Gamma_0^s$ such that the linear tangent tensor $T(\delta\hat{\theta}) \stackrel{\text{def}}{=} S'(\hat{\theta}) \cdot \delta\hat{\theta}$ in the direction $\delta\hat{\theta}$ satisfies the following adjoint identity,

$$\int_{\Omega_0^s} T(\delta\hat{\theta}) \cdot \cdot B + \int_{\Omega_0^s} \delta\hat{\theta} \cdot T_A(B) = \int_{\Gamma_0^s} T_\Gamma(B) \cdot \delta\hat{\theta} \tag{8.4}$$

for any smooth second order tensor B.
The coupling between the solid and the fluid is realized through standard boundary conditions at the fluid-structure interface $\Gamma_0^s$, namely, the kinematic continuity of the velocity and the kinetic continuity of the stress [95]:

$$u = V_{\hat{\theta}} \stackrel{\text{def}}{=} \partial_t \hat{\theta} \circ \hat{\theta}^{-1}, \text{ on } \Gamma_{\hat{\theta}}^s,$$

$$\mathrm{D}\,\hat{\theta} \cdot S(\hat{\theta}) \cdot n_0 = J(\hat{\theta}) \sigma(u, p) ({}^{*}\mathrm{D}\,\hat{\theta})^{-1} \cdot n_0, \text{ on } \Gamma_0^s \tag{8.5}$$

where $n_0$ stands for the unit normal vector on $\Gamma_0^s$ pointing inside $\Omega_0^s$. Moreover, we endow the fluid equations with Dirichlet boundary condition on the far-field boundary $\Gamma_\infty^f$:

$$u = u_\infty, \quad \text{on} \quad \Gamma_\infty^f$$

In summary, the strong coupled problem, with a ALE formulation for the fluid, is given by:

$$
\left\{
\begin{array}{l}
\partial_t u + \mathrm{D}\, u \cdot u - \nu \Delta u + \nabla p = 0, \text{ in } \Omega^f_{\hat{\theta}}, \\[2mm]
\operatorname{div} u = 0, \text{ in } \Omega^f_{\hat{\theta}}, \\[2mm]
u = u_\infty, \text{ on } \Gamma^f_\infty, \\[2mm]
u = \partial_t \hat{\theta} \circ \hat{\theta}^{-1}, \text{ on } \Gamma^s_{\hat{\theta}}, \\[2mm]
\mathrm{D}\,\hat{\theta} \cdot S(\hat{\theta}) \cdot n_0 = J(\hat{\theta})\sigma(u,p)(^*\mathrm{D}\,\hat{\theta})^{-1} \cdot n_0, \text{ on } \Gamma^s_0, \\[2mm]
\partial_{tt}\,\hat{\theta} - \operatorname{div}_0[\mathrm{D}\,\hat{\theta} \cdot S(\hat{\theta})] = f, \text{ in } \Omega^s_0, \\[2mm]
\left(u,\hat{\theta},\partial_t\hat{\theta}\right)\Big|_{t=0} = (u_0,\hat{\theta}_0,\hat{\theta}_1), \text{ in } \Omega^f_0 \times (\Omega^s_0)^2
\end{array}
\right.
\tag{8.6}
$$

Here, $f$ represents the applied body force, $u_0$ the initial fluid velocity and $\hat{\theta}_0$, $\hat{\theta}_1$ the initial displacement and solid velocity, respectively.
The principal result of this chapter reads,
**Main Result:** For smooth $u_\infty \in \mathcal{U}_c$, the following cost functional,

$$
j(u_\infty) = J_{u_\infty}(u(u_\infty), p(u_\infty), \hat{\theta}(u_\infty))
\tag{8.7}
$$

with $(u(u_\infty), p(u_\infty), \hat{\theta}(u_\infty))$ is a weak solution of problem (8.6) and where $J(u,p,\hat{\theta})$ is a real functional of the following form

$$
J_{u_\infty}(u,p,\hat{\theta}) = \frac{\alpha}{2}\int_0^T \int_{\Omega^s_0} |\hat{\theta} - \hat{\theta}_d|^2
\tag{8.8}
$$

admits a gradient given by the following expression,

$$
\nabla j(u_\infty) = \sigma(\varphi,\pi) \cdot n|_{(0,T)\times\Gamma^f_\infty}
\tag{8.9}
$$

with $(\varphi,\pi)$ being a solution of the following backward fluid adjoint system,

$$
\left\{
\begin{array}{ll}
-\partial_t\varphi - \mathrm{D}\,\varphi \cdot u + {}^*\mathrm{D}\,u \cdot \varphi - \nu\Delta\varphi + \nabla\pi = 0, & Q^f \\
\operatorname{div}\varphi = 0, & Q^f \\
\varphi = \hat{\psi} \circ \hat{\theta}^{-1}, & \Sigma^s \\
\varphi = 0, & \Sigma^f_\infty \\
\varphi(T) = 0, & \Omega^f_T
\end{array}
\right.
\tag{8.10}
$$

and $\hat\psi$ being a solution of the following backward solid adjoint system,

$$\begin{cases} \partial_{tt}\,\hat\psi - \mathrm{D}\,\hat\psi \cdot \operatorname{div} S(\hat\theta) - \mathrm{D}^2\,\hat\psi \cdot\!\cdot S(\hat\theta) - T_{\mathrm{A}}({}^*\mathrm{D}\,\hat\theta \cdot \mathrm{D}\,\hat\psi) = \\ \\ \alpha\,(\hat\theta - \hat\theta_d), \hspace{5cm} (0,T)\times\Omega_0^s \\ \\ (\hat\psi, \partial_t\,\hat\psi)_{|t=T} = (0,0), \hspace{3.5cm} \Omega_0^s \end{cases}$$

$$(8.11)$$

together with the following condition,

$$\left[\mathrm{D}\,\hat\psi \cdot {}^*S(\hat\theta)\cdot\hat n + T_\Gamma({}^*\mathrm{D}\,\hat\theta\cdot\mathrm{D}\,\hat\psi)\right]\circ\hat\theta^{-1}\,[J(\hat\theta)|^*(\mathrm{D}\,\hat\theta)^{-1}\cdot\hat n|]^{-1}\circ\hat\theta^{-1} =$$

$$\left[-V_{\hat\theta}\cdot\partial_t\varphi - p\operatorname{div}\varphi - \pi\operatorname{div}V_{\hat\theta} + \nu\,\mathrm{D}\,\varphi\cdot\!\cdot\mathrm{D}\,V_{\hat\theta} - (\operatorname{div}V_{\hat\theta})\,V_{\hat\theta}\cdot\varphi\right.$$

$$\left.- (\mathrm{D}\,\varphi\cdot V_{\hat\theta})\cdot\varphi\right]\cdot n - {}^*\mathrm{D}\,\varphi\cdot\sigma(u,p)\cdot n + {}^*\mathrm{D}\,V_{\hat\theta}\cdot\left[\sigma(\varphi,\pi) - V_{\hat\theta}\cdot\varphi\right]\cdot n$$

$$+ \left[-(\partial_t\mathcal{E})\cdot n + \mathcal{E}\cdot\nabla_\Gamma(V_{\hat\theta}\cdot n) - (\operatorname{div}_\Gamma V_{\hat\theta})\,\mathcal{E}\cdot n - \mathrm{D}\,\mathcal{E}\cdot V_{\hat\theta}\cdot n - \mathcal{E}\cdot\mathrm{D}_\Gamma\,n\cdot V_{\hat\theta}\right]$$

$$\text{on }\Gamma_{\hat\theta} \quad (8.12)$$

with $\mathcal{E} = \sigma(\varphi,\pi) - (V_{\hat\theta}\cdot\varphi)\,\mathrm{I}$.

### 8.2.1  Solid weak state operator

We define the solid state operator,

$$e^s : \hat X^s \longrightarrow (\hat Y^s)^*$$

whose action is defined by the following identity,

$$\langle e^s(\hat\theta), \hat v^s\rangle =$$

$$\int_0^T\int_{\Omega_0^s}[\hat\theta\cdot\partial_{tt}\,\hat v^s + \mathrm{D}\,\hat\theta\cdot S(\hat\theta)\cdot\!\cdot\mathrm{D}\,\hat v^s] + \int_0^T\int_{\Gamma_0^s}[\mathrm{D}\,\hat\theta\cdot S(\hat\theta)\cdot n_0]\cdot\hat v^s$$

$$-\int_0^T\int_{\Omega_0^s}f\cdot\hat v^s + \int_{\Omega_0^s}[\partial_t\,\hat\theta(T)\cdot\hat v^s(T) - \hat\theta(T)\cdot\partial_t\,\hat v^s(T)] + \int_{\Omega_0^s}[-\hat\theta_1\cdot\hat v^s(0) + \hat\theta_0\cdot\partial_t\,\hat v^s(0)]$$

### 8.2.2  Fluid state operator

We define the fluid state spaces,

$$X^f \overset{\text{def}}{=} \left\{u \in H^2(0,T;(H^2(\Omega_t^f)))\right\}$$

$$Z^f \overset{\text{def}}{=} \{p \in H^1(0,T;H^1(D))\}$$

We also need test function spaces that will be useful to define Lagrange multipliers :

$$Y^f \overset{\text{def}}{=} \left\{v^f \in L^2(0,T;H^2(\Omega^f_t))\right\}$$

$$V^f \overset{\text{def}}{=} \left\{q \in H^1(0,T;(H^1(D))^2)\right\}$$

$$W^f_s \overset{\text{def}}{=} \left\{(v^f,\hat{v}^s) \in Y^f \times \hat{Y}^s,\ \hat{v}^f = \hat{v}^s,\ \text{on } \Gamma^s_0\right\}$$

We define the fluid weak state operator,

$$e^f_{u_\infty} : X^f \times Z^f \times U^s \times X^s \longrightarrow (Y^f \times V^f)^*$$

whose action is defined by :

$$\langle e^f_{u_\infty}(u,p,u^s,\hat{\theta}),(v^f,q)\rangle =$$

$$\int_0^T \int_{\Omega^f_t(\hat{\theta})} \left[-u \cdot \partial_t v^f + (\mathrm{D}\, u \cdot u) \cdot v^f - \nu u \cdot \Delta v^f + u \cdot \nabla q - p\, \mathrm{div}\, v^f\right]$$

$$+ \int_0^T \int_{\Gamma^f_\infty} u_\infty \cdot (\sigma(v^f,q)\cdot n) + \int_0^T \int_{\Gamma^s_{\hat{\theta}}} \left[u^s \cdot (\sigma(v^f,q)\cdot n) - (u \cdot v^f)\langle u^s,n\rangle\right]$$

$$- \int_0^T \int_{\Gamma^s_{\hat{\theta}}} v^f \cdot (\sigma(u,p)\cdot n) + \int_{\Omega_T} u(T)\cdot v^f(T) - \int_{\Omega_0} u_0 \cdot v^f(0)$$

$$\forall (v^f,q) \in Y^f \times V^f, \quad \text{with } v^f \overset{\text{def}}{=} \hat{v}^f \circ \theta, \quad \hat{v}^f \in \hat{Y}^f$$

### 8.2.3   Coupled system operator

Our mechanical system consists of a solid part and a fluid part. These subsystems have been represented thanks to a solid and a fluid state operator. It is now possible to couple these two operators in order to build an ad-hoc coupled system operator.

The major point here is to notice that the kinetic continuity of the stress holds on the fluid-structure interface. To achieve the coupling, we need to decide whether or not the fluid and the solid multipliers match at the fluid-solid interface. If not, we have to work with the fluid constraint tensor at the fluid-solid boundary, which may be not convenient due to regularity requirement. Hence, we choose to work with continuous test functions on $\Gamma^s_{\hat{\theta}}$. This means that we shall choose the fluid and solid multipliers inside the space $W^f_s$. We define the coupled system weak state operator as follows,

$$e_{u_\infty} : Y^f \times Z^f \times \hat{X}^s \longrightarrow (W^f_s \times V^f)^*$$

whose action is defined by the following identity,

$$\langle e_{u_\infty}(u,p,\hat\theta),(v^f,q,\hat v^s)\rangle = \langle e^s(\hat\theta),\hat v^s)\rangle + \langle e^f_{u_\infty}(u,p,\partial_t\hat\theta\circ\hat\theta^{-1},\hat\theta),(v^f,q)\rangle$$

$$= \int_0^T\int_{\Omega_t^f(\theta)}\left[-u\cdot\partial_t v^f+(\mathrm{D}\,u\cdot u)\cdot v^f-\nu u\cdot\Delta v^f+u\cdot\nabla q-p\,\mathrm{div}\,v^f\right]$$

$$+\int_0^T\int_{\Gamma_{\hat\theta}^s}\left[(\partial_t\hat\theta\circ\hat\theta^{-1})\cdot(\sigma(v^f,q)\cdot n)-(u\cdot v^f)\langle(\partial_t\hat\theta\circ\hat\theta^{-1}),n\rangle\right]$$

$$+\int_0^T\int_{\Gamma_\infty^f}u_\infty\cdot(\sigma(v^f,q)\cdot n)+\int_0^T\int_{\Omega_0^s}[\hat\theta\cdot\partial_{tt}\hat v^s+\mathrm{D}\,\hat\theta\cdot S(\hat\theta)\cdot\!\cdot\,\mathrm{D}\,\hat v^s]-\int_0^T\int_{\Omega_0^s}f\cdot\hat v^s$$

$$+\int_{\Omega_0^s}[\partial_t\hat\theta(T)\cdot\hat v^s(T)-\hat\theta(T)\cdot\partial_t\hat v^s(T)-\hat\theta_1\cdot\hat v^s(0)+\hat\theta_0\cdot\partial_t\hat v^s(0)]$$

$$+\int_{\Omega_T}u(T)\cdot v^f(T)-\int_{\Omega_0}u_0\cdot v^f(0)$$

$$\forall\,(v^f,\hat v^s,q)\in W_s^f\times V^f$$

### 8.2.4   Min-Max problem

In this section, we introduce the Lagrangian functional associated with problem (8.6) and problem (7.18) :

$$\mathcal{L}_{u_\infty}(u,p,\hat\theta;v^f,q,\hat v^s)\stackrel{\text{def}}{=}J_{u_\infty}(u,p,\hat\theta)-\langle e_{u_\infty}(u,p,\hat\theta),(v^f,q,\hat v^s)\rangle\qquad(8.13)$$

with

$$J_{u_\infty}(u,p,\hat\theta)=\frac{\alpha}{2}\int_0^T\int_{\Omega_0^s}|\hat\theta-\hat\theta_d|^2$$

Using this functional, the cost function can be put in the following form :

$$j(u_\infty)=\min_{(u,p,\hat\theta)\in X^f\times Z^f\times X^s}\ \max_{(v^f,\hat v^s,q)\in W_s^f\times V^f}\ \mathcal{L}_{u_\infty}(u,p,\hat\theta;v^f,q,\hat v^s)\qquad(8.14)$$

### Reduced Gradient

We assume that the conditions to apply the Min-Max principle [40] are fulfilled so we can bypass the derivation with respect to the control variable $u_\infty$ through the min-max subproblem (8.14). It leads to the following result :

**THEOREM 8.1**

For $u_\infty \in \mathcal{U}_c$, and $(u, p, \hat{\theta}, \varphi, \pi, \hat{\psi})$ the unique saddle point of problem (8.14), the gradient of the cost function $j$ at point $u_\infty \in \mathcal{U}_c$ is given by the following expression :

$$\nabla j(u_\infty) = \sigma(\varphi, \pi) \cdot n|_{(0,T) \times \Gamma_\infty^f} + \gamma \left[ {}^*\mathcal{K}\mathcal{K} \right] \cdot u_\infty \qquad (8.15)$$

**PROOF**   Using theorem (3) from [45, 46], we bypass the derivation with respect to $u_\infty$ inside the min-max problem (8.14).
☐

---

## 8.3   KKT optimality conditions

In this section, we are interested in establishing the first order optimality condition for problem (8.14), better known as Karusch-Kuhn-Tucker optimality conditions. This step is crucial, because it leads to the formulation of the adjoint problem satisfied by the Lagrange multipliers $(\varphi, \pi, \hat{\psi})$. We recall the expression of the Lagrangian,

$$\mathcal{L}_{u_\infty}(u, p, \hat{\theta}; v^f, q, \hat{v}^s) \stackrel{\text{def}}{=} J_{u_\infty}(u, p, \hat{\theta}) - \langle e_{u_\infty}(u, p, \hat{\theta}), (v^f, q, \hat{v}^s) \rangle \qquad (8.16)$$

The KKT system will have the following structure :

$$\partial_{(v^f, q, \hat{v}^s)} \mathcal{L}_{u_\infty}(u, p, \hat{\theta}; v^f, q, \hat{v}^s) \cdot (\delta v^f, \delta q, \delta \hat{v}^s) = 0,$$
$$\forall (\delta v^f, \delta \hat{v}^s, \delta q) \in W_f^f \times V^f \to \quad \text{State Equations}$$
$$\partial_{(u, p, \hat{\theta})} \mathcal{L}_{u_\infty}(u, p, \hat{\theta}; v^f, q, \hat{v}^s) \cdot (\delta u, \delta p, \delta \hat{\theta}) = 0,$$
$$\forall (\delta u, \delta p, \delta \hat{\theta}) \in X^f \times Z^f \times X^s \to \quad \text{Adjoint Equations}$$

### 8.3.1   Fluid adjoint system

**LEMMA 8.1**

$$\langle \partial_u \mathcal{L}_{u_\infty}(u, p, \hat{\theta}; v^f, q, \hat{v}^s), \delta u \rangle =$$
$$- \int_0^T \int_{\Omega_{\hat{\theta}}^f} \left[ -\delta u \cdot \partial_t v^f + [\mathrm{D}\, \delta u \cdot u + \mathrm{D}\, u \cdot \delta u] \cdot v^f - \nu \delta u \cdot \Delta v^f + \delta u \cdot \nabla q \right]$$
$$+ \int_0^T \int_{\Gamma_{\hat{\theta}}^s} (\delta u \cdot v^f) \langle \partial_t \hat{\theta} \circ \hat{\theta}^{-1}, n \rangle - \int_{\Omega_T} \delta u(T) \cdot v^f(T) \quad \forall \delta u \in X^f$$

In order to obtain a strong formulation of the fluid adjoint problem, we perform some integration by parts :

**LEMMA 8.2**

$$\int_{\Omega_{\hat{\theta}}^f} (D\,\delta u \cdot u) \cdot v^f = -\int_{\Omega_{\hat{\theta}}^f} \left[ D\,v^f \cdot u + \text{div}(u)\,v^f \right] \cdot \delta u + \int_{\Gamma_{\infty}^f \cup \Gamma_{\hat{\theta}}^s} (\delta u \cdot v^f)\,\langle u, n \rangle$$

It leads to the following identity :

$$\langle \partial_{\hat{u}} \mathcal{L}_{u_\infty}(u, p, \hat{\theta}; \varphi, \pi, \hat{\psi}),\, \delta u \rangle =$$
$$- \int_{Q^f} [-\partial_t \varphi + {}^*D\,u \cdot \varphi - D\,\varphi \cdot u - (\text{div}\,u)\,\varphi - \nu\Delta\varphi + \nabla\pi] \cdot \delta u$$
$$- \int_{\Omega_T^f} \varphi(T) \cdot \delta u(T)$$

**LEMMA 8.3**

$$\langle \partial_p \mathcal{L}_{u_\infty}(u, p, \hat{\theta}; \varphi, \pi, \hat{\psi}),\, \delta p \rangle = \int_0^T \int_{\Omega_{\hat{\theta}}^f} \delta p\,\text{div}\,\varphi, \quad \forall \delta p \in Z^f$$

This leads to the following fluid adjoint strong formulation,

$$\begin{cases} -\partial_t \varphi - D\,\varphi \cdot u + {}^*D\,u \cdot \varphi - \nu\Delta\varphi + \nabla\pi = 0, & Q^f \\ \text{div}\,\varphi = 0, & Q^f \\ \varphi = \hat{\psi} \circ \hat{\theta}^{-1}, & \Sigma^s \\ \varphi = 0, & \Sigma_\infty^f \\ \varphi(T) = 0, & \Omega_T^f \end{cases} \qquad (8.17)$$

### 8.3.2 Solid adjoint system

Now we shall perform the differentiation of the Lagrangian with respect to $\hat{\theta}$. We introduce a perturbation map $\delta\hat{\theta}$ together with a scalar $\rho \geq 0$ and the following perturbed sets,

$$\Omega_{\hat{\theta}+\rho\delta\hat{\theta}}^f \overset{\text{def}}{=} \left[\hat{\theta} + \rho\delta\hat{\theta}\right](\Omega_0^f) = \left[(I + \rho\delta\theta) \circ \hat{\theta}\right](\Omega_0^f)$$
$$\Gamma_{\hat{\theta}+\rho\delta\hat{\theta}}^s \overset{\text{def}}{=} \left[\hat{\theta} + \rho\delta\hat{\theta}\right](\Gamma_0^s) = \left[(I + \rho\delta\theta) \circ \hat{\theta}\right](\Gamma_0^s)$$

with $\delta\theta \overset{\text{def}}{=} \delta\hat{\theta} \circ \hat{\theta}^{-1}$.

**Shape derivative tools**

We shall need the following shape derivative formulas,

**LEMMA 8.4**

$$\frac{d}{d\rho}\left(\int_{\Omega^f_{\hat{\theta}+\rho\delta\hat{\theta}}} G(\rho)\,d\Omega\right)\Bigg|_{\rho=0} = \int_{\Omega^f_{\hat{\theta}}} \partial_\rho G(\rho)|_{\rho=0}\,d\Omega + \int_{\Gamma^s_{\hat{\theta}}} G(\rho=0)\langle\delta\theta,n\rangle\,d\Gamma$$

$$(8.18)$$

**LEMMA 8.5**

$$\frac{d}{d\rho}\left(\int_{\Gamma^s_{\hat{\theta}+\rho\delta\hat{\theta}}} \phi(\rho)\,d\Gamma\right)\Bigg|_{\rho=0} = \int_{\Gamma^s_{\hat{\theta}}} \left[\phi'_\Gamma + H\phi\,\langle\delta\theta,n\rangle\right]\,d\Gamma \qquad (8.19)$$

*where $\phi'_\Gamma$ stands for the tangential shape derivative of $\phi(\rho,.) \in L^1(\Gamma^s_{\hat{\theta}})$*

We recall classical definitions of shape derivative functions :

**DEFINITION 8.1**    *For $\phi(\rho,x) \in C^0((0,\rho_0; C^1(\Gamma^s_{\hat{\theta}+\rho\delta\hat{\theta}}))$, the material derivative is given by the following expression :*

$$\dot{\phi} = \frac{d}{d\rho}\left[\phi(\rho,.) \circ (\mathrm{I}+\rho\delta\theta)\right]\Bigg|_{\rho=0}$$

*then the tangential shape derivative of $\phi$ is given by the following expression,*

$$\phi'_\Gamma \overset{\text{def}}{=} \dot{\phi} - \nabla_\Gamma\phi\cdot\delta\theta$$

**REMARK 8.4**    If $\phi$ is the trace of a vector field $\tilde{\phi}$ defined over $\Omega$, then we have

$$\phi'_\Gamma \overset{\text{def}}{=} \tilde{\phi}'|_{\Gamma^s_{\hat{\theta}}} + \partial_n\tilde{\phi}\,\langle\delta\theta,n\rangle$$

with $\tilde{\phi}'|_{\Gamma^s_{\hat{\theta}}} \overset{\text{def}}{=} \dfrac{d}{d\rho}\left(\tilde{\phi}(\rho,.)\right)\Bigg|_{\rho=0}\Bigg|_{\Gamma^s_{\hat{\theta}}}.$      ⬜

Following this remark, we have

**LEMMA 8.6**

$$\frac{d}{d\rho}\left(\int_{\Gamma^s_{\hat{\theta}+\rho\delta\hat{\theta}}} \tilde{\phi}(\rho,x)\,da\right)\Bigg|_{\rho=0} = \int_{\Gamma^s_{\hat{\theta}}} \left[\tilde{\phi}' + \left[H\tilde{\phi} + \partial_n\tilde{\phi}\right]\langle\delta\theta,n\rangle\right]\,d\Gamma \qquad (8.20)$$

**Derivation of the perturbed Lagrangian**

The map $\mathrm{I}+\rho\delta\theta$ sends respectively the sets $(\Omega^f_{\hat{\theta}}, \Gamma^s_{\hat{\theta}})$ into $(\Omega^f_{\hat{\theta}+\rho\delta\hat{\theta}}, \Gamma^s_{\hat{\theta}+\rho\delta\hat{\theta}})$. Then defining its inverse $\mathcal{R}^t_\rho \overset{\text{def}}{=} (\mathrm{I}+\rho\delta\theta)^{-1}$ we can perturb the integral support inside the Lagrangian and keep functions defined on $\Omega^f_{\hat{\theta}}$ and $\Gamma^s_{\hat{\theta}}$. This leads to the following perturbed Lagrangian,

$$\mathcal{L}_{u_\infty}^\rho \overset{\text{def}}{=} \mathcal{L}(u, p, \hat{\theta}; v^f, q, \hat{v}^s) = J_{u_\infty}(u, p, \hat{\theta} + \rho \delta \hat{\theta})$$

$$- \int_0^T \int_{\Omega^f_{\hat{\theta} + \rho \delta \hat{\theta}}} \left[ -(u \circ \mathcal{R}_\rho^t) \cdot \partial_t (v^f \circ \mathcal{R}_\rho^t) + [\mathrm{D}(u \circ \mathcal{R}_\rho^t) \cdot u \circ \mathcal{R}_\rho^t] \cdot (v^f \circ \mathcal{R}_\rho^t) \right.$$

$$\left. - \nu(u \circ \mathcal{R}_\rho^t) \cdot \Delta(v^f \circ \mathcal{R}_\rho^t) + (u \circ \mathcal{R}_\rho^t) \cdot \nabla q - p \, \mathrm{div}(v^f \circ \mathcal{R}_\rho^t) \right]$$

$$- \int_0^T \int_{\Gamma^s_{\hat{\theta} + \rho \delta \hat{\theta}}} \left\{ [\partial_t (\hat{\theta} + \rho \delta \hat{\theta}) \circ (\hat{\theta} + \rho \delta \hat{\theta})^{-1}] \cdot (\sigma(v^f \circ \mathcal{R}_\rho^t, q) \cdot n^\rho) \right.$$

$$- (u \circ \mathcal{R}_\rho^t) \cdot (v^f \circ \mathcal{R}_\rho^t) \left\langle \partial_t (\hat{\theta} + \rho \delta \hat{\theta}) \circ (\hat{\theta} + \rho \delta \hat{\theta})^{-1}, n^\rho \right\rangle \right\} - \int_0^T \int_{\Gamma^f_\infty} u_\infty \cdot (\sigma(v^f, q) \cdot n)$$

$$- \int_0^T \int_{\Omega^s_0} \left[ (\hat{\theta} + \rho \delta \hat{\theta}) \cdot \partial_{tt} \hat{v}^s + \mathrm{D}(\hat{\theta} + \rho \delta \hat{\theta}) \cdot S(\hat{\theta} + \rho \delta \hat{\theta}) \cdot\cdot \mathrm{D} \, \hat{v}^s \right]$$

$$- \int_{\Omega^s_0} \left[ \partial_t (\hat{\theta} + \rho \delta \hat{\theta})(T) \cdot \hat{v}^s(T) - (\hat{\theta} + \rho \delta \hat{\theta})(T) \cdot \partial_t \hat{v}^s(T) \right.$$

$$\left. - \hat{\theta}_1 \cdot \hat{v}^s(0) + \hat{\theta}_0 \cdot \partial_t \hat{v}^s(0) \right] - \int_{\Omega^{f,\rho}_T} (u \circ \mathcal{R}_\rho^t)(T) \cdot (v^f \circ \mathcal{R}_\rho^t)(T)$$

$$+ \int_{\Omega_0} u_0 \cdot v^f(0) + \int_0^T \int_{\Omega^s_0} f \cdot \hat{v}^s, \quad \forall (v^f, \hat{v}^s, q) \in W^f_s \times V^f$$

**Fluid distributed terms**

We set

$$G(\rho, .) = \left[ -(u \circ \mathcal{R}_\rho^t) \cdot \partial_t (v^f \circ \mathcal{R}_\rho^t) + [\mathrm{D}(u \circ \mathcal{R}_\rho^t) \cdot u \circ \mathcal{R}_\rho^t] \cdot (v^f \circ \mathcal{R}_\rho^t) \right.$$

$$\left. - \nu(u \circ \mathcal{R}_\rho^t) \cdot \Delta(v^f \circ \mathcal{R}_\rho^t) + (u \circ \mathcal{R}_\rho^t) \cdot \nabla q - p \, \mathrm{div}(v^f \circ \mathcal{R}_\rho^t) \right]$$

**LEMMA 8.7**

$$\left( \frac{d\mathcal{R}_\rho^t}{d\rho} \right) \bigg|_{\rho = 0} = -\delta \theta$$

Then we have the following derivative,

$$
\partial_\rho G(\rho,.)|_{\rho=0} = \big[(\mathrm{D}\,u \cdot \delta\theta) \cdot \partial_t v^f + u \cdot \partial_t(\mathrm{D}\,v^f \cdot \delta\theta)
$$
$$
- [(\mathrm{D}(\mathrm{D}\,u \cdot \delta\theta)) \cdot u + \mathrm{D}\,u \cdot (\mathrm{D}\,u \cdot \delta\theta)] \cdot v^f - (\mathrm{D}\,u \cdot u) \cdot (\mathrm{D}\,v^f \cdot \delta\theta)
$$
$$
+\nu(\mathrm{D}\,u \cdot \delta\theta) \cdot \Delta v^f + \nu u \cdot \Delta(\mathrm{D}\,v^f \cdot \delta\theta) + p\,\mathrm{div}(\mathrm{D}\,v^f \cdot \delta\theta) - (\mathrm{D}\,u \cdot \delta\theta) \cdot \nabla q\big]
$$

Then we have an expression for the derivative of fluid distributed terms coming from the Lagrangian with respect to $\rho$,

$$
\frac{d}{d\rho}\left(\int_{\Omega^f_{\hat\theta+\rho\delta\hat\theta}} G(\rho)d\Omega\right)\Bigg|_{\rho=0} = \int_{\Omega^f_{\hat\theta}} \big[(\mathrm{D}\,u \cdot \delta\theta) \cdot \partial_t v^f + u \cdot \partial_t(\mathrm{D}\,v^f \cdot \delta\theta)
$$
$$
- [(\mathrm{D}(\mathrm{D}\,u \cdot \delta\theta)) \cdot u + \mathrm{D}\,u \cdot (\mathrm{D}\,u \cdot \delta\theta)] \cdot v^f - (\mathrm{D}\,u \cdot u) \cdot (\mathrm{D}\,v^f \cdot \delta\theta)
$$
$$
+\nu(\mathrm{D}\,u \cdot \delta\theta) \cdot \Delta v^f + \nu u \cdot \Delta(\mathrm{D}\,v^f \cdot \delta\theta) + p\,\mathrm{div}(\mathrm{D}\,v^f \cdot \delta\theta) - (\mathrm{D}\,u \cdot \delta\theta) \cdot \nabla q\big]
$$
$$
+ \int_{\Gamma^s_{\hat\theta}} \big[-u \cdot \partial_t v^f + (\mathrm{D}u \cdot u) \cdot v^f - \nu u \cdot \Delta v^f + u \cdot \nabla q - p\,\mathrm{div}\,v^f\big] \langle \delta\theta, n \rangle
$$

## Fluid boundary terms

We must now take into account the terms coming from the moving boundary $\Gamma^s_{\hat\theta+\rho\delta\hat\theta}$. Then we set

$$
\phi(\rho,.) = \big[\partial_t\,(\hat\theta+\rho\delta\hat\theta) \circ (\hat\theta+\rho\,\delta\hat\theta)^{-1}\big] \cdot \big[\sigma(v^f \circ \mathcal{R}^t_\rho, q) - (u \circ \mathcal{R}^t_\rho) \cdot (v^f \circ \mathcal{R}^t_\rho)\big] \cdot n^\rho
$$

### *LEMMA 8.8 [53]*

$$
\partial_\rho n^\rho|_{\rho=0} = n'_\Gamma = -\nabla_\Gamma(\delta\theta \cdot n)
$$

### *LEMMA 8.9*

$$
\frac{d}{d\rho}\left(\int_{\Gamma^s_{\hat\theta+\rho\delta\hat\theta}} \langle E(\rho), n^\rho\rangle d\Gamma\right)\Bigg|_{\rho=0} = \int_{\Gamma^s_{\hat\theta}} \langle E'|_{\Gamma_t}, n\rangle + (\mathrm{div}\,E)\langle \delta\theta, n\rangle
$$

**PROOF**  First, we use that

$$
\int_{\Gamma^s_{\hat\theta+\rho\delta\hat\theta}} \langle E(\rho), n^\rho\rangle = \int_{\Omega^f_{\hat\theta+\rho\delta\hat\theta}} \mathrm{div}\,E(\rho)
$$

then we derive this quantity using lemma (8.4),

$$
\frac{d}{d\rho}\left(\int_{\Omega^f_{\hat\theta+\rho\delta\hat\theta}} \mathrm{div}\,E(\rho)\right)\Bigg|_{\rho=0} = \int_{\Omega^f_{\hat\theta}} \mathrm{div}\,E' + \int_{\Gamma^s_{\hat\theta}} (\mathrm{div}\,E)\langle \delta\theta, n\rangle
$$

We conclude using $\int_{\Omega_{\hat\theta}^f} \mathrm{div}\, E' = \int_{\Gamma_{\hat\theta}^s} \langle E', n\rangle.$                    ☐

**LEMMA 8.10**

$$W_{\delta\theta} \overset{\mathrm{def}}{=} \frac{d}{d\rho}\left[\partial_t\,(\hat\theta + \rho\delta\hat\theta)\circ(\hat\theta + \rho\,\delta\hat\theta)^{-1}\right]\Big|_{\rho=0} = \partial_t\,(\delta\theta\circ\hat\theta)\circ\hat\theta^{-1} - \mathrm{D}\,V_{\hat\theta}\cdot\delta\theta$$

(8.21)

*where we recall that* $V_{\hat\theta} \overset{\mathrm{def}}{=} (\partial_t\hat\theta)\circ\hat\theta^{-1}.$

Using the last identities, we shall obtain,

**LEMMA 8.11**

$$E'|_{\Gamma_{\hat\theta}^s} = W_{\delta\theta}\cdot\left[-q\,\mathrm{I} + \nu\,\mathrm{D}\,v^f - u\cdot v^f\right] + V_{\hat\theta}\cdot\left[-\nu\,\mathrm{D}(\mathrm{D}\,v^f\cdot\delta\theta) + (\mathrm{D}\,u\cdot\delta\theta)\cdot v^f\right.$$
$$\left. + u\cdot(\mathrm{D}\,v^f\cdot\delta\theta)\right]$$

It leads to the final expression,

$$\frac{d}{d\rho}\left(\int_{\Gamma_{\hat\theta+\rho\delta\hat\theta}^s}\phi(\rho)\,d\Gamma\right)\Big|_{\rho=0} = \int_{\Gamma_{\hat\theta}^s} W_{\delta\theta}\cdot\left[-q\,\mathrm{I} + \nu\,\mathrm{D}\,v^f - u\cdot v^f\right]\cdot n$$

$$+ \int_{\Gamma_{\hat\theta}^s} V_{\hat\theta}\cdot\left[-\nu\,\mathrm{D}(\mathrm{D}\,v^f\cdot\delta\theta) + (\mathrm{D}\,u\cdot\delta\theta)\cdot v^f + u\cdot(\mathrm{D}\,v^f\cdot\delta\theta)\right]\cdot n$$

$$+ \int_{\Gamma_{\hat\theta}^s}\mathrm{div}(V_{\hat\theta}\cdot\left[-q\,\mathrm{I} + \nu^*\,\mathrm{D}\,v^f - u\cdot v^f\right])\langle\delta\theta, n\rangle$$

**REMARK 8.5**    We recall that

$$\int_{\Gamma_{\hat\theta}^s} V_{\hat\theta}\cdot(\mathrm{D}\,v^f\cdot n) = \int_{\Omega_{\hat\theta}^f}\mathrm{div}(^*\mathrm{D}\,v^f\cdot V_{\hat\theta})$$

$$= \int_{\Omega_{\hat\theta}^f}\mathrm{D}\,v^f\cdot\cdot\mathrm{D}\,V_{\hat\theta} + V_{\hat\theta}\cdot\Delta v^f$$

(8.22)

☐

**Distributed solid terms**

We set

$$m(\rho) = \left[(\hat\theta + \rho\delta\hat\theta)\cdot\partial_{tt}\,\hat v^s + \mathrm{D}(\hat\theta + \rho\delta\hat\theta)\,S(\hat\theta + \rho\delta\hat\theta)\cdot\cdot\mathrm{D}\,\hat v^s\right]$$

### LEMMA 8.12

$$\frac{d}{d\rho}\left(\int_0^T \int_{\Omega_0^s} m(\rho)\right)\Bigg|_{\rho=0} =$$

$$\int_0^T \int_{\Omega_0^s} \left[\delta\hat{\theta}\cdot\partial_{tt}\,\hat{v}^s + \left(\mathrm{D}\,\delta\hat{\theta}\cdot S(\hat{\theta}) + \mathrm{D}\,\hat{\theta}\cdot S'(\hat{\theta})\cdot\delta\hat{\theta}\right)\cdot\cdot\mathrm{D}\,\hat{v}^s\right]$$

### Final time terms

We set

$$\varpi(\rho) = \int_{\Omega_0^s}\left[\partial_t\,(\hat{\theta}+\rho\delta\hat{\theta})(T)\cdot\hat{v}^s(T) - (\hat{\theta}+\rho\delta\hat{\theta})(T)\cdot\partial_t\,\hat{v}^s(T)\right]$$

$$+ \int_{\Omega_T^{f,\rho}}(u\circ\mathcal{R}_\rho^t)(T)\cdot(v^f\circ\mathcal{R}_\rho^t)(T)$$

### LEMMA 8.13

$$\frac{d}{d\rho}\varpi(\rho)\Bigg|_{\rho=0} = \int_{\Omega_0^s}\left[(\partial_t\,\delta\hat{\theta})(T)\cdot\hat{v}^s(T) - \delta\hat{\theta}(T)\cdot\partial_t\,\hat{v}^s(T)\right]$$

$$+ \int_{\Omega_T^f}\left[-(\mathrm{D}\,u\cdot\delta\theta)(T)\cdot v^f(T) - u(T)\cdot(\mathrm{D}\,v^f\cdot\delta\theta)(T)\right] + \int_{\Gamma_T^s}u(T)\cdot v^f(T)\langle\delta\theta(T),n\rangle$$

### Complete derivative

We set

$$\ell(v^f,\hat{v}^s) = \int_{\Omega_0^s}\left[-\hat{\theta}_1\cdot\hat{v}^s(0) + \hat{\theta}_0\cdot\partial_t\,\hat{v}^s(0)\right] - \int_{\Omega_0}u_0\cdot v^f(0) - \int_0^T\int_{\Omega_0^s}f\cdot\hat{v}^s$$

The perturbed Lagrangian has the following form,

$$\mathcal{L}^\rho = J^\rho$$

$$- \left[\int_0^T\int_{\Omega_{\hat{\theta}+\rho\delta\hat{\theta}}^f}G(\rho) + \int_0^T\int_{\Gamma_{\hat{\theta}+\rho\delta\hat{\theta}}^s}\phi(\rho) + \int_0^T\left(\int_{\Omega_0^s}m(\rho)\right) + \varpi(\rho) + \ell\right]$$

Then,

$$\frac{d}{d\rho}J^\rho\Bigg|_{\rho=0} = \alpha\int_0^T\int_{\Omega_0^s}(\hat{\theta}-\hat{\theta}_d)\cdot\delta\hat{\theta}$$

Using the previous identities, we shall furnish the derivative of the perturbed Lagrangian $\mathcal{L}^\rho$ with respect to $\rho$ at point $\rho=0$,

$$\frac{d}{d\rho}\left(\mathcal{L}_{\hat{\theta},\delta\hat{\theta}}^\rho\right)\Bigg|_{\rho=0} = -\left[A_{\Omega_{\hat{\theta}}^f} + B_{\Gamma_{\hat{\theta}}^s} + C_{\Gamma_{\hat{\theta}}^s} + D_{\Omega_0^s} + F_{\Gamma_\infty^f}\right] \qquad (8.23)$$

with

$$A_{\Omega_{\hat{\theta}}^f} = \int_0^T \int_{\Omega_{\hat{\theta}}^f} (\mathrm{D}\,u \cdot \delta\theta) \cdot \partial_t v^f - [(\mathrm{D}(\mathrm{D}\,u \cdot \delta\theta)) \cdot u$$

$$+ \mathrm{D}\,u \cdot (\mathrm{D}\,u \cdot \delta\theta)] \cdot v^f + \nu(\mathrm{D}\,u \cdot \delta\theta) \cdot \Delta v^f - (\mathrm{D}\,u \cdot \delta\theta) \cdot \nabla q$$

$$+ \int_0^T \int_{\Omega_{\hat{\theta}}^f} \left[ u \cdot \partial_t(\mathrm{D}\,v^f \cdot \delta\theta) - (\mathrm{D}\,u \cdot u) \cdot (\mathrm{D}\,v^f \cdot \delta\theta) + \nu u \cdot \Delta(\mathrm{D}\,v^f \cdot \delta\theta) \right.$$

$$+ p\,\mathrm{div}(\mathrm{D}\,v^f \cdot \delta\theta) \Big] + \int_{\Omega_T^f} \left[ -(\mathrm{D}\,u \cdot \delta\theta)(T) \cdot v^f(T) - u(T) \cdot (\mathrm{D}\,v^f \cdot \delta\theta)(T) \right]$$

$$B_{\Gamma_{\hat{\theta}}^s} = \int_0^T \int_{\Gamma_{\hat{\theta}}^s} \left[ -u \cdot \partial_t v^f + (Du \cdot u) \cdot v^f - \nu u \cdot \Delta v^f + u \cdot \nabla q - p\,\mathrm{div}\,v^f \right] \langle \delta\theta, n \rangle$$

$$+ \int_0^T \int_{\Gamma_{\hat{\theta}}^s} (\mathrm{D}\,V_{\hat{\theta}} \cdot \delta\theta) \cdot \left[ -q\,\mathrm{I} + \nu\,\mathrm{D}\,v^f - u \cdot v^f \right] \cdot n$$

$$+ \int_0^T \int_{\Gamma_{\hat{\theta}}^s} V_{\hat{\theta}} \cdot \left[ -\nu\,\mathrm{D}(\mathrm{D}\,v^f \cdot \delta\theta) + (\mathrm{D}\,u \cdot \delta\theta) \cdot v^f + u \cdot (\mathrm{D}\,v^f \cdot \delta\theta) \right] \cdot n$$

$$+ \int_0^T \int_{\Gamma_{\hat{\theta}}^s} \mathrm{div}(V_{\hat{\theta}} \cdot \left[ -q\,\mathrm{I} + \nu^*\,\mathrm{D}\,v^f - u \cdot v^f \right]) \langle \delta\theta, n \rangle$$

$$+ \int_{\Gamma_T^s} u(T) \cdot v^f(T) \langle \delta\theta(T), n \rangle$$

$$C_{\Gamma_{\hat{\theta}}^s} = \int_0^T \int_{\Gamma_{\hat{\theta}}^s} (\partial_t\,\delta\hat{\theta}) \circ \hat{\theta}^{-1} \cdot \left[ -q\,\mathrm{I} + \nu\,\mathrm{D}\,v^f - u \cdot v^f \right] \cdot n$$

$$D_{\Omega_0^s} = \int_0^T \int_{\Omega_0^s} \left[ \delta\hat{\theta} \cdot \partial_{tt}\,\hat{v}^s + \left( \mathrm{D}\,\delta\hat{\theta} \cdot S(\hat{\theta}) + \mathrm{D}\,\hat{\theta} \cdot S'(\hat{\theta}) \cdot \delta\hat{\theta} \right) \cdot\cdot \mathrm{D}\,\hat{v}^s \right]$$

$$- \alpha \int_0^T \int_{\Omega_0^s} (\hat{\theta} - \hat{\theta}_d) \cdot \delta\hat{\theta} + \int_{\Omega_0^s} \left[ (\partial_t\,\delta\hat{\theta})(T) \cdot \hat{v}^s(T) - \delta\hat{\theta}(T) \cdot \partial_t\,\hat{v}^s(T) \right]$$

$$F_{\Gamma_\infty^f} = - \int_0^T \int_{\Gamma_\infty^f} \nu\,u_\infty \cdot \mathrm{D}(\mathrm{D}\,v^f \cdot \delta\theta) \cdot n$$

**The shape derivative kernel identity**

We shall now assume that $(u, p, \varphi, \pi)$ is a saddle point of the Lagrangian functional $\mathcal{L}$. This will help us to simplify several terms involved in the derivative of $\mathcal{L}$ with respect to $\rho$.

*LEMMA 8.14*

$$\int_0^T \int_{\Omega_{\hat{\theta}}^f} \{ [(D\,u \cdot \delta\theta) \cdot \partial_t\, \varphi - [(D(D\,u \cdot \delta\theta) \cdot u)$$

$$+ D\,u \cdot (D\,u \cdot \delta\theta)] \cdot \varphi + \nu(D\,u \cdot \delta\theta) \cdot \Delta\varphi - (D\,u \cdot \delta\theta) \cdot \nabla\,\pi]$$

$$+ [u \cdot \partial_t(D\,\varphi \cdot \delta\theta) - (D\,u \cdot u) \cdot D\,\varphi \cdot \delta\theta + \nu u \cdot \Delta(D\,\varphi \cdot \delta\theta) + p\,\mathrm{div}(D\,\varphi \cdot \delta\theta)] \}$$

$$+ \int_0^T \int_{\Gamma_{\hat{\theta}}^s} \nu(\varphi - \hat{v}^s \circ \hat{\theta}^{-1}) \cdot (D(D\,u \cdot \delta\theta) \cdot n) + (D\,\varphi \cdot \delta\theta) \cdot (-p\,n + \nu\,D\,u \cdot n)$$

$$- \int_0^T \int_{\Gamma_{\hat{\theta}}^s} V_{\hat{\theta}} \cdot [\nu\,D(D\,\varphi \cdot \delta\theta) - (D\,u \cdot \delta\theta) \cdot \varphi - u \cdot (D\,\varphi \cdot \delta\theta)] \cdot n$$

$$- \int_0^T \int_{\Gamma_\infty^f} \nu u_\infty \cdot (D(D\,\varphi \cdot \delta\theta) \cdot n) - \int_{\Omega_T} (D\,u \cdot \delta\theta)(T) \cdot \varphi(T) + u(T) \cdot (D\,\varphi \cdot \delta\theta)(T) = 0,$$

$$\forall \delta\hat{\theta} \in \hat{Y}^s$$

**PROOF**    We define a new Lagrangian, where we take into account the normal stress on the moving interface by choosing test functions discontinuous at the fluid-structure interface, i.e., $(v^f, \hat{v}^s) \in Y^f \times \hat{Y}^s$,

$$
\mathcal{L}^d_{u_\infty}(u, p, \hat{\theta}; v^f, q, \hat{v}^s) = J_{u_\infty}(u, p, \hat{\theta})
$$

$$
- \int_0^T \int_{\Omega^f_{\hat{\theta}}} \left[ -u \cdot \partial_t v^f + (\mathrm{D}\, u \cdot u) \cdot v^f - \nu u \cdot \Delta v^f + u \cdot \nabla q - p \,\mathrm{div}\, v^f \right]
$$

$$
- \int_0^T \int_{\Gamma^s_{\hat{\theta}}} V_{\hat{\theta}} \cdot \left[ \sigma(v^f, q) - (u \cdot v^f) \right] \cdot n + (v^f - \hat{v}^s \circ \hat{\theta}^{-1}) \cdot (\sigma(u, p) \cdot n)
$$

$$
- \int_0^T \int_{\Gamma^f_\infty} u_\infty \cdot (\sigma(v^f, q) \cdot n) - \int_0^T \int_{\Omega^s_0} [\hat{\theta} \cdot \partial_{tt}\, \hat{v}^s + \mathrm{D}\,\hat{\theta}) \cdot S(\hat{\theta}) \cdot\cdot \mathrm{D}\,\hat{v}^s] + \int_0^T \int_{\Omega^s_0} f \cdot \hat{v}^s
$$

$$
- \int_{\Omega^s_0} [\partial_t \hat{\theta}(T) \cdot \hat{v}^s(T) - \hat{\theta}(T) \cdot \partial_t \hat{v}^s(T) - \hat{\theta}_1 \cdot \hat{v}^s(0) + \hat{\theta}_0 \cdot \partial_t \hat{v}^s(0)]
$$

$$
- \int_{\Omega_T} u(T) \cdot v^f(T) + \int_{\Omega_0} u_0 \cdot v^f(0)
$$

$$
\forall (v^f, \hat{v}^s, q) \in Y^f \times \hat{Y}^s \times V^f
$$

Let us differentiate the above functional with respect to $(u, v)$,

$$
\partial_{(u, v^f)} \mathcal{L}^d_{u_\infty} \cdot (\delta u, \delta v^f)
$$
$$
= \int_0^T \int_{\Omega^f_{\hat{\theta}}} \left[ \delta u \cdot \partial_t v^f - [(\mathrm{D}\, \delta u \cdot u) + (\mathrm{D}\, u \cdot \delta u)] \cdot v^f + \nu \delta u \cdot \Delta v^f - \delta u \cdot \nabla q \right]
$$

$$
+ \int_0^T \int_{\Omega^f_{\hat{\theta}}} \left[ u \cdot \partial_t \delta v^f - (\mathrm{D}\, u \cdot u) \cdot \delta v^f + \nu u \cdot \Delta \delta v^f + p \,\mathrm{div}\, \delta v^f \right]
$$

$$
- \int_0^T \int_{\Gamma^f_\infty} \nu u_\infty \cdot (\mathrm{D}\, \delta v^f \cdot n) + \int_0^T \int_{\Gamma^s_{\hat{\theta}}} \nu (v^f - \hat{v}^s \circ \hat{\theta}^{-1}) \cdot (\mathrm{D}\, \delta u \cdot n) + \delta v^f \cdot (-p\, n + \nu\, \mathrm{D}\, u \cdot n)
$$

$$- \int_0^T \int_{\Gamma_{\hat{\theta}}^s} V_{\hat{\theta}} \cdot \left[ \nu \, D \, \delta v^f - \delta u \cdot v^f - u \cdot \delta v^f \right] \cdot n$$

$$- \int_{\Omega_T} \delta u(T) \cdot v^f(T) + u(T) \cdot \delta v^f(T)$$

$$\forall \, (v^f, \hat{v}^s, q) \in \ Y^f \times \hat{Y}^s \times V^f$$

The first order optimality conditions corresponding to the saddle point formulation with respect to $(u, v^f)$ writes

$$\partial_{(u,v^f)} \mathcal{L}_{u_\infty}^d \big|_{(u,\varphi,\hat{\theta})} \cdot (\delta u, \delta v^f) = 0, \quad \forall \, (v^f, \hat{v}^s, q) \in \ Y^f \times \hat{Y}^s \times V^f$$

We choose specific perturbations,

$$\delta u = D \, u \cdot \delta\theta, \qquad \delta v^f = D \, v^f \cdot \delta\theta$$

☐

## Solid adjoint equation

Using the shape derivative kernel identity, we simplify the Lagrangian derivative at the saddle point $(u, p, \varphi, \pi, \hat{\psi})$ and the distributed $A_{\Omega_{\hat{\theta}}^f}$ only involves boundary terms,

$$A_{\Omega_{\hat{\theta}}^f} =$$

$$\int_0^T \int_{\Gamma_{\hat{\theta}}^s} V_{\hat{\theta}} \cdot [\nu \, D(D \, \varphi \cdot \delta\theta) - (D \, u \cdot \delta\theta) \cdot \varphi - u \cdot (D \, \varphi \cdot \delta\theta)] \cdot n$$

$$- \int_0^T \int_{\Gamma_{\hat{\theta}}^s} \nu(\varphi - \hat{\psi} \circ \hat{\theta}^{-1}) \cdot (D(D \, u \cdot \delta\theta) \cdot n) + (D \, \varphi \cdot \delta\theta) \cdot (-p \, n + \nu \, D \, u \cdot n)$$

$$+ \int_0^T \int_{\Gamma_\infty^f} \nu u_\infty \cdot (D(D \, \varphi \cdot \delta\theta) \cdot n)$$

Using the fluid adjoint system (8.17), we have

$$\varphi = \hat{\psi} \circ \hat{\theta}^{-1}, \qquad \text{on } \Gamma_{\hat{\theta}}^s$$

We have

$$A_{\Omega_{\hat{\theta}}^f} + F_{\Gamma_\infty^f} = \int_0^T \int_{\Gamma_{\hat{\theta}}^s} V_{\hat{\theta}} \cdot [\nu \, D(D \, \varphi \cdot \delta\theta) - (D \, u \cdot \delta\theta) \cdot \varphi - u \cdot (D \, \varphi \cdot \delta\theta)] \cdot n$$

$$- \int_0^T \int_{\Gamma_{\hat{\theta}}^s} (D \, \varphi \cdot \delta\theta) \cdot \sigma(u, p) \cdot n$$

**LEMMA 8.15**

*The following identity holds,*

$$\operatorname{div}(V_{\hat{\theta}} \cdot [-\pi\,\mathrm{I} + \nu^* \,\mathrm{D}\,\varphi - u \cdot \varphi]) = -\pi \,\operatorname{div} V_{\hat{\theta}} - V_{\hat{\theta}} \cdot \nabla \pi + \nu \,\mathrm{D}\,\varphi \cdot\cdot\, \mathrm{D}\, V_{\hat{\theta}}$$

$$+ \nu\, V_{\hat{\theta}} \Delta\varphi - (\operatorname{div} V_{\hat{\theta}})\, u \cdot \varphi - (\mathrm{D}\, u \cdot V_{\hat{\theta}}) \cdot \varphi - (\mathrm{D}\,\varphi \cdot u) \cdot \varphi$$

Using the above identity, we get

$$B_{\Gamma^s_{\hat{\theta}}} =$$

$$\int_0^T \int_{\Gamma^s_{\hat{\theta}}} \big[ -V_{\hat{\theta}} \cdot \partial_t \varphi + (Du \cdot V_{\hat{\theta}}) \cdot \varphi - \nu V_{\hat{\theta}} \cdot \Delta\varphi + V_{\hat{\theta}} \cdot \nabla\pi - p \,\operatorname{div}\varphi \big] \langle \delta\theta, n \rangle$$

$$+ \int_0^T \int_{\Gamma^s_{\hat{\theta}}} (\mathrm{D}\, V_{\hat{\theta}} \cdot \delta\theta) \cdot [-\pi\,\mathrm{I} + \nu\,\mathrm{D}\,\varphi - u \cdot \varphi] \cdot n$$

$$+ \int_0^T \int_{\Gamma^s_{\hat{\theta}}} V_{\hat{\theta}} \cdot [-\nu\,\mathrm{D}(\mathrm{D}\,\varphi \cdot \delta\theta) + (\mathrm{D}\, u \cdot \delta\theta) \cdot \varphi + u \cdot (\mathrm{D}\,\varphi \cdot \delta\theta)] \cdot n$$

$$+ \int_0^T \int_{\Gamma^s_{\hat{\theta}}} \big[ -\pi \,\operatorname{div} V_{\hat{\theta}} - V_{\hat{\theta}} \cdot \nabla\pi + \nu\,\mathrm{D}\,\varphi \cdot\cdot\, \mathrm{D}\, V_{\hat{\theta}} + \nu\, V_{\hat{\theta}} \Delta\varphi - (\operatorname{div} V_{\hat{\theta}})\, u \cdot \varphi$$

$$- (\mathrm{D}\, u \cdot V_{\hat{\theta}}) \cdot \varphi - (\mathrm{D}\,\varphi \cdot V_{\hat{\theta}}) \cdot \varphi \big] \langle \delta\theta, n \rangle + \int_{\Gamma^s_T} u(T) \cdot \varphi(T) \langle \delta\theta(T), n \rangle$$

Several terms can be eliminated inside the above expressions. Furthermore, adding the terms $A_{\Omega_{\hat\theta}^f} + F_{\Gamma_\infty^f}$, we obtain

$$A_{\Omega_{\hat\theta}^f} + B_{\Gamma_{\hat\theta}^s} + F_{\Gamma_\infty^f} =$$

$$\int_0^T \int_{\Gamma_{\hat\theta}^s} \left[ - V_{\hat\theta} \cdot \partial_t\varphi - p \operatorname{div}\varphi - \pi \operatorname{div} V_{\hat\theta} + \nu \operatorname{D}\varphi \cdot\cdot \operatorname{D} V_{\hat\theta} \right.$$

$$\left. - (\operatorname{div} V_{\hat\theta}) \, u \cdot \varphi - (\operatorname{D}\varphi \cdot V_{\hat\theta}) \cdot \varphi \right] \langle \delta\theta, n \rangle - (\operatorname{D}\varphi \cdot \delta\theta) \cdot (-p\,n + \nu \operatorname{D} u \cdot n)$$

$$+ \int_0^T \int_{\Gamma_{\hat\theta}^s} (\operatorname{D} V_{\hat\theta} \cdot \delta\theta) \cdot \left[ -\pi \operatorname{I} + \nu \operatorname{D}\varphi - u \cdot \varphi \right] \cdot n$$

$$+ \int_{\Gamma_T^s} u(T) \cdot \varphi(T) \langle \delta\theta(T), n \rangle$$

We now turn to the analysis of the term $C_{\Gamma_{\hat\theta}^s}$,

$$C_{\Gamma_{\hat\theta}^s} = \int_0^T \int_{\Gamma_{\hat\theta}^s} (\partial_t \, \delta\hat\theta) \circ \hat\theta^{-1} \cdot \left[ \sigma(\varphi, \pi) - V_{\hat\theta} \cdot \varphi \right] \cdot n$$

We shall use the following fundamental adjoint identity,

**LEMMA 8.16**

*For any smooth $E$ defined in the hold-all domain $D$,*

$$\int_0^T \int_{\Gamma_{\hat\theta}} \partial_t(\delta\hat\theta) \circ \hat\theta^{-1} \cdot E = \int_0^T \int_{\Gamma_{\hat\theta}} \left[ -\partial_t E - (\operatorname{div}_\Gamma V_{\hat\theta}) E - \operatorname{D} E \cdot V_{\hat\theta} \right] \cdot \delta\theta$$

$$+ \left[ \int_{\Gamma_{\hat\theta}^s} E \cdot \delta\theta \right]_0^T$$

$$(8.24)$$

Let us apply this lemma, with $(E)_i = \mathcal{E}_{i,j}\, n_j$

$$\int_0^T \int_{\Gamma_{\hat{\theta}}} \partial_t(\delta\hat{\theta}) \circ \hat{\theta}^{-1} \cdot (\mathcal{E} \cdot n) =$$

$$\int_0^T \int_{\Gamma_{\hat{\theta}}} \left[ -(\partial_t \mathcal{E}) \cdot n + \mathcal{E} \cdot \nabla_\Gamma (V_{\hat{\theta}} \cdot n) - (\mathrm{div}_\Gamma\, V_{\hat{\theta}})\, \mathcal{E} \cdot n \right.$$

$$\left. - \mathrm{D}\,\mathcal{E} \cdot V_{\hat{\theta}} \cdot n - \mathcal{E} \cdot \mathrm{D}_\Gamma\, n \cdot V_{\hat{\theta}} \right] \cdot \delta\theta + \left[ \int_{\Gamma_{\hat{\theta}}^s} (\mathcal{E} \cdot n) \cdot \delta\theta \right]_0^T$$

Using the previous identity, we get

$$C_{\Gamma_{\hat{\theta}}^s} = \int_0^T \int_{\Gamma_{\hat{\theta}}} \left[ -(\partial_t \mathcal{E}) \cdot n + \mathcal{E} \cdot \nabla_\Gamma (V_{\hat{\theta}} \cdot n) - (\mathrm{div}_\Gamma\, V_{\hat{\theta}})\, \mathcal{E} \cdot n \right.$$

$$\left. - \mathrm{D}\,\mathcal{E} \cdot V_{\hat{\theta}} \cdot n - \mathcal{E} \cdot \mathrm{D}_\Gamma\, n \cdot V_{\hat{\theta}} \right] \cdot \delta\theta + \int_{\Gamma_{\hat{\theta}}^s} (\mathcal{E} \cdot n)(T) \cdot \delta\theta(T)$$

with $\mathcal{E} = \sigma(\varphi, \pi) - (V_{\hat{\theta}} \cdot \varphi)\, I$ and $\delta\theta(0) = 0$.
Let us now analyse the solid distributed term,

$$D_{\Omega_0^s} =$$

$$\int_0^T \int_{\Omega_0^s} \left[ \delta\hat{\theta} \cdot \partial_{tt}\, \hat{\psi} + \left( \mathrm{D}\, \delta\hat{\theta} \cdot S(\hat{\theta}) + \mathrm{D}\, \hat{\theta} \cdot S'(\hat{\theta}) \cdot \delta\hat{\theta} \right) \cdot\cdot \mathrm{D}\, \hat{\psi} \right]$$

$$- \alpha \int_0^T \int_{\Omega_0^s} (\hat{\theta} - \hat{\theta}_d) \cdot \delta\hat{\theta} + \int_{\Omega_0^s} \left[ (\partial_t\, \delta\hat{\theta})(T) \cdot \hat{\psi}(T) - \delta\hat{\theta}(T) \cdot \partial_t\, \hat{\psi}(T) \right]$$

The Green formula leads to the following identity,

$$\int_{\Omega_0^s} [\mathrm{D}\, \delta\hat{\theta} \cdot S(\hat{\theta})] \cdot\cdot \mathrm{D}\, \hat{\psi} = - \int_{\Omega_0^s} \left[ \mathrm{D}\, \hat{\psi} \cdot \mathrm{div}\, S(\hat{\theta}) + \mathrm{D}^2\, \hat{\psi} \cdot\cdot S(\hat{\theta}) \right] \cdot \delta\hat{\theta}$$

$$+ \int_{\Gamma_0^s} \left[ \mathrm{D}\, \hat{\psi} \cdot {}^*S(\hat{\theta}) \cdot n \right] \cdot \delta\hat{\theta}$$

We use the adjoint identity (8.4) with $B = {}^*\mathrm{D}\,\hat{\theta} \cdot \mathrm{D}\,\hat{\psi}$ and we get

$$\int_{\Omega_0^s} T(\delta\theta) \cdot\cdot ({}^*\mathrm{D}\,\hat{\theta} \cdot \mathrm{D}\,\hat{\psi}) = - \int_{\Omega_0^s} \delta\theta \cdot T_A({}^*\mathrm{D}\,\hat{\theta} \cdot \mathrm{D}\,\hat{\psi}) + \int_{\Gamma_0^s} T_\Gamma({}^*\mathrm{D}\,\hat{\theta} \cdot \mathrm{D}\,\hat{\psi}) \cdot \delta\hat{\theta}$$

$$(8.25)$$

This leads to the following expression,

$$D_{\Omega_0^s} =$$

$$\int_0^T \int_{\Omega_0^s} \left[ \partial_{tt}\hat{\psi} - D\hat{\psi} \cdot \operatorname{div} S(\hat{\theta}) - D^2\hat{\psi} \cdot \cdot S(\hat{\theta}) - T_A(^*D\hat{\theta} \cdot D\hat{\psi}) - \alpha\,(\hat{\theta} - \hat{\theta}_d) \right] \cdot \delta\hat{\theta}$$

$$+ \int_0^T \int_{\Gamma_0^s} \left[ D\hat{\psi} \cdot {}^*S(\hat{\theta}) \cdot n + T_\Gamma(^*D\hat{\theta} \cdot D\hat{\psi}) \right] \cdot \delta\hat{\theta} + \int_{\Omega_0^s} \left[ (\partial_t\,\delta\hat{\theta})(T) \cdot \hat{\psi}(T) \right.$$

$$\left. - \delta\hat{\theta}(T) \cdot \partial_t\,\hat{\psi}(T) \right]$$

Using the optimality condition (8.23), we shall deduce that the adjoint solid state is the solution of the following backward second order system,

$$\begin{cases} \partial_{tt}\hat{\psi} - D\hat{\psi} \cdot \operatorname{div} S(\hat{\theta}) - D^2\hat{\psi} \cdot \cdot S(\hat{\theta}) - T_A(^*D\hat{\theta} \cdot D\hat{\psi}) = \\[2mm] \alpha\,(\hat{\theta} - \hat{\theta}_d), & (0,T) \times \Omega_0^s \\[2mm] (\hat{\psi}, \partial_t\,\hat{\psi})|_{t=T} = (0,0), & \Omega_0^s \end{cases}$$

$$(8.26)$$

Furthermore, the following identity holds on the moving boundary $\Gamma_{\hat{\theta}}$,

$$0 = \left[ -V_{\hat{\theta}} \cdot \partial_t\varphi - p\,\operatorname{div}\varphi - \pi\,\operatorname{div} V_{\hat{\theta}} + \nu\,D\varphi \cdot \cdot D V_{\hat{\theta}} - (\operatorname{div} V_{\hat{\theta}})\,u \cdot \varphi \right.$$

$$- (D\varphi \cdot V_{\hat{\theta}}) \cdot \varphi] \cdot n - {}^*D\varphi \cdot \sigma(u,p) \cdot n + {}^*D V_{\hat{\theta}} \cdot [\sigma(\varphi,\pi) - u \cdot \varphi] \cdot n$$

$$+ \left[ -(\partial_t\mathcal{E}) \cdot n + \mathcal{E} \cdot \nabla_\Gamma(V_{\hat{\theta}} \cdot n) - (\operatorname{div}_\Gamma V_{\hat{\theta}})\,\mathcal{E} \cdot n - D\mathcal{E} \cdot V_{\hat{\theta}} \cdot n - \mathcal{E} \cdot D_\Gamma n \cdot V_{\hat{\theta}} \right]$$

$$+ \left[ D\hat{\psi} \cdot {}^*S(\hat{\theta}) \cdot \hat{n} + T_\Gamma(^*D\hat{\theta} \cdot D\hat{\psi}) \right] \circ \hat{\theta}^{-1}\,[J(\hat{\theta})|^*(D\hat{\theta})^{-1} \cdot \hat{n}|]^{-1} \circ \hat{\theta}^{-1}$$

with $\mathcal{E} = \sigma(\varphi,\pi) - (V_{\hat{\theta}} \cdot \varphi)\,I$.
This last identity ends the proof of the main result of this chapter.

# Appendix A

## Functional spaces and regularity of domains

### A.1 Classical functions

Let $\Omega$ be an open subset of $\mathbb{R}^d$, and we denote by $\mathcal{C}^0(\Omega)$ the space of continuous functions from $\Omega$ to $\mathbb{R}$. For $k \in \mathbb{N}^*$,

$$\mathcal{C}^k(\Omega) \stackrel{\text{def}}{=} \left\{ f \in \mathcal{C}^{k-1}(\Omega) : \mathrm{D}^\alpha f \in \mathcal{C}^0(\Omega), \quad \forall \alpha, |\alpha| = k \right\} \tag{A.1}$$

where $\alpha = (\alpha_1, \ldots, \alpha_d) \in \mathbb{N}^d$ is a multi-index. $|\alpha| = \displaystyle\sum_{1 \leq i \leq d} \alpha_i$ is the order of derivative and

$$\mathrm{D}^\alpha f \stackrel{\text{def}}{=} \prod_{1 \leq i \leq d} \mathrm{D}_i^{\alpha_i} f = \frac{\partial^{|\alpha|} f}{\partial x_1^{\alpha_1} \ldots x_d^{\alpha_d}} \tag{A.2}$$

**REMARK A.1** The spaces $\mathcal{C}^0(\Omega)$ and $\mathcal{C}^k(\Omega)$ for $k \in \mathbb{N}^*$ are not Banach[1] spaces. However if $\{K_n\}_{n \in \mathbb{N}}$ is a sequence of compact sets of $\Omega$ such that

$$\Omega = \bigcup_{n \in \mathbb{N}} K_n,$$

we can define a sequence of semi-norms

$$d_n(x) = \sup_{x \in K_n} |u(x)|$$

---

[1] A complete normed space.
A norm on a vector space $E$ is a real-valued functional $m$ on $E$ such that

- $m(x) \geq 0, \quad \forall\, x \in E$ with equality iff $x = 0$,
- $m(c\,x) = |c|\, m(x), \quad \forall\, x \in E, \quad c \in \mathbb{C}$,
- $m(x + y) \leq m(x) + m(y), \quad \forall\, x, y \in E$

A normed space is a vector space which is provided with a norm.
A Cauchy sequence in a metric space is a sequence $\{x_n\}_{n \geq 0}$ which satisfies $\lim_{n, p \to \infty} d(x_n, x_p) = 0$.
A metric space is said to be complete in a metric space iff every Cauchy squence is convergent.

and a distance

$$d(u, v) = \sum_{n \in \mathbb{N}} a_n \frac{d_n(u - v)}{1 + d_n(u - v)}$$

with $a_n \geq 0$ such that $\sum_{n \in \mathbb{N}} a_n = 1$. Then $\left(\mathcal{C}^0(\Omega), d(.,.)\right)$ is a Fréchet space.[2]

In a similar way, we can define on $\mathcal{C}^k(\Omega)$ a Fréchet space structure with the following sequence of semi-norms

$$d_n^k(u) = \sum_{|\alpha| \leq k} d_n(\mathrm{D}^\alpha u)$$

⬚

Since $\Omega$ is open, functions in $\mathcal{C}^k(\Omega)$ can be unbounded. Thus, let us denote by $\mathcal{B}^0(\Omega)$ the space of bounded continuous functions. For $k \in \mathbb{N}^*$, we define

$$\mathcal{B}^k(\Omega) = \left\{ f \in \mathcal{B}^{k-1}(\Omega) : \mathrm{D}^\alpha f \in \mathcal{B}^0(\Omega), \quad \forall \alpha, \, |\alpha| = k \right\}$$

Endowed with the norm

$$\|f\|_{\mathcal{C}^k(\Omega)} \overset{\text{def}}{=} \max_{0 \leq |\alpha| \leq k} \sup_{x \in \Omega} |\mathrm{D}^\alpha f(x)| \tag{A.3}$$

$\mathcal{B}^k(\Omega)$ is a Banach space.

If a function $f$ is bounded and uniformly continuous[3] on $\Omega$, it has a unique, continuous extension to the closure $\overline{\Omega}$. Hence we introduce the space

$$\mathcal{C}^k(\overline{\Omega}) = \left\{ f \in \mathcal{B}^k(\Omega) : \mathrm{D}^\alpha f \text{ uniformly continuous on } \Omega, \quad \forall \, 0 \leq |\alpha| \leq k \right\}$$

The space $\left(\mathcal{C}^k(\overline{\Omega}), \|.\|_{\mathcal{C}^k(\Omega)}\right)$ is a Banach space. Given $0 < \ell \leq 1$, a function $f$ is $(0, \ell)$-Hölder continuous in $\Omega$ if

$$\exists\, c > 0, \quad |f(y) - f(x)| \leq c|y - x|^\ell, \quad \forall x, y \in \Omega$$

When $\ell = 1$, we say that $f$ is Lipschitz. Similarly for $k \geq 1$, $f$ is $(k, \ell)$-Hölder continuous in $\Omega$ if

$$\forall \alpha, 0 \leq |\alpha| \leq k, \exists\, c > 0, |\mathrm{D}^\alpha f(y) - \mathrm{D}^\alpha f(x)| \leq c|y - x|^\ell, \quad \forall x, y \in \Omega$$

---

[2]A Fréchet space is complete metrizable locally convex topological vector space. $E$ is a topological vector space if $E$ is furnished with a topology compatible with the structure of the vector space, i.e., such that the vector space operations of addition and scalar multiplication are continuous. It is locally convex if each neighborhood of the origin in $E$ contains a convex neighborhood.

A topological space is said to be metrizable if its topology can be defined by a metric.

[3]A function $f : \Omega \to \mathbb{R}$ is uniformly continuous if for each $\varepsilon > 0$ there exists $\delta > 0$ such that

$$|f(x) - f(y)| < \varepsilon, \quad \forall x, y \in \Omega \text{ with } |x - y| < \delta$$

We denote by $\mathcal{C}^{k,\ell}(\Omega)$ the space of all $(k,\ell)$-Hölder continuous functions in $\Omega$. We also define $\mathcal{C}^{k,\ell}(\overline{\Omega}) \overset{\text{def}}{=} \mathcal{C}^{k,\ell}(\Omega) \cup \mathcal{C}^k(\overline{\Omega})$. Endowed with the norm

$$\|f\|_{\mathcal{C}^{k,\ell}(\Omega)} \overset{\text{def}}{=} \|f\|_{\mathcal{C}^k(\Omega)} + \max_{0 \leq |\alpha| \leq k} \sup_{\substack{x,y \in \Omega \\ x \neq y}} \frac{|\mathrm{D}^\alpha f(y) - \mathrm{D}^\alpha f(x)|}{|y-x|^\ell} \tag{A.4}$$

$\mathcal{C}^{k,\ell}(\overline{\Omega})$ is a Banach space.
The spaces $\mathcal{D}^k(\Omega)$ for $k \in \mathbb{N}^*$ and $\mathcal{D}(\Omega)$ contain respectively all the functions of class $\mathcal{C}^k$ and infinitely continuously differentiable with compact support.

---

## A.2   Lebesgue spaces

Let $\Omega$ be any open set in $\mathbb{R}^d$. The space $L^p(\Omega)$, $1 \leq p < +\infty$ is the set of real functions defined on $\Omega$ that have finite norm

$$\|u\|_{L^p(\Omega)} \overset{\text{def}}{=} \left( \int_\Omega |u(x)|^p \, \mathrm{d}x \right)^{1/p} \tag{A.5}$$

We note that we identify functions that are different only on a set of Lebesgue measure zero.
The space $L^\infty(\Omega)$ is the set of functions on $\Omega$ that are measurable with respect to the Lebesgue measure $\mathrm{d}x$ and have a finite norm

$$\|u\|_{L^\infty(\Omega)} = \inf_{\mathrm{d}x-\mathrm{meas}\{x \in \Omega: |u(x)| > C\} = 0} C \tag{A.6}$$

### A.2.1   Various inequalities

#### LEMMA A.1 Hölder inequality

$$\left| \int_\Omega u(x)\, v(x)\, \mathrm{d}x \right| \leq \|u\|_{L^p(\Omega)} \|v\|_{L^q(\Omega)}, \quad 1/p + 1/q = 1, \tag{A.7}$$

*for $u \in L^p(\Omega)$ , $v \in L^q(\Omega)$.*

**REMARK A.2**   This identity is a consequence of the Young inequality,

$$a\,b \leq 1/p\, a^p + 1/q\, b^q, \quad 1/p + 1/q = 1, \tag{A.8}$$

for $a \geq 0, b \geq 0$.  ◻

**REMARK A.3**    For $p = q = 2$, this inequality is referred to as the Cauchy inequality.                                                                          ⬚

**LEMMA A.2 Interpolation inequality**

$$\|u\|_{L^r(\Omega)} \leq \|u\|_{L^p(\Omega)}^{\alpha} \|u\|_{L^q(\Omega)}^{1-\alpha}, \quad 1/r = \alpha/p + (1-\alpha)/q, \qquad (A.9)$$

*for $u \in L^p(\Omega) \cap L^q(\Omega)$.*

## A.2.2    Completeness, separability and reflexivity

**LEMMA A.3**
*For $1 \leq p \leq +\infty$, the space $L^p(\Omega)$ is a Banach space[4].*
*For $1 < p < +\infty$ the space $L^p(\Omega)$ is a separable reflexive space and its dual space can be identified with $L^q(\Omega)$ with $1/p + 1/q = 1$.*
*The space $L^1(\Omega)$ is separable and its dual can be identified with $L^\infty(\Omega)$.*

## A.2.3    Convolution

**THEOREM A.1**
*Let $u \in L^1(\mathbb{R}^d)$ and $v \in L^p(\mathbb{R}^d)$ with $1 \leq p \leq \infty$. Then $y \mapsto u(x-y)\,v(y)$ is summable over $\mathbb{R}^d$ for a.e. $x \in \mathbb{R}^d$. We define their convolution product by*

$$u \star v(x) = \int_{\mathbb{R}^d} u(x-y)\,v(y)\,\mathrm{d}y \qquad (A.10)$$

*which satisfies $u \star v \in L^p(\mathbb{R}^d)$ and*

$$\|u \star v\|_{L^p} \leq \|u\|_{L^1}\|v\|_{L^p} \qquad (A.11)$$

The convolution can be extended to a function $u \in L^p(\mathbb{R}^d)$ thanks to the following theorem,

**THEOREM A.2 Young inequality**
*Let $u \in L^p(\mathbb{R}^d)$ and $v \in L^q(\mathbb{R}^d)$ with*

$$1 \leq p \leq \infty,\ 1 \leq q \leq \infty,\ 1/r = 1/p + 1/q - 1 \geq 0 \qquad (A.12)$$

*then*

$$u \star v \in L^r(\mathbb{R}^d) \quad and \quad \|u \star v\|_{L^r} \leq \|u\|_{L^p}\|v\|_{L^q} \qquad (A.13)$$

---

[4]Fischer-Riez theorem [17] Theorem IV. 8, p. 57

The following regularizing property will be useful to state a density property of regular functions space in $L^p$,

**THEOREM A.3**
*Let $g \in \mathcal{D}^k(\mathbb{R}^d)$ for $k \in \mathbb{N}$ and $u \in L^1_{loc}(\mathbb{R}^d)$, then*

$$g \star u \in \mathcal{C}^k(\mathbb{R}^d) \quad and \quad \mathrm{D}^\alpha(g \star u) = (\mathrm{D}^\alpha g) \star u, \quad |\alpha| \leq k \qquad (A.14)$$

**REMARK A.4**  As an application of the previous theorem, we deduce that if $g \in \mathcal{D}(\mathbb{R}^d)$ and $u \in L^1_{loc}(\mathbb{R}^d)$ then $g \star u \in \mathcal{C}^\infty(\mathbb{R}^d)$. ⬜

## A.2.4  Density property

**THEOREM A.4 [44]**
*The space $\mathcal{D}(\Omega)$ is dense in $L^p(\Omega)$ for $1 \leq p < +\infty$ and every convergent sequence in $\mathcal{D}(\Omega)$ converges in $L^p(\Omega)$.*

**REMARK A.5**  This theorem can be established using the *truncation + regularization* procedure introduced by Leray and Friedrichs that will be introduced in the next section. ⬜

---

## A.3  Smooth domains and boundary measure

Let $\Omega$ be an open set of $\mathbb{R}^d$ and let its associated boundary $\Gamma \stackrel{\text{def}}{=} \partial\Omega$ be a manifold in $\mathbb{R}^d$ of codimension 1. The smoothness of $\Gamma$ is classically characterized by introducing at each point of $\Gamma$ a local diffeomorphism that locally flattens the boundary. In order to make the purpose more precise, let us introduce some notations and definitions.
Let $\{e_i\}_{1 \leq i \leq d}$ be the canonical unit orthogonal basis in $\mathbb{R}^d$. We use the notation $\zeta = (\zeta', \zeta_d)$ for a point $\zeta \in \mathbb{R}^d$ with $\zeta' = (\zeta_1, \ldots, \zeta_{d-1})$. Let $B$ be the unit open ball in $\mathbb{R}^d$ and let us define the following sets,

$$B_0 \stackrel{\text{def}}{=} \{\zeta \in B : \zeta_d = 0\} \qquad (A.15)$$

$$B_+ \stackrel{\text{def}}{=} \{\zeta \in B : \zeta_d > 0\}, \, B_- \stackrel{\text{def}}{=} \{\zeta \in B : \zeta_d < 0\} \qquad (A.16)$$

**DEFINITION A.1**  *Let $\Omega$ be a subset of $\mathbb{R}^d$ such that $\Gamma \stackrel{\text{def}}{=} \partial\Omega \neq \emptyset$.*

a) $\Omega$ *is said to be of class* $\mathcal{C}^k, 0 \le k \le +\infty$, *if for each point* $x \in \Gamma$ *there exist*

- *a neighbourhood* $U(x) \subset \mathbb{R}^d$ *of* $x$,
- *a bijective map* $g_x : U(x) \to B$ *with the following properties :*

$$g_x \in \mathcal{C}^k(U(x), B),\ h_x \stackrel{\text{def}}{=} g_x^{-1} \in \mathcal{C}^k(B, U(x)) \qquad (A.17)$$

$$\text{int}[\Omega \cap U(x)] = h_x(B^+) \qquad (A.18)$$

$$\Gamma_x \stackrel{\text{def}}{=} \Gamma \cap U(x) = h_x(B_0),\ B_0 = g_x(\Gamma_x) \qquad (A.19)$$

b) $\Omega$ *is said to be of class* $\mathcal{C}^{k,\ell}$, $k \ge 0$, $0 < \ell \le 1$, *if the above conditions are satisfied with a map* $g_x \in C^{k,\ell}(U(x), B)$ *together with its inverse* $h_x \in C^{k,\ell}(B, U(x))$.

For sets of class $\mathcal{C}^1$, the unit exterior normal $n$ to the boundary $\Gamma$ can be characterized through the Jacobian matrices of $g_x$ and $h_x$.

### LEMMA A.4

*The outward unit normal field* $n(y)$ *at point* $y \in \Gamma_x$ *is given by the following expressions,*

$$n(y) = -\frac{{}^*(\mathrm{D}\, h_x)^{-1}(\zeta', 0) \cdot e_d}{|{}^*(\mathrm{D}\, h_x)^{-1}(\zeta', 0)e_d|}, \forall\, h_x(\zeta', 0) = y \in \Gamma_x \qquad (A.20)$$

$$= -\frac{{}^*\mathrm{D}\, g_x(y)e_d}{|{}^*\mathrm{D}\, g_x(y) \cdot e_d|}, \qquad \forall\, y \in \Gamma_x \qquad (A.21)$$

*with* $(\mathrm{D}\, g)_{i,j} = \partial_j g_i$.

The family of neighbourhoods $U(x)$ associated with all the points $x$ of $\Gamma$ is an open cover of $\Gamma$.

### LEMMA A.5

*Let* $\Omega \subset \mathbb{R}^d$ *be a bounded domain with compact boundary* $\Gamma$. *There exists a finite subcover, i.e., there exists a finite sequence of points* $\{x_i\}_{1 \le i \le m}$ *of* $\Gamma$ *such that* $\Gamma \subset U_1 \cup \ldots \cup U_m$, *where* $U_i = U(x_i)$.
*Furthermore there exists a partition of unity* $\{r_i\}_{1 \le i \le m}$ *associated to the family of open neighbourhoods* $\{U_i\}_{1 \le i \le m}$ *of* $\Gamma$ *such that :*

$$\begin{cases} r_i \in \mathcal{D}(U_i), & 0 \le r_i(x) \le 1, \\ \sum_{1 \le i \le m} r_i(x) = 1, & \forall\, x \in U \end{cases} \qquad (A.22)$$

with $\bar{U} \subset \bigcup_{1 \leq i \leq m} U_i$.

Let us define the boundary integral of a continuous function $f$ on $\Gamma$ :

### LEMMA A.6
*For $f \in \mathcal{C}^0(\Gamma)$, its integral over $\Gamma$ is defined as follows*

$$\int_\Gamma f d\Gamma \overset{\text{def}}{=} \sum_{1 \leq i \leq m} \int_{\Gamma_i} f r_i d\Gamma_i \qquad \text{(A.23)}$$

*where*

$$\int_{\Gamma_i} f r_i d\Gamma_i \overset{\text{def}}{=} \int_{B_0} (f r_i) \circ h_i(\zeta', 0) \omega_i(\zeta') d\zeta'$$

*with $\Gamma_i = U(x_i) \cap \Gamma$ and $\omega_i = \Omega_{x_i}$ and*

$$\omega_x(\zeta') = |m_x(h_x(\zeta', 0))| |\det \mathrm{D}\, h_x(\zeta', 0)|$$

*with*

$$m_x(y) = -^* \mathrm{D}\, g_x(y) e_d$$

# Appendix B

## Distribution spaces

### B.1 The space $\mathcal{D}(\Omega)$

Let $\Omega$ be an open set in $\mathbb{R}^d$. The spaces $\mathcal{D}^k(\Omega)$ for $k \in \mathbb{N}^*$ and $\mathcal{D}(\Omega)$ contain respectively all the functions of class $\mathcal{C}^k$ and infinitely continuously differentiable with compact support[1] in $\Omega$. For any open set $\mathcal{O} \subset \mathbb{R}^d$ such that $\overline{\Omega} \subset \mathcal{O}$, we define

$$\mathcal{D}(\overline{\Omega}) = \{\phi|_\Omega : \phi \in \mathcal{D}(\mathcal{O})\} \tag{B.1}$$

The space $\mathcal{D}(\Omega)$ does not reduce to the null function. Indeed let $\omega : \mathbb{R}^d \to \mathbb{R}$ defined as follows :

$$\omega(x) = \begin{cases} e^{1/(|x|^2 - 1)}, & |x| < 1 \\ 0, & |x| \geq 1 \end{cases}$$

Let $B_r(a)$ be an open ball contained in $\Omega$, then the function

$$\varphi_{a,r}(x) = \omega(\frac{x - a}{r})$$

belongs to $\mathcal{D}(\Omega)$. There exists a general procedure for the construction of functions in $\mathcal{D}(\Omega)$ based on the previous particular function and the notion of convolution product.

**DEFINITION B.1 mollifier**   *We call a family of function $\{\rho_\varepsilon\}_{\varepsilon \geq 0}$ a mollifier if*

- *$\rho_\varepsilon \in \mathcal{D}(\mathbb{R}^d)$,*

- *Supp $\rho_\varepsilon \subset B(0, \varepsilon)$,*

- *$\rho_\varepsilon \geq 0$ and $\int_{\mathbb{R}^d} \rho_\varepsilon = 1$.*

---

[1] For $f : \Omega \to \mathbb{R}$, the support of $f$ is defined as follows

$$\text{Supp } f = \overline{\{x \in \Omega : f(x) \neq 0\}}$$

An example of a mollifier is furnished by

$$\rho_\varepsilon(x) = \varepsilon^{-d}\,\varphi_{0,1}(x/\varepsilon) \tag{B.2}$$

### LEMMA B.1
*Let $u$ be a function locally integrable in $\Omega$ with compact support subset of $\Omega$. Then there exists $\varepsilon_0 > 0$ such that*

$$u_\varepsilon = \rho_\varepsilon \star u \in \mathcal{D}(\Omega)$$

*for $\varepsilon < \varepsilon_0$.*

This lemma shows that the space $\mathcal{D}(\Omega)$ is sufficiently rich and we shall see that it has density properties in numerous common functional spaces.
The space $\mathcal{D}(\Omega)$ is neither a Banach space nor a Fréchet space, but the inductive limit of Fréchet spaces which makes it a locally convex topological vector space. But the corresponding inductive limit topology is not easy to manipulate. Fortunately, the notion of convergence of sequences will be sufficient for our purpose.

### DEFINITION B.2 
*Let $\{\varphi_k\}_{k\geq 0}$ be sequence of elements of $\mathcal{D}(\Omega)$. It converges to $\varphi \in \mathcal{D}(\Omega)$ if there exists a compact subset $K \subset \Omega$ such that*

$$\begin{cases} \text{Supp } \varphi_k \subset K, & \forall\, k \geq 0 \\ D^\alpha\,\varphi_k \text{ converges uniformly on } K \text{ to } D^\alpha\varphi \text{ as } k \to \infty \end{cases}$$

### LEMMA B.2 Generalized Urysohn
*Let $K$ be a compact set in $\mathbb{R}^d$ and $\varepsilon > 0$ arbitrary. Then there exists a function $\psi_\varepsilon \in \mathcal{D}(\mathbb{R}^d)$ such that*

- $\psi_\varepsilon(x) = 1, \quad \forall\, x \in K$ *and* $\psi_\varepsilon \geq 0$,

- *Supp* $\psi_\varepsilon \subset K_\varepsilon = \cup_{x \in K} B(x, \varepsilon)$.

With the above definition, we can state a useful density property for $\mathcal{D}(\mathbb{R}^d)$,

### THEOREM B.1
*The space $\mathcal{D}(\mathbb{R}^d)$ is contained and dense in $\mathcal{C}^k(\mathbb{R}^d)$ for $0 \leq k \leq \infty$ and every convergent sequence in $\mathcal{D}(\mathbb{R}^d)$ converges in $\mathcal{C}^k(\mathbb{R}^d)$.*

### REMARK B.1 
The proof consists of two steps :

- Truncation :
  We prove that $\mathcal{D}^k$ is dense in $\mathcal{C}^k$ using the family $u_p(x) = \psi(x/p)\,u(x) \in \mathcal{D}^k$ with $\psi$ given by lemma B.2 with $K = \overline{B(0,1)}$.

- Regularization :
  We show that $\mathcal{D}$ is dense in $\mathcal{D}^k$ using the sequence $u_p = u \star \rho_p \in \mathcal{D}$ with $\rho_p = \rho_{\varepsilon=1/p}$ and $\rho_\varepsilon$ a mollifying family.

  □

## B.2 The space of distributions $\mathcal{D}'(\Omega)$

**DEFINITION B.3** *A continuous linear form on $\mathcal{D}(\Omega)$ is called a distribution on $\Omega$. We denote by $\mathcal{D}'(\Omega)$ the space of distributions on $\Omega$. If $T \in \mathcal{D}'(\Omega)$, we note*

$$\langle T, \varphi \rangle = T(\varphi), \quad \forall \varphi \in \mathcal{D}(\Omega)$$

*Here $T$ continuous[2] on $\mathcal{D}(\Omega)$ means*

$$\lim_{k \to \infty} \langle T, \varphi_k \rangle = \langle T, \varphi \rangle, \quad \forall \{\varphi_k\}_{k \in \mathbb{N}} \in \mathcal{D}(\Omega) \text{ with } \lim_{k \to \infty} \varphi_k = \varphi \text{ in } \mathcal{D}(\Omega)$$

The space $\mathcal{D}'(\Omega)$ is a vector space and is in fact the topological dual space of $\mathcal{D}(\Omega)$. As a dual space, $\mathcal{D}'(\Omega)$ is equipped with the weak-star topology $\sigma(\mathcal{D}'(\Omega), \mathcal{D}(\Omega))$, defined by the family of semi-norms

$$\theta_\varphi : \mathcal{D}'(\Omega) \to \mathbb{R}$$
$$T \mapsto \theta_\varphi(T) = T(\varphi) = \langle T, \varphi \rangle$$

for $\varphi \in \mathcal{D}(\Omega)$ and where $\langle .,. \rangle$ stands for the duality pairing between $\mathcal{D}'(\Omega)$ and $\mathcal{D}(\Omega)$.
In this topology, which is again not metrizable, a sequence of distributions $T_k$ converges to $T$ if and only if

$$\langle T_k, \varphi \rangle \to_{k \to \infty} \langle T, \varphi \rangle, \quad \forall \varphi \in \mathcal{D}(\Omega)$$

**THEOREM B.2**
*If $\{T_k\}_{k \in \mathbb{N}}$ is a sequence of elements in $\mathcal{D}'(\Omega)$ such that $\langle T_k, \varphi \rangle$ is convergent for all $\varphi \in \mathcal{D}(\Omega)$ as $k \to \infty$, then the linear form*

$$\varphi \mapsto \lim_{k \to \infty} \langle T_k, \varphi \rangle$$

---

[2]Equivalently, given any compact subset $K \subset \Omega$, there exist a constant $C(K)$ and an integer $m(K) \geq 0$ such that

$$|T(\varphi)| \leq C(K) \max_{|\alpha| \leq m(K)} \sup_{x \in \Omega} |D^\alpha \varphi(x)|, \quad \forall \varphi \in \mathcal{D}(\Omega) \text{ with Supp } \varphi \subset K$$

When the integer $m$ can be chosen independently on $K$, the distribution $T$ is said to be of finite order and its order is defined as the smallest suitable integer $m$.

belongs to $\mathcal{D}'(\Omega)$.

When $f$ is locally integrable on $\Omega$, then $f$ can be identified with a distribution by

$$\langle f, \varphi \rangle = \int_\Omega f(x)\,\varphi(x)\,\mathrm{d}x, \quad \forall\, \varphi \in \mathcal{D}(\Omega)$$

Hence the duality pairing $\langle .,. \rangle$ is an extension of the scalar product of the Hilbert space $(L^2(\Omega), \|.\|_{L^2(\Omega)})$ .
The differentiation of distributions (which coincides with the usual differentiation when applied to continuously differentiable functions) is the essential property which justifies the introduction of distributions.

**DEFINITION B.4**    *Letting $T \in \mathcal{D}'(\Omega)$ , the derivative $\mathrm{D}^\alpha T$, $\alpha \in \mathbb{N}^d$ is the distribution on $\Omega$ defined by*

$$\langle \mathrm{D}^\alpha T, \varphi \rangle = (-1)^{|\alpha|} \langle T, \mathrm{D}^\alpha \varphi \rangle, \quad \forall\, \varphi \in \mathcal{D}(\Omega)$$

**THEOREM B.3**
*Let $\{T_k\}_{k\in\mathbb{N}} \in \mathcal{D}'(\Omega)$ be a convergent sequence of distributions towards $T \in \mathcal{D}'(\Omega)$. Then the derivative of $T$ has the following property*

$$\lim_{k\to\infty} \mathrm{D}^\alpha T_k = \mathrm{D}^\alpha T \ \text{in } \mathcal{D}'(\Omega), \quad \forall\, \alpha \in \mathbb{N}^d$$

**THEOREM B.4**
*Let $\Omega$ be an open set of $\mathbb{R}^d$ and $E$ a locally convex topological vector space of functions on $\Omega$ satisfying :*

- *$\mathcal{D}(\Omega)$ is contained and dense in $E$,*

- *every convergent sequence in $\mathcal{D}(\Omega)$ converges in $E$.*

*Then the continuous linear forms on $E$ are identified with distributions ( i.e., elements of $\mathcal{D}'(\Omega)$).*

**REMARK B.2**    The identification is made in the following manner : let $T$ be a continuous linear form on $E$, then the restriction of $T$ to $\mathcal{D}(\Omega)$ is the distribution we are looking for.                                                                    □

## B.3   Examples of distributions

### B.3.1   Measures on $\Omega$

**DEFINITION B.5**   *We call every continuous linear form on the space* $\mathcal{D}^0(\Omega)$ *of continuous functions with compact support a measure on* $\Omega$. *The space of measures is denoted* $\mathcal{M}(\Omega)$.

Using that $\mathcal{D}(\Omega) \hookrightarrow \mathcal{D}^0(\Omega)$, we can state that $\mu \in \mathcal{M}(\Omega)$ defines an element of $\mathcal{D}'(\Omega)$.

We usually note $\mu(\varphi) = \langle \mu, \varphi \rangle$ the value of the linear form $\mu$ on $\varphi \in \mathcal{D}^0(\Omega)$. We also say that $\mu(\varphi)$ is the integral of the function $\varphi$ relative to the measure $\mu$ and we write

$$\mu(\varphi) = \int_\Omega \varphi(x) d\mu$$

We can extend the definition of $\mu(\varphi)$ for measurable functions that do not belong to $\mathcal{D}^0(\Omega)$. These are called summable functions.

### B.3.2   Regular distributions

We recall that $L^1_{loc}(\Omega)$ is the set of functions integrable on every compact set in $\Omega$.

**DEFINITION B.6**   *Let* $f \in L^1_{loc}(\Omega)$; *we define a corresponding measure* $\mu_f$ *by*

$$\mu_f(\varphi) = \int_\Omega f(x)\varphi(x)\, dx$$

The application

$$[.] : L^1_{loc}(\Omega) \to \mathcal{M}(\Omega)$$
$$f \mapsto [f] = \mu_f$$

is an injection and its image is called the set of absolutely continuous measures in $\Omega$.

$f$ is called the density of $\mu_f$ with respect to the Lebesgue measure $dx$ ( $d\mu_f = f(x)\,dx$). It is usual to identify $\mu_f$ with its density $f$. Furthermore $\mu_f$ defines a distribution and we say that this distribution is a function $f$.

**REMARK B.3**   As $L^p(\Omega) \hookrightarrow L^1_{loc}(\Omega)$ with $1 \le p \le +\infty$, every element of $L^p(\Omega)$ defines a distribution on $\Omega$ ( which is a measure on $\Omega$). ⬚

### B.3.3   The Dirac measure

**DEFINITION B.7 Dirac measure**     *Let $x_0 \in \Omega$. The mapping*

$$\delta_{x_0} : \mathcal{D}^0(\Omega) \to \mathbb{R}$$
$$\varphi \mapsto \varphi(x_0)$$

*defines a measure in $\mathcal{M}(\Omega)$ called the Dirac measure at point $x_0$.*

**REMARK B.4**     The Dirac measure is the simplest example of measure which is not a function. However we shall abuse the notation and write sometimes $\delta(x - x_0)$ for $\delta_{x_0}$. Using integral notation, we have the following properties,

$$\delta_{x_0}(\varphi) = \varphi(x_0) = \int_{\mathbb{R}^d} \varphi(x)\delta(x - x_0)\,\mathrm{d}x$$

$$\int_{\mathbb{R}^d} \delta(x - x_0)\,\mathrm{d}x = 1$$

▯

The Dirac function can also be characterized as the derivative in the sense of distributions of a particular function,

**DEFINITION B.8 Heaviside function**     *Let*

$$Y : \mathbb{R} \to \mathbb{R}$$
$$x \mapsto Y(x) : \begin{cases} +1, & x > 0 \\ 0, & x < 0 \end{cases}$$

Obviously $Y \in L^1_{loc}(\mathbb{R})$. Then it defines a distribution. We can then compute its derivation in $\mathcal{D}'(\mathbb{R})$,

$$\langle \mathrm{D}\,Y, \varphi \rangle = - \int_0^{+\infty} \varphi'(x)\,\mathrm{d}x = \varphi(0)$$

then

$$\mathrm{D}\,Y = \delta, \quad \text{in } \mathcal{D}'(\mathbb{R})$$

### B.3.4   Derivative of the Dirac measure

Let us compute the derivative of the Dirac measure in the sense of distributions,

$$\langle \mathrm{D}^m\,\delta, \varphi \rangle = (-1)^m \langle \delta, \mathrm{D}^m\,\varphi \rangle = (-1)^m \mathrm{D}^m\,\varphi(0)$$

Hence to define the distributions $D^m \delta$, we need to consider test functions which belong at least to the class $C^m$. We thus conclude that $D^m \delta$ is not a measure for $m \geq 1$.

**REMARK B.5**   In general, a derivative in the sense of distributions of a measure on $\Omega$ defines a distribution which is not a measure except in the case where the measure is a sufficiently regular distribution.   ⬚

## B.3.5   The Dirac comb

*DEFINITION B.9*   *The application*

$$S : \mathcal{D}(\mathbb{R}) \to \mathbb{R}$$
$$\varphi \mapsto \sum_{k \in \mathbb{Z}} \varphi(k)$$

*defines a distribution called Dirac comb.*

## B.3.6   Principal value of $1/x$

Let us consider the function,

$$\mathbb{R} \to \mathbb{R}$$
$$x \mapsto \log(|x|)$$

By a direction computation, it can be stated that its derivative in the sense of distributions is given by the following identity,

$$\langle D \log(|x|), \varphi \rangle = \lim_{\varepsilon \to 0} \int_\varepsilon^{+\infty} \frac{\varphi(x) - \varphi(-x)}{x} \, dx$$

The right hand side is referred to as the Cauchy principal value of $\int_{-\infty}^{+\infty} \frac{\varphi(x)}{x} \, dx$ and is denoted $\mathrm{v.\,p} \int_{-\infty}^{+\infty} \frac{\varphi(x)}{x} \, dx$.
It can be checked that the mapping

$$\mathcal{D}(\mathbb{R}) \to \mathbb{R}$$
$$\varphi \mapsto \mathrm{v.\,p} \int_{-\infty}^{+\infty} \frac{\varphi(x)}{x} \, dx$$

is linear and continuous on $\mathcal{D}(\mathbb{R})$ (but not in $\mathcal{D}^0(\mathbb{R})$). Then it defines a distribution (which is not a measure) which we denote by

$$\mathrm{v.\,p} \, \frac{1}{x}$$

It is sometimes called the pseudo-function $\frac{1}{x}$ and it is the derivative of the $\log(|x|)$ in the sense of distributions.

**REMARK B.6**    $\frac{1}{x}$ is not locally summable and there exists no associated regular distribution.                                                                          ⬚

# Appendix C

## The Fourier transform

### C.1 The case of $L^1$ functions and extension to $L^2$

**DEFINITION C.1** Let $u \in L^1(\mathbb{R}^d)$. We define the Fourier transform as follows

$$\hat{u}(\xi) = (\mathscr{F}u)(\xi) \overset{\text{def}}{=} (2\pi)^{-d/2} \int_{\mathbb{R}_d} e^{-i\,x\cdot\xi}\, u(x)\, dx, \quad \forall\, \xi \in \mathbb{R}^d \qquad (C.1)$$

**THEOREM C.1**
The Fourier transform belongs to $\mathscr{L}(L^1(\mathbb{R}^d), \mathcal{B}_0(\mathbb{R}^d_\xi))$.

**REMARK C.1** The space $\mathcal{B}_0(\mathbb{R}^d)$ stands for the space of bounded continuous functions on $\mathbb{R}^d$ which tend to zero at infinity. Provided with the norm $\|u\|_\infty = \sup_{x\in\mathbb{R}^d} |u(x)|$, $\mathcal{B}_0(\mathbb{R}^d)$ is a Banach space. $\quad\Box$

**DEFINITION C.2** For $u \in L^1(\mathbb{R}^d)$, we define the Fourier co-transform as follows

$$\bar{\hat{u}}(\xi) = (\overline{\mathscr{F}}u)(\xi) \overset{\text{def}}{=} (2\pi)^{-d/2} \int_{\mathbb{R}_d} e^{i\,x\cdot\xi}\, u(x)\, dx, \quad \forall\, \xi \in \mathbb{R}^d \qquad (C.2)$$

**DEFINITION C.3** The translate of the function $f$ of amplitude $a \in \mathbb{R}^d$ is defined by
$$\tau_a f(x) = f(x - a), \quad \forall\, x \in \mathbb{R}^d \qquad (C.3)$$

and its symmetriser by

$$\varsigma f(x) = f(-x), \quad \forall\, x \in \mathbb{R}^d \qquad (C.4)$$

**THEOREM C.2**
For $u, v \in L^1(\mathbb{R}^d)$, the following properties hold

*i)* $\bar{\hat{u}} = \overline{\mathscr{F}(u)} = \mathscr{F}(\bar{u}),$     $\overline{\mathscr{F}(\varsigma\,u)} = \mathscr{F}u = \hat{u},$     $\varsigma(\mathscr{F}u) = \mathscr{F}\bar{u},$

*ii)* $\displaystyle\int_{\mathbb{R}^d} \hat{u}(x)\,v(x)\,\mathrm{d}x = \int_{\mathbb{R}^d} u(y)\,\hat{v}(y)\,\mathrm{d}y,$

*iii)* $\mathscr{F}(\tau_a u)(\xi) = e^{-i\,a\cdot\xi}\hat{u}(\xi),$     $\mathscr{F}(e^{-i\,a\cdot x}\,u) = \tau_{-a}\hat{u}.$

**REMARK C.2**   The property (iii) expresses that the Fourier transform exchanges translation and multiplication by a complex exponential.   ◻

### THEOREM C.3

*i)* If $x^\alpha\,u \in L^1(\mathbb{R}^d)$ for $|\alpha| \le k$, then $\hat{u} \in \mathcal{C}^k(\mathbb{R}^d)$ and we have

$$\mathrm{D}^\alpha\,\hat{u} = (-i)^{|\alpha|}\mathscr{F}(x^\alpha\,u) \tag{C.5}$$

*ii)* If $u \in \mathcal{C}^k(\mathbb{R}^d)$ with $\mathrm{D}^\beta\,u \in L^1(\mathbb{R}^d)$ for $|\beta| \le k$, then $\xi^\beta\,\hat{u} \in L^\infty$ and

$$i^{|\beta|}\xi^\beta\,\hat{u} = \mathscr{F}(\mathrm{D}^\beta\,u) \tag{C.6}$$

**REMARK C.3**   The two last properties assert that the Fourier transform exchanges reciprocally differentiation of order $|\alpha|$ with multiplication by $x^\alpha$ which suggests a kind of stability that will be established for a subspace of $L^1$.   ◻

### THEOREM C.4
Let $u \in L^1(\mathbb{R}^d)$ and real number $\lambda > 0$. We have

$$\mathscr{F}(f(\lambda.))(x) = \lambda^{-d}\hat{f}(\xi/\lambda) \tag{C.7}$$

### THEOREM C.5
Let $u, v \in L^1(\mathbb{R}^d)$. Then the Fourier transform of the convolution product is given by the following identity

$$\mathscr{F}(u \star v) = (\mathscr{F}u)\,(\mathscr{F}v) \tag{C.8}$$

The next theorem is a fundamental result justifying the power of the Fourier transform,

### THEOREM C.6 Fourier inversion
Let $u \in L^1(\mathbb{R}^d)$ such that $\hat{f} \in L^1(\mathbb{R}^d_\xi)$. Then

$$f(x) = \overline{\mathscr{F}}(\hat{f})(x) \tag{C.9}$$

**REMARK C.4**   This result is a consequence of both the following formula

$$\mathscr{F}(e^{-a|.|^2})(\xi) = (2\,a)^{-d/2}e^{-|\xi|^2/4a} \tag{C.10}$$

and the regularizing property of convolution with Gaussian functions.   ⬜

**REMARK C.5**   The inversion formula gives $f = \mathscr{F}(g)$ with $g = \hat{f} \in L^1$. This implies that $f \in \mathcal{B}_0(\mathbb{R}^d)$ which is not convenient for the applications. A great improvement is the extension of the Fourier transform to square summable functions furnished by the next theorem.   ⬜

**THEOREM C.7 Fourier-Plancherel**
*The Fourier transform $\mathscr{F} : L^1(\mathbb{R}^d) \cup L^2(\mathbb{R}^d) \to L^2(\mathbb{R}^d)$ can be continuously extended into a surjective isometry $\mathscr{F} : L^2(\mathbb{R}^d) \to L^2(\mathbb{R}^d)$.*

**REMARK C.6**   This result is a consequence of the Fourier inversion Theorem and the identity $(ii)$ in Theorem (C.2).   ⬜

---

## C.2   The space $\mathscr{S}(\mathbb{R}^d)$

We define the space of $\mathcal{C}^\infty$ functions with rapid decay at infinity[1],

$$\mathscr{S}(\mathbb{R}^d) \overset{\text{def}}{=} \left\{ u \in \mathcal{C}^\infty(\mathbb{R}^d), \quad \lim_{|x| \to +\infty} x^\alpha D^\beta u(x) = 0, \quad \forall \alpha, \beta \in \mathbb{N}^d \right\} \tag{C.11}$$

**REMARK C.7**   $\mathscr{S}(\mathbb{R}^d)$ can be endowed with the structure of a Fréchet space with the family of semi-norms,

$$\mathcal{N}_k(\varphi) = \sum_{|\alpha|,|\beta| \le k} \|x^\alpha D^\beta \varphi\|_{L^\infty}$$

$\forall k \in \mathbb{N}$.   ⬜

**THEOREM C.8**
*The space $\mathcal{D}(\mathbb{R}^d)$ is contained with dense and continuous injection in the Fréchet space $\mathscr{S}(\mathbb{R}^d)$.*

---

[1]It can be equivalently defined as follows

$$\mathscr{S}(\mathbb{R}^d) \overset{\text{def}}{=} \left\{ u \in \mathcal{C}^\infty(\mathbb{R}^d), \quad x^\alpha D^\beta u(x) \in L^2(\mathbb{R}^d), \quad \forall \alpha, \beta \in \mathbb{N}^d \right\}$$

**THEOREM C.9**
*The Fourier transform $\mathscr{F}$ is an isometry from $\mathscr{S}$ into itself when $\mathscr{S}$ is endowed with the topology induced by $L^2(\mathbb{R}^d)$. Its inverse is given by*

$$\mathscr{F}^{-1} = \overline{\mathscr{F}}$$

*and the following identities hold true,*

$$(u, v)_{L^2} = (\hat{u}, \hat{v})_{L^2}, \quad Parseval \tag{C.12}$$
$$\|u\|_{L^2} = \|\hat{u}\|_{L^2}, \quad Plancherel \tag{C.13}$$

**REMARK C.8**   Using the density of $\mathscr{S}$ in $L^2$, the Fourier transform $\mathscr{F} \in \text{ISOM}(\mathscr{S})$ can be extended in a unique manner as $\mathscr{F} \in \text{ISOM}(L^2)$. This furnishes an alternative way of defining the Fourier transform in $L^2$.   ▯

---

## C.3   The space of tempered distributions

**DEFINITION C.4**   *The space $\mathscr{S}'(\mathbb{R}^d)$ is defined as the space of continuous linear forms on the Fréchet space $\mathscr{S}(\mathbb{R}^d)$.*

**REMARK C.9**   Using the continuous and dense injection of $\mathcal{D}(\mathbb{R}^d) \hookrightarrow \mathscr{S}(\mathbb{R}^d)$ (Theorem (C.8)) and Theorem (B.4), we can identifiy elements of $\mathscr{S}'(\mathbb{R}^d)$ with distributions in $\mathcal{D}'(\mathbb{R}^d)$. The distributions in $\mathscr{S}'(\mathbb{R}^d)$ are called tempered distributions.   ▯

**REMARK C.10**   The elements of $L^p$ for $1 \leq p \leq \infty$ can be identified with tempered distributions.   ▯

A sequence $\{T_k\}_{k \in \mathbb{N}}$ of tempered distributions is convergent in $\mathscr{S}'(\mathbb{R}^d)$ if

$$\lim_{k \to \infty} \langle T_k, \varphi \rangle = \langle T, \varphi \rangle, \quad \forall \varphi \in \mathscr{S}(\mathbb{R}^d)$$

**THEOREM C.10**
*If $T \in \mathscr{S}'(\mathbb{R}^d)$, all its derivatives belong to $\mathscr{S}'(\mathbb{R}^d)$. Furthermore, let $\{T_k\}_{k \in \mathbb{N}} \in \mathscr{S}'(\mathbb{R}^d)$ be a convergent sequence of tempered distributions towards $T \in \mathscr{S}'(\Omega)$. Then the derivative of $T_k$ has the following property*

$$\lim_{k \to \infty} \mathrm{D}^\alpha T_k = \mathrm{D}^\alpha T \text{ in } \mathscr{S}'(\Omega), \quad \forall \alpha \in \mathbb{N}^d$$

**DEFINITION C.5**   *Let $\mathscr{O}_M(\mathbb{R}^d)$ be the space of infinitely differentiable functions of slow growth in $\mathbb{R}^d$, as well as all their derivatives, i.e.,*

$$\mathscr{O}_M(\mathbb{R}^d) \stackrel{\text{def}}{=} \left\{ \varphi \in \mathscr{E}(\mathbb{R}^d) : \forall \beta \in \mathbb{N}^d \; \exists C_\beta, m_\beta > 0; |\, D^\beta \, \varphi| \leq C_\beta (1 + |x|)^{m_\beta} \right\}$$

**THEOREM C.11**
*Let $f \in \mathscr{O}_M(\mathbb{R}^d)$, then we have*

(i) $f \varphi \in \mathscr{S}(\mathbb{R}^d)$, $\quad \forall \varphi \in \mathscr{S}(\mathbb{R}^d)$.

(ii) $f T \in \mathscr{S}'(\mathbb{R}^d)$, $\quad \forall T \in \mathscr{S}'(\mathbb{R}^d)$.

(ii) $\lim_{k \to \infty} f T_k = f T$, $\mathscr{S}'(\mathbb{R}^d)$ *if* $\lim_{k \to \infty} T_k = T$, $\mathscr{S}'(\mathbb{R}^d)$.

**REMARK C.11**   Hence the space $\mathscr{O}_M$ leaves stable by multiplication the space $\mathscr{S}$ and $\mathscr{S}'$, whence the notation $"M"$ for multiplicator. ▯

---

## C.4   Fourier transform in $\mathscr{S}'$

**DEFINITION C.6**   *Let $T \in \mathscr{S}'(\mathbb{R}^d)$. Its Fourier transform $\mathscr{F}T \in \mathscr{S}'(\mathbb{R}^d)$ is the tempered distribution defined by*

$$\langle \mathscr{F}T, \varphi \rangle \stackrel{\text{def}}{=} \langle T, \mathscr{F}\varphi \rangle, \quad \forall \varphi \in \mathscr{S}(\mathbb{R}^d) \tag{C.14}$$

**THEOREM C.12**
*The Fourier transform is such that*

$$\mathscr{F} \in \mathrm{ISOM}(\mathscr{S}'(\mathbb{R}^d))$$

*with inverse $\mathscr{F}^{-1} = \overline{\mathscr{F}}$ and the following properties,*

i) $\mathscr{F}(\tau_a T) = e^{-i\,a\cdot\cdot}\mathscr{F}T$, $\qquad \mathscr{F}(e^{-i\,a\cdot\cdot}\,T) = \tau_{-a}\mathscr{F}T$.

ii) *For* $\alpha \in \mathbb{N}^d$, $\mathscr{F}(D^\alpha\,T) = (i\cdot.)^\alpha \mathscr{F}T$.

iii) *For* $\beta \in \mathbb{N}^d$, $D^\beta\,\mathscr{F}(T) = \mathscr{F}((-i\cdot.)^\beta T)$.

Let us consider an automorphism $\theta$ of $\mathbb{R}^d$. We denote by ${}^{\mathrm{T}}\theta$ its transpose, i.e.,

$${}^{\mathrm{T}}\theta(x) \cdot y = x \cdot \theta(y), \quad \forall\, x, y \in \mathbb{R}^d$$

For $T \in \mathscr{S}'(\mathbb{R}^d)$, we define its direct image $\theta_* T \in \mathscr{S}'(\mathbb{R}^d)$ by $\theta$,

$$\langle \theta_* T, \varphi \rangle = \langle T, \varphi \circ \theta \rangle, \quad \forall \varphi \in \mathscr{S}(\mathbb{R}^d) \tag{C.15}$$

and its inverse image $^{\mathrm{T}}\theta^* T \overset{\text{def}}{=} |\det \mathrm{D}\,^{\mathrm{T}}\theta|^{-1} (^{\mathrm{T}}\theta^{-1})_* T$.

### THEOREM C.13
*Let $T \in \mathscr{S}'(\mathbb{R}^d)$, then we have*

$$\mathscr{F}(\theta_* T) = {}^{\mathrm{T}}\theta^* \mathscr{F}(T) \tag{C.16}$$

In the special case of distributions of compact support, we have the following result,

### THEOREM C.14
*Let $T \in \mathscr{E}'(\mathbb{R}^d)$. Then its Fourier transform $\mathscr{F}T \in \mathscr{O}_M(\mathbb{R}^d)$ is given by the following identity,*

$$(\mathscr{F}T)(\xi) = \langle T(x), e^{-i x \cdot \xi} \rangle \tag{C.17}$$

**REMARK C.12**    This theorem furnishes an explicit calculus of the Fourier transform of a distribution, when its support is compact. As an application, we can easily compute the Fourier transform of the Dirac measure, since its support is reduced to a singleton

$$\mathscr{F}(\delta_a) = e^{i\,a\cdot}.$$

⬜

---

## C.5    Fourier transform and convolution

### THEOREM C.15
*Let $\varphi, \psi \in \mathscr{S}(\mathbb{R}^d)$. We define their convolution product $\varphi \star \psi \in \mathscr{S}(\mathbb{R}^d)$ by*

$$(\varphi \star \psi)(y) = \int_{\mathbb{R}^d} \varphi(y - x)\,\psi(x)\,\mathrm{d}x \tag{C.18}$$

*and whose Fourier transform satisfies*

$$\mathscr{F}(\varphi \star \psi) = (2\pi)^{d/2} \mathscr{F}(\varphi)\,\mathscr{F}(\psi) \tag{C.19}$$

**DEFINITION C.7** *We introduce the space of convolutors of $\mathscr{S}(\mathbb{R}^d)$,*

$$\mathcal{O}'(\mathbb{R}^d) = \left\{ T \in \mathscr{S}'(\mathbb{R}^d) : T \star \varphi \in \mathscr{S}(\mathbb{R}^d), \quad \forall \varphi \in \mathscr{S}(\mathbb{R}^d) \right\}$$

**REMARK C.13** This space can be characterized as the space of distributions with rapid decay at infinity. $\quad\Box$

**THEOREM C.16**

(i) $\mathscr{F}(\mathcal{O}'_M(\mathbb{R}^d)) = \mathcal{O}'_c(\mathbb{R}^d)$ *and* $\mathscr{F}(\mathcal{O}'_c(\mathbb{R}^d)) = \mathcal{O}'_M(\mathbb{R}^d)$.

(ii) *If* $S \in \mathscr{S}'(\mathbb{R}^d)$, $T \in \mathcal{O}'_c(\mathbb{R}^d)$ *and* $\alpha \in \mathcal{O}'_M(\mathbb{R}^d)$, *we have*

$$\mathscr{F}(T \star S) = (2\pi)^{d/2} \mathscr{F}(T) \mathscr{F}(S) \tag{C.20}$$

$$\mathscr{F}(\alpha S) = (2\pi)^{d/2} \mathscr{F}(\alpha) \star \mathscr{F}(S) \tag{C.21}$$

---

## C.6 Partial Fourier transform

**THEOREM C.17**
*Let* $\varphi \in \mathscr{S}(\mathbb{R} \times \mathbb{R}^d)$. *Its partial Fourier transform* $\mathscr{F}_x$ *with respect to* $x \in \mathbb{R}^d$ *is defined by*

$$(\mathscr{F}_x \varphi)(t, \xi) = \int_{\mathbb{R}^d} e^{-ix \cdot \xi} \varphi(t, x) \, dx \tag{C.22}$$

*Furthermore* $\mathscr{F}_x \in \mathrm{ISOM}(\mathscr{S}(\mathbb{R} \times \mathbb{R}^d))$ *with inverse* $\mathscr{F}_x^{-1} = \overline{\mathscr{F}_x}$.

**REMARK C.14** By transposition we define $\mathscr{F}_x$ in $\mathscr{S}'(\mathbb{R}^d)$ and we have $\mathscr{F}_x \in \mathrm{ISOM}(\mathscr{S}'(\mathbb{R} \times \mathbb{R}^d))$. $\quad\Box$

---

## C.7 Vector valued distributions

Let $X$ be a Banach space and $]a, b[$ an open set of $\mathbb{R}$. The measure $dt$ stands for the Lebesgue measure on $]a, b[$.

**DEFINITION C.8** *The space of linear continuous mapping of* $\mathcal{D}(]a, b[)$ *into* $X$, *is called the space of vectorial distributions over* $]a, b[$ *with values in*

*X and we note*

$$\mathcal{D}'(]a, b[; X) \overset{\text{def}}{=} \mathscr{L}(\mathcal{D}(]a, b[); X) \tag{C.23}$$

*It is endowed with the topology of the uniform convergence on bounded sets in* $\mathcal{D}(]a, b[)$.

Hence, if $f \in \mathcal{D}'(]a, b[; X)$, then the application

$$\mathcal{D}(]a, b[) \to X$$
$$\varphi \mapsto \langle f, \varphi \rangle$$

is continuous.

**REMARK C.15**    This definition can be extended to any open set $\Omega$ of $\mathbb{R}^d$.                                                                     ⬚

As in the case of scalar distributions, we can define derivatives in the sense of distributions,

**DEFINITION C.9**    *For $f \in \mathcal{D}'(]a, b[; X)$, we define its first order derivative in the sense of distributions as the unique element $\frac{d}{dt} f \in \mathcal{D}'(]a, b[; X)$ such that*

$$\langle \frac{d}{dt} f, \varphi \rangle = -\langle f, \frac{d}{dt} \varphi \rangle, \quad in \; X \tag{C.24}$$

$\forall \varphi \in \mathcal{D}(]a, b[)$

**REMARK C.16**    The application

$$\mathcal{D}'(]a, b[; X) \to \mathcal{D}'(]a, b[; X)$$
$$f \mapsto \frac{d}{dt} f$$

is continuous.                                                                                          ⬚

**REMARK C.17**    This definition can be extended to the higher order derivatives, with

$$\langle \frac{d^i}{dt^i} f, \varphi \rangle = -\langle f, \frac{d^i}{dt^i} \varphi \rangle, \quad in \; X \tag{C.25}$$

$\forall \varphi \in \mathcal{D}(]a, b[)$, and $i \in \mathbb{N}$.                                              ⬚

## C.8  Vector valued Lebesgue spaces

Let $X$ be a Banach space and $]a, b[$ an open set of $\mathbb{R}$. The measure $dt$ stands for the Lebesgue measure on $]a, b[$.

**DEFINITION C.10**  *We denote by $L^p(a, b; X)$ for $1 \leq p < +\infty$ the space of functions*

$$]a, b[ \rightarrow X$$
$$t \mapsto f(t)$$

*such that*

*(i) $f$ is (strongly) dt-measurable on $]a, b[$,*

*(ii)*

$$\|f\|_{L^p(a,b;X)} \overset{\text{def}}{=} \left( \int_a^b \|f(t)\|_X^p \, dt \right)^{1/p} < +\infty \qquad (C.26)$$

**DEFINITION C.11**  *We denote by $L^\infty(a, b; X)$, the space of functions*

$$]a, b[ \rightarrow X$$
$$t \mapsto f(t)$$

*such that*

*(i) $f$ is (strongly) dt-measurable on $]a, b[$,*

*(ii) $f$ is bounded almost everywhere over $]a, b[$ and we set*

$$\|f\|_{L^\infty(a,b;X)} = \inf_{dt-\text{meas}\{t \in ]a,b[: \|f(t)\|_X > C\}=0} C \qquad (C.27)$$

**THEOREM C.18 [16]**
For $1 \leq p \leq \infty$, the space $L^p(a, b; X)$ is a Banach space if $X$ is a Banach space. It is a Hilbert space if $X$ is a Hilbert space.

In the applications, the following lemmas will be of great use,

**LEMMA C.1**
Let $X, Y$ be two Banach spaces and $u \in L^1(a, b; X)$ and $\mathscr{A} \in \mathscr{L}(X, Y)$, then

1. $\mathscr{A} u \in L^1(a, b; Y)$,

2. $\int_a^b \mathscr{A}\,u(t)\,\mathrm{d}t = \mathscr{A}\left(\int_a^b u(t)\,\mathrm{d}t\right).$

**LEMMA C.2**

1. *Let* $]a,b[$ *be an open set of* $\mathbb{R}$, $u \in L^1(a,b;X)$ *and* $\phi \in X'$, *then*

$$\left\langle \phi, \int_a^b u(t)\,\mathrm{d}t \right\rangle = \int_a^b \langle \phi, u(t)\rangle\,\mathrm{d}t \qquad (\mathrm{C}.28)$$

2. *For* $|a|, |b| < \infty$, *the equality* (C.28) *is valid for* $u \in L^p(a,b;X)$ *and* $p \geq 1$.

# Appendix D

## Sobolev spaces

### D.1  Spaces $H^m(\Omega)$

Let $\Omega$ be an open set in $\mathbb{R}^d$. We introduce the Sobolev space of integer regularity $m \in \mathbb{N}$,

$$H^m(\Omega) \stackrel{\text{def}}{=} \left\{ u \in \mathcal{D}'(\Omega) : D^\alpha u \in L^2(\Omega), \quad |\alpha| \leq m \right\} \qquad (D.1)$$

**THEOREM D.1**
*The space $H^m(\Omega)$ endowed with the scalar product*

$$(u, v)_{H^m(\Omega)} \stackrel{\text{def}}{=} \sum_{|\alpha| \leq m} \int_\Omega D^\alpha u(x) \overline{D^\alpha v(x)} \, dx \qquad (D.2)$$

*and the associated norm*

$$\|u\|_{H^m(\Omega)} = \left( \sum_{|\alpha| \leq m} \| D^\alpha u \|_{L^2(\Omega)}^2 \right)^{1/2} \qquad (D.3)$$

*is a Hilbert space.*

**THEOREM D.2**
*If $m_1 > m_2$ the space $H^{m_1}(\Omega) \hookrightarrow H^{m_2}(\Omega)$ with continuous injection.*

In the case $\Omega = \mathbb{R}^d$, it is possible to define equivalently the space $H^m(\mathbb{R}^d)$ thanks to the Fourier transform.

**THEOREM D.3**
*Let $m \in \mathbb{N}$, then*

*(i) $H^m(\mathbb{R}^d) \subset \mathscr{S}'(\mathbb{R}^d)$.*

*(ii) The space $H^m(\mathbb{R}^d)$ coincides with following space*

$$\left\{ u \in \mathscr{S}'(\mathbb{R}^d) : (1 + |\xi|^2)^{m/2} \hat{u} \in L^2(\mathbb{R}^d) \right\}$$

*where $\hat{u}$ is the Fourier transform of $u$ defined in section 1.4.4.*

*(iii) The norm $\|.\|_{H^m(\mathbb{R}^d)}$ is equivalent to the following norm,*

$$\left( \int_{\mathbb{R}^d} (1 + |\xi|^2)^m |\hat{u}(\xi)|^2 d\xi \right)^{1/2}$$

**REMARK D.1**    The proof is based on the identity $\mathscr{F}(D^\alpha u) = (i\xi)^\alpha \mathscr{F} u$ and the following inequality that holds $\forall 0 \leq |\alpha| \leq m$,

$$\sum_{1 \leq j \leq m} |\xi_j|^{2\alpha_j} \leq \left( 1 + \sum_{1 \leq j \leq m} |\xi_j|^2 \right)^m \leq C \left( 1 + \sum_{0 < |\alpha| \leq m} \prod_{1 \leq j \leq m} |\xi_j|^{2\alpha_j} \right)$$

□

This theorem furnishes a new definition of the space $H^m(\mathbb{R}^d)$ that we shall extend to real parameter $m$ in the next section.

## D.2    The space $H^s(\mathbb{R}^d)$

For $s \in \mathbb{R}$ we define the following Sobolev space,

$$H^s(\mathbb{R}^d) \stackrel{\text{def}}{=} \left\{ u \in \mathscr{S}'(\mathbb{R}^d) : (1 + |\xi|^2)^{s/2} \hat{u}(\xi) \in L^2(\mathbb{R}^d_\xi) \right\} \tag{D.4}$$

endowed with the following scalar product

$$(u, v)_{H^s(\mathbb{R}^d)} \stackrel{\text{def}}{=} \int_{\mathbb{R}^d} (1 + |\xi|^2)^s \hat{u}(\xi) \, \overline{\hat{v}}(\xi) \, d\xi \tag{D.5}$$

and the associated norm,

$$\|u\|_{H^s(\mathbb{R}^d)} = \left( \int_{\mathbb{R}^d} (1 + |\xi|^2)^s |\hat{u}(\xi)|^2 \, dxi \right)^{1/2} \tag{D.6}$$

**THEOREM D.4**
*For $s \in \mathbb{R}$, the space $(H^s(\mathbb{R}^d), (.,.)_{H^s(\mathbb{R}^d)})$ is a Hilbert space. Furthermore the continuous injection $H^{s_1}(R^d) \hookrightarrow H^{s_2}(R^d)$ holds for $s_1 \geq s_2$.*

**REMARK D.2**    The proof of completeness uses the fact that $\mathscr{F} \in$ ISOM($\mathscr{S}'(\mathbb{R}^d)$) and the completeness of $L^2(\mathbb{R}^d)$. The injection property is an easy consequence of the inequality

$$(1 + \|\xi\|^2)^{s_2} \leq (1 + \|\xi\|^2)^{s_1}, \quad s_1 \geq s_2$$

⬜

**THEOREM D.5**
*The space $H^s(\mathbb{R}^d)$ is a normal space of distributions[1].*

---

## D.3    The topological dual of $H^s(\mathbb{R}^d)$

In this section we shall characterize the space $(H^s(\mathbb{R}^d))'$ of continuous linear forms on $H^s(\mathbb{R}^d)$. We already have the following inclusions

$$(H^s(\mathbb{R}^d))' \subset \mathscr{S}'(\mathbb{R}^d) \subset \mathcal{D}'(\mathbb{R}^d)$$

since $\mathcal{D}(\mathbb{R}^d)$ is dense in $\mathscr{S}(\mathbb{R}^d)$ which is dense in $H^s(\mathbb{R}^d)$.

**THEOREM D.6**
*$\forall s \in \mathbb{R}$, the dual space of $H^s(\mathbb{R}^d)$ coincides (algebraically and topologically )
in $\mathcal{D}'(\mathbb{R}^d)$ with $H^{-s}(\mathbb{R}^d)$.*

---

## D.4    Sobolev embedding theorems

**THEOREM D.7**
*Let $s \in \mathbb{R}$ and $k \in \mathbb{N}$ satisfying $s > d/2 + k$. Then the continuous injection*

$$H^s(\mathbb{R}^d) \hookrightarrow \mathcal{B}_0(\mathbb{R}^d) \tag{D.7}$$

*holds true.*

**REMARK D.3**    The proof is based on Theorem (C.1) and $(1 + |\xi|^2)^{s/2} \in L^2(\mathbb{R}^d_\xi)$ for $s > d/2$.

⬜

**REMARK D.4**    The condition $s > d/2 + k$ is optimal. ⬜

---

[1] i.e., $\mathcal{D}(\mathbb{R}^d)$ is dense in $H^s(\mathbb{R}^d)$.

**THEOREM D.8**
*For $s > d/2$, the Hilbert space $H^s(\mathbb{R}^d)$ is an algebra.*

**REMARK D.5**   Using Theorem(A.1) we have

$$\mathscr{F}(u\,v) = \mathscr{F}u \star \mathscr{F}v$$

and using $w(\xi) = (1 + |\xi|^2)^s \leq 4^s \left[(1 + |\xi - \eta|^2)^s + (1 + |\eta|^2)^s\right]$, for $s \geq 0$, we get

$$(1 + |\xi|^2)^s|\mathscr{F}(u\,v)(\xi)| \leq C\,|(w\,|\mathscr{F}u|) \star v| + C\,|\mathscr{F}u| \star (w\,\mathscr{F}v) \in L^2(\mathbb{R}^d)$$

for $s \geq d/2$. 

**THEOREM D.9**
*For $s \in \mathbb{R}$, if $u \in H^s(\mathbb{R}^d)$ and $\varphi \in \mathscr{S}(\mathbb{R}^d)$ then $\varphi\,u \in H^s(\mathbb{R}^d)$.*

**REMARK D.6**   Again we use

$$\mathscr{F}(u\,\varphi) = \mathscr{F}u \star \mathscr{F}\varphi$$

and the inequality $(1 + |\xi|^2)^s \leq 2^{|s|}(1 + |\eta|^2)^s(1 + |\xi - \eta|^2)^{|s|}$. 

**THEOREM D.10**
*The continuous injection*

$$H^s(\mathbb{R}^d) \hookrightarrow L^p(\mathbb{R}^d), \quad \forall 2 \leq p \leq 2d/(d - 2s) \tag{D.8}$$

*holds for $s < d/2$.*

**REMARK D.7**   The Lebesgue indice can be deduced from homogeneity arguments. We set $v_\lambda(x) = v(\lambda\,x)$ and we have

$$\|v_\lambda\|_{L^p} = \lambda^{-d/p}\|v\|_{L^p}$$

$$|v_\lambda|_{H^s} = \lambda^{(-d+2s)/2}|v|_{H^s}$$

with $|v|_{H^s} \overset{\text{def}}{=} \left(\int_{\mathbb{R}^d} |\xi|^{2s}|\hat{v}(\xi)|^2\,d\xi\right)^{1/2}$, from which we deduce $1/p = 1/2 - s/d$. 

**REMARK D.8**   The proof is based on a decomposition of the function $u \in H^s$ as

$$u = u_{L,A} + u_{H,A}$$

with the low frequency part $u_{L,A} = \mathscr{F}^{-1}(\chi_{B(0,A)}\hat{u})$ and the high frequency part $u_{H,A} = \mathscr{F}^{-1}(\chi_{\complement B(0,A)}\hat{u})$. We use also the characterization

$$\|u\|_{L^p}^p = p\int_0^\infty \lambda^{p-1}\operatorname{meas}(|u| > \lambda)\,d\lambda$$

and the Bienaymé-Tchebytchev inequality

$$\operatorname{meas}(|u| > \lambda) \leq \lambda^{-1}\|u\|_{L^2}^2$$

□

### THEOREM D.11

*The following continuous injection*

$$H^s(\mathbb{R}^d) \hookrightarrow L^p(\mathbb{R}^d), \quad 2 \leq p < 2d/(d-2) \tag{D.9}$$

*holds for $s = d/2$.*

---

## D.5 Density properties

We already saw in Theorem (D.5) that $\mathcal{D}(\mathbb{R}^d)$ is dense in $H^s(\mathbb{R}^d)$ for $s \in \mathbb{R}$. In this section, we deal with the case where $\Omega \subsetneq \mathbb{R}^d$.

### THEOREM D.12

*Let $k \in \mathbb{N}^*$ and $\Omega$ a domain of class $\mathcal{C}^k$. The space $\mathcal{D}(\overline{\Omega})$ is dense in $H^k(\mathbb{R}^d)$.*

**REMARK D.9**  The proof is first performed for $\mathbb{R}^d_+$ by truncation and regularization. In the general case, we may assume that $\overline{\Omega}$ is compact and use local maps (Definition(A.1)) and partition of unity. □

We define the space

$$H_0^m(\Omega) = \overline{\mathcal{D}(\Omega)}^{H^m(\Omega)} \tag{D.10}$$

## D.6   Trace theorem for $H^s(\mathbb{R}^d)$

**DEFINITION D.1**   *The linear application*

$$\gamma_0 : \mathscr{S}(\overline{\mathbb{R}^d}) \to \mathscr{S}(\mathbb{R}^{d-1})$$
$$u \mapsto x' \mapsto \gamma_0 u(x') = u(x', x_d = 0)$$

*stands for the trace operator of order 0.*

**THEOREM D.13**
*For $s > 1/2$,*

(i) *The linear mapping $\gamma_0$ is continuous from $\mathscr{S}(\overline{\mathbb{R}^d})$ endowed with the topology of $H^s(\mathbb{R}^d)$ into $\mathscr{S}(\mathbb{R}^{d-1})$ endowed with the topology of $H^{s-1/2}(\mathbb{R}^{d-1})$.*

(ii) *This mapping can be extended to a continuous linear mapping from $H^s(\mathbb{R}^d)$ into $H^{s-1/2}(\mathbb{R}^{d-1})$, again denoted $\gamma_0$. Furthermore $\gamma_0$ is surjective.*

**REMARK D.10**   We use the partial Fourier transform $(\mathscr{F}_{x'}u)(\xi', x_d)$ of $u(x', x_d)$ in the direction $x' \in \mathbb{R}^{d-1}$. ☐

## D.7   Trace theorem for $H^m(\mathbb{R}^d_+)$

**DEFINITION D.2**   *The linear application,*

$$\gamma_j : \mathcal{D}(\overline{\mathbb{R}^d_+}) \to \mathcal{D}(\mathbb{R}^{d-1})$$
$$u \mapsto x' \mapsto (\gamma_j u)(x') = \partial_{x_d}^j u(x', x_d)|_{x_d=0}$$

*stands for the trace operator of order $j$.*

**THEOREM D.14**
*The following trace mapping can be defined on $\mathcal{D}(\overline{\mathbb{R}^d_+})$ and extended by density to a continuous linear surjective application.*

$$\gamma_m : H^m(\mathbb{R}^d_+) \to \prod_{0 \leq j \leq m-1} H^{m-j-1/2}(\mathbb{R}^{d-1})$$
$$u \mapsto (\gamma_j u)_{1 \leq j \leq m-1}$$

*and* $\ker \gamma_m = H_0^m(\mathbb{R}_+^d)$.

---

## D.8    The space $H^s(\Gamma)$

There exists different methods to define Sobolev spaces on submanifolds of codimension 1 in $\mathbb{R}^d$. Here we may consider a bounded domain $\Omega \subset \mathbb{R}^d$ with compact boundary $\Gamma$ of class $\mathcal{C}^\infty$.

- Definition by localisation and diffeomorphism :
  We use local maps $(U_j, h_j)_{1 \le j \le m}$ and partition of unity $\{r_j\}_{1 \le j \le m}$ to define functional spaces on $\Gamma$. Using the result of Lemma (A.6), we define the function

$$(\theta_j u)(\zeta') = \begin{cases} (r_j u) \circ h_j(\zeta', 0), & \zeta' \in B_0 \\ 0, & \zeta' \in \complement B_0 \end{cases} \tag{D.11}$$

**DEFINITION D.3**    *We define the Sobolev space of fractional order on the $\mathcal{C}^\infty$ manifold $\Gamma$ by*

$$H^s(\Gamma) \overset{\text{def}}{=} \left\{ u \in \mathcal{D}'(\Gamma) : \theta_j u \in H^s(\mathbb{R}^{d-1}), \quad 1 \le j \le m \right\} \tag{D.12}$$

**THEOREM D.15**

1. *The space $H^s(\Gamma)$ provided with the norm,*

$$\|u\|_{H^s(\Gamma)} = \left( \sum_{1 \le j \le m} \|\theta_j u\|^2_{H^s(\mathbb{R}^{d-1})} \right)^{1/2} \tag{D.13}$$

   *is a Hilbert space.*

2. *The space $\mathcal{D}(\Gamma)$ is dense in $H^s(\Gamma)$ for $s \ge 0$.*

3. *Using the space $L^2(\Gamma) \overset{\text{def}}{=} H^0(\Gamma)$ as the pivot space, we have the following identification,*

$$(H^s(\Gamma))' \equiv H^{-s}(\Gamma) \tag{D.14}$$

**REMARK D.11**    The definition of the space $H^s(\Gamma)$ depends neither on the atlas $(U_j, h_j)_{1 \le j \le m}$ nor on the associated partition of unity $(r_j)_{1 \le j \le m}$. The associated norm does depend on the choice of the atlas, but defines a class of equivalent norms.    ⬚

- Definition as the domain of tangential operator :
  We consider the Laplace-Beltrami operator $\Delta_\Gamma$ on the manifold $\Gamma$. Considered as a positive definite self-adjoint operator in $L^2(\Omega)$ with domain $D(-\Delta_\Gamma)$, we define the space

  $$H^s(\Gamma) \stackrel{\text{def}}{=} D((-\Delta_\Gamma)^{s/2}), \quad s \in \mathbb{R} \tag{D.15}$$

  provided with the norm of the graph

  $$\|v\|_{H^s(\Gamma)} = (\|v\|^2_{L^2(\Gamma)} + \|(-\Delta_\Gamma)^{s/2}v\|^2_{L^2(\Gamma)})^{1/2} \tag{D.16}$$

---

## D.9 Trace theorem in $H^m(\Omega)$

**THEOREM D.16**
*Let $\Omega$ be a bounded domain in $\mathbb{R}^d$ with compact boundary $\Gamma$ of class $\mathcal{C}^k$. Let $m \leq k$, then the mapping $\gamma_m$ defined on $\mathcal{D}(\overline{\Omega})$ with values in $[\mathcal{D}(\Gamma)]^m$ can be extended by density to a continuous linear surjection :*

$$\gamma_m : H^m(\Omega) \to \prod_{0 \leq j \leq m-1} H^{m-j-1/2}(\Gamma)$$

$$u \mapsto (\gamma_j u)_{1 \leq j \leq m-1}$$

*and $\ker \gamma_m = H_0^m(\Omega)$.*

---

## D.10 Extension theorems

**DEFINITION D.4** *An open set $\Omega \subset \mathbb{R}^d$ has the m-extension property for $m \in \mathbb{N}$, if there exists an operator $P$ such that*

*(i) $P \in \mathscr{L}(H^m(\Omega), H^m(\mathbb{R}^d))$,*

*(ii) $Pu = u$, a.e. in $\Omega$, $\forall u \in H^m(\Omega)$.*

We may use a general procedure that may allow to define m-extension operators by density,

**LEMMA D.1**
*Let $V \subset H^m(\Omega)$ be a dense subspace and an operator $P$ such that*

*(i)* $P \in \mathscr{L}(V, H^m(\mathbb{R}^d))$,

*(ii)* $Pu = u$, *a.e. in* $\Omega$, $\forall u \in V$.

*Then* $P$ *has a unique extension into an operator of m-extension from* $H^m(\Omega)$ *into* $H^m(\mathbb{R}^d)$.

### LEMMA D.2
*Let* $\Omega, \Omega_1, \Omega_2 \subset \mathbb{R}^d$ *and* $S \overset{\text{def}}{=} \overline{\Omega_1} \cap \overline{\Omega_2}$ *of class* $\mathcal{C}^{m+1}$ *such that*

$$\Omega = \Omega_1 \cup S \cup \Omega_2$$

*Let* $u \in \mathcal{C}^m(\Omega)$ *such that*

$$u|_{\Omega_i} \in \mathcal{D}(\overline{\Omega_i}), \quad i \in \{1,2\}$$

*then*

$$u \in H^{m+1}(\Omega)$$

**REMARK D.12**  If we set $v_i|_{\Omega_j} = \partial_{x_i}(u|_{\Omega_j})$ for $1 \le i \le d$ and $j \in \{1,2\}$, the following identity holds

$$\langle v_i, \varphi \rangle = \int_S [u]\, \varphi \cdot n_i - \int_\Omega u\, \partial_{x_i}\varphi, \quad \forall \varphi \in \mathcal{D}(\Omega)$$

where $[u]$ is the saltus of $u$ across $S$. ⬜

### THEOREM D.17
*The open set* $\mathbb{R}^d_+$ *possesses the m-extension property for* $m \in \mathbb{N}$.

**REMARK D.13**  For $m = 0$, we only use the extension by zero. For $m \ge 1$, we use the extension by reflexion (Babitch extension),

$$(Pu)(x', x_d) = \begin{cases} u(x', x_d), & x_d \ge 0 \\ \sum_{1 \le k \le m} \alpha_k\, u(x', -k\, x_d), & x_d < 0 \end{cases}$$

where we choose $(\alpha_k)_{1 \le k \le m}$ such that $Pu \in \mathcal{C}^{m-1}(\Omega)$ which impose continuity of derivatives with respect to $x_d$ up to order $m - 1$. This implies $\sum_{1 \le k \le m} (-k)^j \alpha_k = 1$, $0 \le j \le m - 1$ which is a unique solvable Van der Monde system. We conclude using Lemma (D.2). ⬜

### THEOREM D.18
*Let* $\Omega \subset \mathbb{R}^d$ *be a bounded open set with boundary* $\Gamma$ *of class* $\mathcal{C}^k$. *Then* $\Omega$ *possesses the m-extension property with* $0 \le m \le k$.

**DEFINITION D.5**   *For $s \in \mathbb{R}$, we define the space*

$$\mathscr{H}^s(\Omega) \stackrel{\text{def}}{=} \left\{ u \in \mathcal{D}'(\Omega), \exists U \in H^s(\mathbb{R}^d) \text{ such that } u = U|_\Omega \right\} \tag{D.17}$$

*provided with the quotient norm*

$$\|u\|_{\mathscr{H}^s(\Omega)} \stackrel{\text{def}}{=} \inf_{\substack{U \in H^s(\mathbb{R}^d) \\ U|_\Omega = u}} \|U\|_{H^s(\mathbb{R}^d)} \tag{D.18}$$

**THEOREM D.19**
*If $\Omega$ has the m-extension property with $m \in \mathbb{N}$, then*

$$\mathscr{H}^s(\Omega) \equiv H^m(\Omega)$$

---

## D.11   The space $H^{-m}(\Omega)$

**DEFINITION D.6**   *For $m \in \mathbb{N}$, we define the space $H^{-m}(\Omega)$ as the dual space of $H_0^m(\Omega)$.*

**THEOREM D.20**

(i) *The space $H^{-m}(\Omega)$ endowed with the norm*

$$\|F\|_{H^{-m}(\Omega)} \stackrel{\text{def}}{=} \sup_{\substack{\|u\|_{H^m(\Omega)} \le 1 \\ u \neq 0}} \frac{|\langle F, u \rangle|}{\|u\|_{H^m(\Omega)}}$$

*is a Hilbert space.*

(ii) *For $m_1 > m_2$, $H^{-m_2}(\Omega) \hookrightarrow H^{-m_1}(\Omega)$.*

**DEFINITION D.7**   *For $m \in \mathbb{N}$, we define the space of distributions*

$$E_m \stackrel{\text{def}}{=} \left\{ T \in \mathcal{D}'(\Omega) : T = \sum_{|\alpha| \le m} D^\alpha f_\alpha, \quad f_\alpha \in L^2(\Omega) \right\} \tag{D.19}$$

*endowed with the topology induced by*

$$\|T\|^2 = \inf_{\substack{f_\alpha \in L^2(\Omega) \\ T = \sum_{|\alpha| \le m} D^\alpha f_\alpha}} \sum_{|\alpha| \le m} \|f_\alpha\|^2$$

**THEOREM D.21**
*The space $E_m$ coincides algebraically and topologically with $H^{-m}(\Omega)$.*

**REMARK D.14**    This is a consequence of the Hahn-Banach theorem allowing to extend linear forms on a sub-space into the whole Banach space and the Riesz representation theorem in Hilbert spaces.    ▯

**THEOREM D.22**
*The operator $-\Delta + \mathrm{I} \in \mathrm{ISOM}(H_0^1(\Omega), H^{-1}(\Omega))$.*

---

## D.12    Compact embeddings

**THEOREM D.23 Rellich**
*Let $\Omega \subset \mathbb{R}^d$ be a bounded open set. Then for $m \in \mathbb{N}$, the injection*

$$H_0^{m+1}(\Omega) \hookrightarrow H_0^m(\Omega)$$

*is compact.*

**THEOREM D.24**
*Let $\Omega \subset \mathbb{R}^d$ be a bounded open set possessing the (m+1)-extension property. Then the injection*

$$H^{m+1}(\Omega) \hookrightarrow H^m(\Omega)$$

*is compact.*

---

## D.13    Poincaré inequalities

**DEFINITION D.8**    *Let $\zeta \in \mathbb{R}^d$ with $|\zeta| = 1$, $]a, b[ \subset \mathbb{R}$ and $0 < e < \infty$ such that $\mathrm{meas}(]a, b[) = e$. We define the band of $\mathbb{R}^d$ of width $e$ in the direction $\zeta$ by*

$$B_e(\zeta) \stackrel{\text{def}}{=} \left\{ x \in \mathbb{R}^d : a < x \cdot \zeta < b \right\} \tag{D.20}$$

**THEOREM D.25 Poincaré inequality in $H_0^1(\Omega)$**
*Let $\Omega \subset \mathbb{R}^d$ be an open set bounded in some direction, i.e., such that there*

*exists a band* $B_e(\zeta)$ *with*

$$\Omega \subset B_e(\zeta)$$

*Then*

$$\|u\|_{L^2(\Omega)}^2 \le e^2/2 \|\nabla u\|_{L^2(\Omega)}^2, \quad \forall u \in H_0^1(\Omega) \tag{D.21}$$

### THEOREM D.26 Poincaré inequality in $H^1(\Omega)$
*Let* $\Omega \subset \mathbb{R}^d$ *be a bounded open set such that the following compact injection*

$$H^1(\Omega) \hookrightarrow L^2(\Omega)$$

*holds. Then we have*

$$\|u\|_{L^2(\Omega)}^2 - (\text{meas }\Omega)^{-1} \left| \int_\Omega u(x)\,\mathrm{d}x \right|^2 \le P(\Omega) \|\nabla u\|_{L^2(\Omega)}^2 \tag{D.22}$$

**REMARK D.15**  If $\Omega$ has the 1-extension property then the compact injection holds. We call Nikodym open sets the sets for which the Poincaré inequality holds.  ▯

**REMARK D.16**  The Poincaré constant $P(\Omega) = 1/\lambda_2$ where $\lambda_2$ is the first non-zero eigenvalue of the Neumann problem,

$$\begin{cases} -\Delta u = \lambda u, & \Omega \\ \partial_n u = 0, & \Gamma \end{cases}$$

▯

### THEOREM D.27
*Let* $\Omega \subset \mathbb{R}^d$ *be a bounded open set with regular boundary* $\Gamma$. *Then there exists* $C > 0$ *such that*

$$\|u\|_{L^2(\Omega)}^2 \le C \left( \|\nabla u\|_{L^2(\Omega)}^2 + \|\gamma_0 u\|_{L^2(\Gamma)}^2 \right), \quad \forall u \in H^1(\Omega) \tag{D.23}$$

# References

[1] F. Abergel and R. Temam. On Some Control Problems in Fluid Mechanics. *Theoretical and Computational Fluid Dynamics*, 1:303–325, 1990.

[2] P. Acquistapace, F. Flandoli, and B. Terreni. Initial boundary value problems and optimal control for nonautonomous parabolic systems. *SIAM J. Control Optimization*, 29(1):89–118, 1991.

[3] J.-C. Aguilar and J.-P. Zolésio. Coque fluide intrinsèque sans approximation géométrique. *C. R. Acad. Sci. Paris Sér. I Math.*, 326(11):1341–1346, 1998.

[4] H. Amann. Nonhomogeneous Navier-Stokes equations in spaces of low regularity. *Quaderni di matematica*, IX(In print), 2002.

[5] L. Ambrosio. Transport equation and Cauchy problem for BV vector fields. CVGMT Preprint Server - http://cvgmt.sns.it/papers/luia/.

[6] L. Ambrosio. Variational problems in SBV and image segmentation. *Acta Appl. Math.*, 17(1):1–40, 1989.

[7] J.-L. Armand. Minimum-mass design of a plate-like structure for specified fundamental frequency. *AIAA J.*, 9:1739–1745, 1971.

[8] J-P. Aubin. *Mutational and morphological analysis - Tools for shape evolution and morphogenesis*. Birkhäuser Boston Inc., Boston, MA, 1999.

[9] H. Beirão da Veiga. On the existence of strong solutions to a coupled fluid-structure evolution problem. *J. Math. Fluid. Mech.*, 6(1):21–52, 2004.

[10] A. Bensoussan, J.-L. Lions, and G. C. Papanicolaou. Homogenization in deterministic and stochastic problems. In *Stochastic problems in dynamics (Sympos., Univ. Southampton, Southampton, 1976)*, pages 106–115. Pitman, London, 1977.

[11] M. L. Bernardi, G. A. Pozzi, and G. Savaré. Variational equations of Schroedinger-type in non-cylindrical domains. *J. Differ. Equations*, 171(1):63–87, 2001.

[12] W. A. Blankinship. The curtain rod problem. *Amer. Math. Monthly*, 50:186–189, 1943.

[13] S. Boisgérault. *Optimisation de forme : systèmes nonlinéaires et mécanique des fluides.* PhD thesis, Ecole des Mines de Paris - Informatique Temps réel, Robotique, Automatique, 2000.

[14] S. Boisgérault and J.-P. Zolésio. Boundary variations in the Navier-Stokes equations and Lagrangian functionals. In *Cagnol, John et al., Shape optimization and optimal design. Proceedings of the IFIP conference. Marcel Dekker. Lect. Notes Pure Appl. Math. 216, 7-26.* 2001.

[15] M. Boulakia. Existence of weak solutions for the motion of an elastic structure in an incompressible viscous fluid. *C. R. Math. Acad. Sci. Paris*, 336(12):985–990, 2003.

[16] N. Bourbaki. *Elements of mathematics. Integration I: Chapters 1–6.* Berlin: Springer. xv, 2004.

[17] H. Brezis. *Functional analysis. Theory and applications.* Masson, "collection mathématiques appliquées pour la maitrise (in french)" edition, 1994.

[18] D. Bucur and J.-P. Zolésio. Optimisation de forme sous contrainte capacitaire. *C. R. Acad. Sci. Paris Sér. I Math.*, 318(9):795–800, 1994.

[19] D. Bucur and J.-P. Zolésio. $N$-dimensional shape optimization under capacitary constraint. *J. Differential Equations*, 123(2):504–522, 1995.

[20] D. Bucur and J.-P. Zolésio. Pseudo-courbure dans l'optimisation de forme. *C. R. Acad. Sci. Paris Sér. I Math.*, 321(3):387–390, 1995.

[21] D. Bucur and J.-P. Zolésio. Boundary optimization under pseudo curvature constraint. *Ann. Scuola Norm. Sup. Pisa Cl. Sci. (4)*, 23(4):681–699 (1997), 1996.

[22] D. Bucur and J.-P. Zolésio. Free boundary problems and density perimeter. *J. Differential Equations*, 126(2):224–243, 1996.

[23] D. Bucur and J.-P. Zolesio. Wiener's criterion and shape continuity for the Dirichlet problem. *Boll. Un. Mat. Ital. B (7)*, 11(4):757–771, 1997.

[24] J. Cagnol and J.-P. Zolésio. Shape derivative in the wave equation with Dirichlet boundary conditions. *J. Differential Equations*, 158(2):175–210, 1999.

[25] J. Cagnol and J.-P. Zolésio. Intrinsic geometric model for the vibration of a constrained shell. In *Differential geometric methods in the control of partial differential equations (Boulder, CO, 1999)*, volume 268 of *Contemp. Math.*, pages 23–39. Amer. Math. Soc., Providence, RI, 2000.

[26] J. Cagnol and J.-P. Zolésio. Shape analysis in membrane vibration. *Math. Methods Appl. Sci.*, 23(11):985–1010, 2000.

[27] J. Cagnol and J.-P. Zolésio. Vibration of a pre-constrained elastic thin shell. I. Modeling and regularity of the solutions. *C. R. Math. Acad. Sci. Paris*, 334(2):161–166, 2002.

[28] J. Cagnol and J.-P. Zolésio. Vibration of a pre-constrained elastic thin shell. II. Intrinsic exact model. *C. R. Math. Acad. Sci. Paris*, 334(3):251–256, 2002.

[29] P. Cannarsa, G. Da Prato, and J.-P. Zolésio. Evolution equations in non-cylindrical domains. *Atti Accad. Naz. Lincei, VIII. Ser., Rend., Cl. Sci. Fis. Mat. Nat.*, 83:73–77, 1989.

[30] P. Cannarsa, G. Da Prato, and J.-P. Zolésio. The damped wave equation in a moving domain. *J. Differ. Equations*, 85(1):1–16, 1990.

[31] J. Céa, A. Gioan, and J. Michel. Quelques résultats sur l'identification de domaines. *Calcolo*, 10:207–232, 1973.

[32] G. Chavent. Sur une méthode de résolution du problème inverse dans les équations aux dérivées partielles paraboliques. *C. R. Acad. Sci. Paris Sér. A-B*, 269:A1135–A1138, 1969.

[33] G. Chavent. *Analyse fonctionnelle et identification de coefficients répartis dans les équations aux dérivées partielles*. PhD thesis, Université Paris VI, 1971.

[34] G. Chavent. Identification of functional parameters in partial differential equations. In *Identification of parameters in distributed systems (Joint Automat. Control Conf., Univ. Texas, Austin, Tex., 1974)*, pages 31–48. Amer. Soc. Mech. Engrs., New York, 1974.

[35] C. F. Chen and J. H. Clarke. Body under lifting wing. *J. Aerospace Sci.*, 28:547–562, 1961.

[36] P.G Ciarlet. *Mathematical Elasticity : Volume I,II* . North-Holland - Studies in Mathematics and its Applications, 1997.

[37] S.S Collis, K. Ghayour, M. Heinkenschloss, M. Ulbrich, and S. Ulbrich. Numerical solution of optimal control problems governed by the compressible Navier-Stokes equations. *Optimal Control of Complex Structures; K.-H. Hoffmann and I. Lasiecka, G. Leugering, J. Sprekels, F. Trltzsch (eds.), Birkhäuser Verlag, International Series of Numerical Mathematics*, 139:43–55, 2001.

[38] C. Conca, J. Planchard, B. Thomas, and M. Vanninathan. *Problèmes mathématiques en couplage Fluide-Structure*. Collection de la Direction des Etudes et de la Recherche - EDF, 1994.

[39] C. Conca, J.A. San Martín, and M. Tucsnak. Existence of solutions for the equations modelling the motion of a rigid body in a viscous fluid. *Commun. Partial Differ. Equations*, 25(5-6):1019–1042, 2000.

[40] R. Correa and A. Seeger. Directional derivative of a minimax function. *Nonlinear Anal., Theory Methods Appl.*, 9:13–22, 1985.

[41] D. Coutand and S. Shkoller. On the motion of an elastic solid inside of an incompressible viscous fluid. *Arch. for Rat. Mech. Anal.*, 2004.

[42] M. Cuer and J.-P. Zolésio. Control of singular problem via differentiation of a min-max. *Systems Control Lett.*, 11(2):151–158, 1988.

[43] G. Da Prato and J.-P. Zolésio. An optimal control problem for a parabolic equation in non-cylindrical domains. *Syst. Control Lett.*, 11(1):73–77, 1988.

[44] R. Dautray and Lions J-L. *Mathematical analysis and numerical methods for science and technology*. Springer - Volume 5, 1990.

[45] M.C. Delfour and J.-P. Zolésio. Shape sensitivity analysis via min max differentiability. *SIAM J. Control Optimization*, 26(4):834–862, 1988.

[46] M.C. Delfour and J.-P. Zolésio. Further developments in the application of min-max differentiability to shape sensitivity analysis. *Control of partial differential equations, Lect. Notes Control Inf. Sci.* , 114:108–119, 1989.

[47] M.C. Delfour and J.-P. Zolésio. Shape analysis via oriented distance functions. *J. Funct. Anal.*, 123(1):129–201, 1994.

[48] M.C. Delfour and J.-P. Zolésio. A boundary differential equation for thin shells. *J. Differential Equations*, 119(2):426–449, 1995.

[49] M.C. Delfour and J.-P. Zolésio. Tangential differential equations for dynamical thin/shallow shells. *J. Differential Equations*, 128(1):125–167, 1996.

[50] M.C. Delfour and J.-P. Zolésio. Shape analysis via distance functions: local theory. In *Boundaries, interfaces, and transitions (Banff, AB, 1995)*, volume 13 of *CRM Proc. Lecture Notes*, pages 91–123. Amer. Math. Soc., Providence, RI, 1998.

[51] M.C. Delfour and J.-P. Zolésio. *Shapes and Geometries - Analysis, Differential Calculus and Optimization*. Advances in Design and Control - SIAM, 2001.

[52] M.C. Delfour and J.-P. Zolésio. Velocity method and Courant metric topologies in shape analysis of partial differential equations. In *Control of nonlinear distributed parameter systems (College Station, TX, 1999)*, volume 218 of *Lecture Notes in Pure and Appl. Math.*, pages 45–68. Dekker, New York, 2001.

[53] F.R. Desaint and J.-P. Zolésio. Manifold derivative in the Laplace-Beltrami equation. *J. Funct. Anal.*, 151(1):234–269, 1997.

[54] B. Desjardins and M.J. Esteban. Existence of solutions for a Model of Fluid-Rigid Structure Interaction. *Arch. for Rat. Mech. Anal.*, 146, 1999.

[55] J. Donea, S. Giuliani, and J.P. Halleux. An arbitrary Lagrangian-Eulerian finite element method for transient dynamic fluid-structure interactions. *Comput. Methods Appl. Mech. Eng.*, 33:689–723, 1982.

[56] R. Dziri. *Problèmes de frontière libre en fluides visqueux.* PhD thesis, Ecole des Mines de Paris - Informatique Temps réel, Robotique, Automatique, 1995.

[57] R. Dziri, M. Moubachir, and J.-P. Zolésio. Dynamical shape gradient for the Navier-Stokes system. *C. R. Acad. Sci. Paris Ser. I*, 338:183–186, 2004.

[58] R. Dziri and J.-P. Zolésio. Dynamical shape control in non-cylindrical Navier-Stokes equations. *J. Convex Anal.*, 6(2):293–318, 1999.

[59] R. Dziri and J.-P. Zolésio. Eulerian derivative for non-cylindrical functionals. *Cagnol, John et al., Shape optimization and optimal design. Lect. Notes Pure Appl. Math*, 216:87–107, 2001.

[60] R. Dziri and J.-P. Zolésio. Eulerian derivative for non-cylindrical functionals. In *Shape optimization and optimal design (Cambridge, 1999)*, volume 216 of *Lecture Notes in Pure and Appl. Math.*, pages 87–107. Dekker, New York, 2001.

[61] A. El Badia and F. Moutazaim. A one-phase inverse Stefan problem. *Inverse Probl.*, 15(6):1507–1522, 1999.

[62] J. Escher and G. Simonett. Moving surfaces and abstract parabolic evolution equations. *Prog. Nonlinear Differ. Equ. Appl.*, 35:183–212, 1999.

[63] T. Fanion, M.A. Fernández, and P. Le Tallec. Deriving adequate formulations for fluid-structure interactions problems: from ALE to transpiration. *Rév. Européenne Élém. Finis*, 9(6-7):681–708, 2000.

[64] M.A. Fernández and P. Le Tallec. Linear stability analysis in fluid-structure interaction with transpiration. I. Formulation and mathematical analysis. *Comput. Methods Appl. Mech. Engrg.*, 192(43):4805–4835, 2003.

[65] M.A. Fernández and P. Le Tallec. Linear stability analysis in fluid-structure interaction with transpiration. II. Numerical analysis and applications. *Comput. Methods Appl. Mech. Engrg.*, 192(43):4837–4873, 2003.

[66] M.A. Fernández and M. Moubachir. Sensitivity analysis for an incompressible aeroelastic system. *Math. Models Methods Appl. Sci.*, 12(8):1109–1130, 2002.

[67] M.A. Fernández and M. Moubachir. An exact block-Newton algorithm for solving fluid-structure interaction problems. *C. R. Math. Acad. Sci. Paris*, 336(8):681–686, 2003.

[68] F. Flori and P. Orenga. Analysis of a nonlinear fluid-structure interaction problem in velocity-displacement formulation. *Nonlinear Analysis*, 35:561–587, 1999.

[69] G. Fourestey and M. Moubachir. Optimal control of Navier-Stokes equations using Lagrange-Galerkin methods. Technical report, INRIA, RR-4609, 2002.

[70] A. Friedman. Free boundary problems in science and technology. *Notices Am. Math. Soc.*, 47(8):854–861, 2000.

[71] A. Fursikov, M. Gunzburger, and L. Hou. Trace theorems for three-dimensional, time-dependent solenoidal vector fields and their applications. *Trans. Am. Math. Soc.*, 354(3):1079–1116, 2002.

[72] A.V. Fursikov, M.D. Gunzburger, and L.S. Hou. Boundary value problems and optimal boundary control for the Navier-Stokes system: The two-dimensional case. *SIAM J. Control Optimization*, 36(3):852–894, 1998.

[73] A.R Galper and T. Miloh. Motion stability of deformable body in an ideal fluid with applications to the N spheres problem. *Physics of Fluid*, 10(1):119–130, 1992.

[74] P. R. Garabedian and M. Schiffer. Identities in the theory of conformal mapping. *Trans. Amer. Math. Soc.*, 65:187–238, 1949.

[75] P. R. Garabedian and M. Schiffer. Convexity of domain functionals. *J. Analyse Math.*, 2:281–368, 1953.

[76] A. Gioan. *Une méthode d'approximations successives pour l'identification de domaines.* PhD thesis, Université de Nice, 1974.

[77] E. Giusti. *Minimal surfaces and functions of bounded variation.* Monographs in Mathematics, Vol. 80. Boston-Basel-Stuttgart: Birkhäuser. XII, 1984.

[78] R. Glowinski and O. Pironneau. Towards the computation of minimum drag profiles in viscous laminar flow. *Appl. Math. Modelling*, 1(2):58–66, 1976/77.

[79] N. Gomez and J.-P. Zolésio. Shape sensitivity and large deformation of the domain for Norton-Hoff flows. In *Optimal control of partial differential equations (Chemnitz, 1998)*, volume 133 of *Internat. Ser. Numer. Math.*, pages 167–176. Birkhäuser, Basel, 1999.

[80] C. Grandmont and Y. Maday. Fluid-structure interaction: A theoretical point of view. *Rev. Européenne Élém. Finis*, 9(6-7):633–653, 2001.

[81] M.D Gunzburger, L. Hou, and T.P Svobodny. Boundary velocity control of incompressible flow with application to viscous drag reduction. *SIAM Journal of Control and Optimization*, 30(1):167–181, 1992.

[82] M.D. Gunzburger, H.C. Lee, and G.A. Seregin. Global existence of weak solutions for viscous incompressible flows around a moving rigid body in three dimensions. *J. Math. Fluid. Mech*, 2(3):219–266, 2000.

[83] E. J. Haug. Distributed parameter optimal design. In *Transactions of the Seventeenth Conference of Army Mathematicians (U. S. Army Missile Command, Redstone Arsenal, Ala., 1971)*, pages 859–879. U. S. Army Research Office (Durham, N. C.), Report No. 72–1. U. S. Army Res. Office-Durham, Durham, N. C., 1972.

[84] E. J. Haug, K. C. Pan, and T. D. Streeter. A computational method for optimal structural design. I. Piecewise uniform structures. *Internat. J. Numer. Methods Engrg.*, 5(2):171–184, 1972.

[85] E.J. Haug and J. Céa, editors. *Optimization of distributed parameter structures. Vol. I-II*, volume 50 of *NATO Advanced Study Institute Series E: Applied Sciences*, The Hague, 1981. Martinus Nijhoff Publishers.

[86] M. Hinze. *Optimal and instantaneous control of the instationary Navier-Stokes equations - Habilitation thesis*. PhD thesis, Fachbereich Mathematik, Technische Universitt Berlin, 2000.

[87] K.-H. Hoffmann and V. N. Starovoitov. On a motion of a solid body in a viscous fluid. Two-dimensional case. *Adv. Math. Sci. Appl.*, 9(2):633–648, 1999.

[88] S. Hofmann and J.L. Lewis. The $L^p$ regularity problem for the heat equation in non-cylindrical domains. *Ill. J. Math.*, 43(4):752–769, 1999.

[89] T.J. Hughes, W.K Liu, and T.K Zimmermann. Lagrangian-Eulerian finite element formulation for incompressible viscous flows. *Computer Methods in Applied Mechanics and Engineering*, 29:329–349, 1981.

[90] J. Jaffre. *Analyse numérique de deux problèmes de domaine optimal et de contrôle ponctuel*. PhD thesis, Paris VI, 1974.

[91] D.D. Joseph. Parameter and domain dependence of eigenvalues of elliptic partial differential equations. *Arch. Rational Mech. Anal.*, 24:325–351, 1967.

[92] D.D. Joseph. Domain perturbations: the higher order theory of infinitesimal water waves. *Arch. Rational Mech. Anal.*, 51:295–303, 1973.

[93] B. Kawohl, O. Pironneau, L. Tartar, and J.-P. Zolésio. *Optimal shape design*, volume 1740 of *Lecture Notes in Mathematics*. Springer-Verlag, Berlin, 2000. Lectures given at the Joint C.I.M./C.I.M.E. Summer

School held in Tróia, June 1–6, 1998, Edited by A. Cellina and A. Ornelas, Fondazione C.I.M.E.[C.I.M.E. Foundation].

[94] P. Le Tallec. Numerical methods for nonlinear three-dimensional elasticity. In *Handbook of numerical analysis, Vol. III*, pages 465–622. North-Holland, Amsterdam, 1994.

[95] P. Le Tallec and J. Mouro. Fluid structure interaction with large structural displacements. *Comput. Methods Appl. Mech. Engrg.*, 190(24-25):3039–3067, 2001.

[96] J-L. Lions. Sur les problèmes mixtes pour certains systèmes paraboliques dans des ouverts non cylindriques. *Ann. Inst. Fourier*, 7:143–182, 1957.

[97] J-L. Lions. *Quelques méthodes de résolution des problèmes aux limites non linéaires.* Etudes mathématiques. Paris: Dunod; Paris: Gauthier-Villars., 1969.

[98] J-L. Lions. *Optimal control of systems governed by partial differential equations.* Springer-Verlag, 1971.

[99] J-L. Lions and E. Magenes. *Non-homogeneous boundary value problems and applications. Vol. I, Vol. II.* Die Grundlehren der mathematischen Wissenschaften. Band 182. Springer-Verlag, 1972.

[100] G. Lumer and R. Schnaubelt. Time-dependent parabolic problems on non-cylindrical domains with inhomogeneous boundary conditions. *J. Evol. Equ.*, 1(3):291–309, 2001.

[101] M. Masmoudi. *Outils pour la conception optimale de formes.* PhD thesis, Université de Nice - Doctorat ès Sciences Mathématiques, 1987.

[102] B. R. McCart, E. J. Haug, and T. D. Streeter. Optimal design of structures with constraints on natural frequency. *AIAA J.*, 8:1012–1019, 1970.

[103] J. Michel. *Une méthode de gradient pour l'identification de domaines.* PhD thesis, Université de Nice, 1974.

[104] A. M. Micheletti. Metrica per famiglie di domini limitati e proprietà generiche degli autovalori. *Ann. Scuola Norm. Sup. Pisa (3)*, 26:683–694, 1972.

[105] A. M. Micheletti. Perturbazione dello spettro dell'operatore di Laplace, in relazione ad una variazione del campo. *Ann. Scuola Norm. Sup. Pisa (3)*, 26:151–169, 1972.

[106] F. Mignot, F. Murat, and J.-P. Puel. Variation d'un point de retournement par rapport au domaine. *Comm. Partial Differential Equations*, 4(11):1263–1297, 1979.

[107] Ph. Morice. Une méthode d'optimisation de forme de domaine. Application à l'écoulement stationnaire à travers une digue poreuse. In *Control theory, numerical methods and computer systems modelling (Internat. Sympos., IRIA LABORIA, Rocquencourt, 1974)*, pages 454–467. Lecture Notes in Econom. and Math. Systems, Vol. 107. Springer, Berlin, 1975.

[108] M. Moubachir. *Control of fluid-structure interaction phenomena, application to the aeroelastic stability*. PhD thesis, Ecole Nationale des Ponts et Chaussées, 2002.

[109] M. Moubachir and J.-P. Zolésio. Cost function gradient for a coupled fluid-solid system. *Internal Report INRIA*, 2004.

[110] Z. Mróz. Limit analysis of plastic structures subject to boundary variations. *Arch. Mech. Stos.*, 15:63–76, 1963.

[111] F. Murat. Un contre-exemple pour le problème du contrôle dans les coefficients. *C. R. Acad. Sci. Paris Sér. A-B*, 273:A708–A711, 1971.

[112] F. Murat. Théorèmes de non-existence pour des problèmes de contrôle dans les coefficients. *C. R. Acad. Sci. Paris Sér. A-B*, 274:A395–A398, 1972.

[113] F. Murat. *Sur le contrôle d'un système par les coefficients de l'opérateur ou par un domaine géométrique*. PhD thesis, Université Paris VI, 1976.

[114] F. Murat. Contre-exemples pour divers problèmes où le contrôle intervient dans les coefficients. *Ann. Mat. Pura Appl. (4)*, 112:49–68, 1977.

[115] F. Murat and J. Simon. Étude de problèmes d'optimal design. In *Optim. Tech., Part 2, Proc. 7th IFIP Conf., Nice 1975, Lect. Notes Comput. Sci. 41, 54-62*. 1976.

[116] F. Murat and J. Simon. Sur le contrôle par un domaine géométrique. Technical report, Université Paris VI - 76 015, 1976.

[117] J. Nečas. *Les méthodes directes en théorie des équations elliptiques*. Masson et Cie, Éditeurs, Paris, 1967.

[118] T. Nomura and T.J. Hughes. An arbitrary Lagrangian-Eulerian finite element method for interaction of fluid and rigid body. *Computer Methods in Applied Mechanics and Engineering*, 95:115–138, 1992.

[119] B. Palmerio. A two-dimensional FEM adaptive moving-node method for steady Euler Flow Simulations. *Computer Methods in Applied Mechanics and Engineering*, 71:315–340, 1988.

[120] L. Passeron, C. Truchi, and J.-P. Zolésio. Dynamic modeling, control theory and stabilization for flexible structures: Industrial applications

at Aerospatiale, Cannes. In *Stabilization of flexible structures, Proc. ComCon Workshop, Montpellier/Fr. 1987, 183-216.* 1988.

[121] S. Piperno and P.-E. Bournet. Numerical simulations of wind effects on flexible civil engineering structures. *Rev. Eur. Élém. Finis,* 1999.

[122] S. Piperno and C. Farhat. Design of efficient partitioned procedures for the transient solution of aeroelastic problems. *Rev. Eur. Élém. Finis,* 9(6-7):655–680, 2001.

[123] O. Pironneau. On optimum profiles in Stokes flow. *J. Fluid Mech.,* 59:117–128, 1973.

[124] O. Pironneau. On optimum design in fluid mechanics. *J. Fluid Mech.,* 64:97–110, 1974.

[125] O. Pironneau. *Optimal shape design for elliptic systems.* Springer Series in Computational Physics. New York: Springer-Verlag. XII, 1984.

[126] A. Piskorek. Propriétés d'une intégrale de l'équation parabolique dans un domaine non cylindrique. *Ann. Pol. Math.,* 8:125–137, 1960.

[127] H. Poncin. Sur les conditions de stabilité d'une discontinuité dans un milieu continu. *Acta Math.,* 71:1–62, 1939.

[128] G. Prokert. On evolution equations for moving domains. *Z. Anal. Anwend.,* 18(1):67–95, 1999.

[129] C. Ramananjaona, M. Lambert, D. Lesselier, and J.-P. Zolésio. Shape reconstruction of buried obstacles by controlled evolution of a level set: from a min-max formulation to numerical experimentation. *Inverse Problems,* 17(4):1087–1111, 2001. Special issue to celebrate Pierre Sabatier's 65th birthday (Montpellier, 2000).

[130] R.M. Russell, H. Wei, and G.M. Lieberman. Weak solutions of parabolic equations in non-cylindrical domains. *Proc. Am. Math. Soc.,* 125(6):1785–1792, 1997.

[131] J.A. San Martín, V. Starovoitov, and M. Tucsnak. Global weak solutions for the two-dimensional motion of several rigid bodies in an incompressible viscous fluid. *Arch. Ration. Mech. Anal.,* 161(2):113–147, 2002.

[132] G. Savaré. Parabolic problems with mixed variable lateral conditions: an abstract approach. *J. Math. Pures Appl.,* 76:321–351, 1997.

[133] D. Serre. Chute libre d'un solide dans un fluide visqueux incompressible. Existence. (Free falling body in a viscous incompressible fluid. Existence). *Japan J. Appl. Math.,* 4:99–110, 1987.

[134] J. Simon. Differentiation with respect to the domain in boundary value problems. *Numer. Funct. Anal. Optim.,* 2(7-8):649–687, 1980.

[135] J. Sokolowski and J.-P. Zolésio. *Introduction to shape optimization: shape sensitivity analysis.*, volume 16. Springer Series in Computational Mathematics, 1992.

[136] V. Šverák. On optimal shape design. *C. R. Acad. Sci. Paris Sér. I Math.*, 315(5):545–549, 1992.

[137] V. Šverák. On optimal shape design. *J. Math. Pures Appl. (9)*, 72(6):537–551, 1993.

[138] T. Takahashi and M. Tucsnak. Global strong solutions for the two-dimensional motion of an infinite cylinder in a viscous fluid. *J. Math. Fluid Mech.*, 6(1):53–77, 2004.

[139] R. Temam. *Navier-Stokes Equations.* North-Holland, Studies in Mathematics and its Applications edition, 1984.

[140] C. Truchi. *Stabilisation par variation du domaine.* PhD thesis, Université de Nice - Spécialité Mathématiques, 1987.

[141] C. Truchi and J.-P. Zolésio. Wave equation in time periodical domain. In *Stabilization of flexible structures, Proc. ComCon Workshop, Montpellier/Fr. 1987, 282-294.* 1988.

[142] P. K. C Wang. Stabilization and control of distributed systems with time-dependent spatial domains. *J. Optimization Theory Appl.*, 65(2):331–362, 1990.

[143] K. Washizu. Complementary variational principles in elasticity and plasticity. In *Duality and complementarity in mechanics of solids (School-Conf. Duality Problems of Mech. Deformable Bodies, Jabłonna, 1977)*, pages 7–93. Ossolineum, Wrocław, 1979.

[144] G.Z. Yang and N. Zabaras. An adjoint method for the inverse design of solidification processes with natural convection. *Int. J. Numer. Methods Eng.*, 42(6):1121–1144, 1998.

[145] J.-P. Zolésio. Un résultat d'existence de vitesse convergente dans des problèmes d'identification de domaine. *C. R. Acad. Sci. Paris Sér. A-B*, 283(11):Aiii, A855–A858, 1976.

[146] J.-P. Zolésio. An optimal design procedure for optimal control support. In *Convex analysis and its applications (Proc. Conf., Muret-le-Quaire, 1976)*, pages 207–219. Lecture Notes in Econom. and Math. Systems, Vol. 144. Springer, Berlin, 1977.

[147] J.-P. Zolésio. *Identification de domaines par déformations.* PhD thesis, Université de Nice - Doctorat d'Etat en Mathématiques, 1979.

[148] J.-P. Zolésio. Solution variationnelle d'un problème de valeur propre non linéaire et frontière libre en physique des plasmas. *C. R. Acad. Sci. Paris Sér. A-B*, 288(19):A911–A913, 1979.

[149] J.-P. Zolésio. Domain variational formulation for free boundary problems. In *Optimization of distributed parameter structures, Vol. II (Iowa City, Iowa, 1980)*, volume 50 of *NATO Adv. Study Inst. Ser. E: Appl. Sci.*, pages 1152–1194. Nijhoff, The Hague, 1981.

[150] J.-P. Zolésio. The material derivative (or speed) method for shape optimization. In *Optimization of distributed parameter structures, Vol. II (Iowa City, Iowa, 1980)*, volume 50 of *NATO Adv. Study Inst. Ser. E: Appl. Sci.*, pages 1089–1151. Nijhoff, The Hague, 1981.

[151] J.-P. Zolésio. Shape stabilization of flexible structure. *Lect. Notes Control Inf. Sci.*, 75:446–460, 1985.

[152] J.-P. Zolésio. Galerkine approximation for wave equation in moving domain. *Lect. Notes Control Inf. Sci.*, 147:191–225, 1990.

[153] J.-P. Zolésio. Weak shape formulation of free boundary problems. *Ann. Scuola Norm. Sup. Pisa Cl. Sci. (4)*, 21(1):11–44, 1994.

[154] J.-P. Zolésio. Shape differential equation with a non-smooth field. In *Computational methods for optimal design and control (Arlington, VA, 1997)*, volume 24 of *Progr. Systems Control Theory*, pages 427–460. Birkhäuser Boston, Boston, MA, 1998.

[155] J.-P. Zolésio. Variational formulation for incompressible Euler equation by weak shape evolution. In *Optimal control of partial differential equations (Chemnitz, 1998)*, volume 133 of *Internat. Ser. Numer. Math.*, pages 309–323. Birkhäuser, Basel, 1999.

[156] J.-P. Zolésio. Shape analysis and weak flow. *Lect. Notes Math.*, 1740:157–341, 2000.

[157] J.-P. Zolésio. Weak set evolution and variational applications. *Lect. Notes Pure Appl. Math.*, 216:415–439, 2001.

[158] J.-P. Zolésio and C. Truchi. Shape stabilization of wave equation. In *Boundary control and boundary variations (Nice, 1986)*, volume 100 of *Lecture Notes in Comput. Sci.*, pages 372–398. Springer, Berlin, 1988.

# *Index*

Aeroelastic stability, 5

Boundary
    control, 9
    Dirichlet, 8
    free, 4, 33
    kinematic continuity, 187, 218
    kinetic continuity, 218
    noncylindrical, 19
    value problem, 1

Condition
    transpiration, 178
    viability, 14, 50
Control
    active, 9
    feedback law, 9
    optimal, 180
    passive, 1, 9
    piezoelectrical device, 4

Domain
    deformation, 2
    moving, 2, 3
    perturbation, 13, 38
    tube, 19, 35, 38, 60
    weak convection, 49

Equation
    adjoint elastodynamic, 220
    adjoint Navier-Stokes, 11, 139,
        188, 198, 219
    adjoint ODE, 199
    Euler-Lagrange, 103
    Hamilton-Jacobi, 18, 101
    Laplace, 90
    linearized Navier-Stokes, 11

Navier-Stokes, 4, 8, 14, 110, 178,
    187, 216
    noncylindrical heat, 35
    noncylindrical wave, 98
    nonlinear elastodynamic, 217
    ODE, 187
    partial differential, 3
    strong convection, 50
    weak convection, 51

Fluid
    drag, 9, 14
    incompressible, 4, 110
Function
    characteristic, 3, 50
    cost, 10
    cost gradient, 11
    Gâteaux differentiable, 24
    minimization, 10
    oriented distance, 2, 119
    without steps, 102
Functional
    lagrangian, 12, 144, 159, 195, 222
    tracking, 18

Geometry
    additive curvature, 120
    curve, 77
    density perimeter, 61
    intrinsic, 2
    mean curvature, 17, 72, 119
    moving, 4
    perimeter, 23, 55
    perturbation, 3
    tangential calculus, 119

Interface
    motion, 4, 5, 33, 216

Milton Keynes UK
Ingram Content Group UK Ltd.
UKHW021619071024
449327UK00020BA/1110